AHUWHENUA

First published in 2022 by Huia Publishers
39 Pipitea Street, PO Box 12–280
Wellington, Aotearoa New Zealand
www.huia.co.nz

ISBN 978-1-77550-700-0

Copyright © The Ahuwhenua Trophy Management Committee 2022

Cover image adapted from images courtesy of John Cowpland
and Ahuwhenua Archives

This book is copyright. Apart from fair dealing for the purpose of private study, research, criticism or review, as permitted under the Copyright Act, no part may be reproduced by any process without the prior permission of the publisher.

A catalogue record for this book is available from the National Library of New Zealand.

AHUWHENUA

Celebrating 90 Years of Māori Farming

REVISED EDITION

Danny Keenan

Contents

PART ONE – CELEBRATING 80 YEARS OF MĀORI FARMING

viii	——	**Foreword** The Ahuwhenua Legacy – Ka tu Rangatira ai te Tangata
xiv	——	**Introduction** The Ahuwhenua Trophy
2	——	**Chapter One** Retaining the Land – to 1933
26	——	**Chapter Two** The Early Years of the Ahuwhenua Trophy – 1933–1939
44	——	**Chapter Three** The War Years – 1940–1945
58	——	**Chapter Four** Māori 'Economic Advancement' and Farming – 1945–1961
96	——	**Chapter Five** The 'Spirit of Friendly Rivalry' – 1962–1972
120	——	**Chapter Six** Challenges – 1973–2002
142	——	**Chapter Seven** Ahuwhenua Continuing – 2003–2013
180	——	**Conclusion to Part One** 'Administering the Policy Effectively'

PART TWO – AHUWHENUA TROPHY COMPETITIONS CONTINUE

190 — **Introduction**
Growing the Trophy – 2014–2021

196 — **Chapter Eight**
Origins of Modern Māori Horticulture – 1840–1860

208 — **Chapter Nine**
Dispossession of Māori Lands – 1858–1912

226 — **Chapter Ten**
Revisiting the Early Ahuwhenua Farmers – 1933–1963

238 — **Chapter Eleven**
Ahuwhenua Trophy Competition Continues – 2014–2021

284 — **Chapter Twelve**
Looking to the Future – Young Māori Farmers and Growers

294 — **Acknowledgements**

296 — **Appendix 1**
Ahuwhenua Trophy Judging Criteria 2013

298 — **Appendix 2**
Winners of the Ahuwhenua Trophy 1933–2021

308 — **Appendix 3**
Winners of the Young Māori Farmer/Grower of the Year 2012–2021

310 — **Appendix 4**
Judges of the Ahuwhenua Trophy Competition 1933–2022

316 — **Endnotes**

326 — **Bibliography**

330 — **Index**

1933–2013

Part One

Celebrating 80 Years of Māori Farming

Foreword

The Ahuwhenua Legacy – Ka tu Rangatira ai te Tangata

Few events in the history of Aotearoa have had such a profound effect on the economic and social wellbeing of Māori as the Ahuwhenua Trophy competition.

The year 2023 marks ninety years since the great Māori leader and visionary Sir Apirana Ngata and his friend Lord Bledisloe – the Governor-General of New Zealand at the time – created this competition. Both were genuine people of the land and visionary political leaders who had the ability to see potential and set up a process – in this case a competition – that had longevity and legacy. A competition that would grow from its humble beginnings in 1933 to become the biggest and most prestigious event on the agricultural calendar.

The Ahuwhenua Trophy was born out of necessity. Māori had had their best lands confiscated by trickery, treachery and other dubious means, and in many instances were left with land that, even today, can be challenging to farm and located in very remote parts of the country. With the arrival of climate change and the volatility of weather patterns, Māori farms in some hill country regions have become disproportionately disadvantaged with pastures and infrastructure now highly susceptible to damage. In conjunction with this, Māori are still significantly hindered in terms of their culture, education, health and economic wellbeing.

Against all these odds, Māori have not lost heart. Inspired by the vision of Ngata and Bledisloe, the Māori men and women of the land have worked increasingly hard to develop farms and orchards that sit amongst the best of all farms and orchards in Aotearoa. There is no doubt that the Ahuwhenua competition has played an integral part in the rise of Māori agribusiness.

While Māori have successfully incorporated modern farming practices into their operations, they have overlaid this with their own unique values of kaitiakitanga of our whenua. As a result of the advocacy of the Kāhui Wai Māori Advisory Group, the intrinsic values of te ao Māori now underpin the Jacinda Ardern-led sixth Labour Government's Essential Freshwater and Resource Management Reforms. Giving effect to Te Mana te Wai and to Te Oranga o te Taiao will create generational change and build more sustainable and resilient farming practices. The 'industry-led' He Waka Eke Noa, a programme aimed at dealing with climate change, has also seen Māori landowners at the forefront of these changes.

One of the aspects of the Ngata/Bledisloe vision was the establishment of a dynamic competition that functioned at a level ahead of its time. Those responsible for running the event over the years have ensured the realisation of this vision. The Ahuwhenua Trophy logically began as one for all Māori farms, but in 1954 the decision was made to separate sheep and beef and dairy so there were two beautiful Bledisloe Cups. In 2020, a third trophy was added upon the decision to provide a competition solely dedicated to horticulture. The inclusion of horticulture reflected Māori involvement and excellence in this field. Historically, the first exports by Māori were vegetables. In purely economic terms, Māori are a significant producer of a range of horticultural products and, like sheep and beef and dairy, their orchards and commercial growing operations benchmark well against all others.

One of the exciting developments in the history of Ahuwhenua was the creation of the Young Māori Farmer of the Year award in 2012. This award is designed to reward and encourage young Māori and set them on a pathway for future leadership roles. Already, the benefits of this event are evident.

Today, we live in a world of uncertainty caused by the coronavirus pandemic and additional geopolitical events outside our control. This is not entirely unusual in the sense that, in the ninety years that the Ahuwhenua Trophy has existed, there has been a world war and a range of other catastrophic events. Through all these events, Māori, with the support of Ahuwhenua, have been able to navigate a path through adversity and remain focused on and committed to the visions of Ngata and Bledisloe. This updated book tells a remarkable story of commitment and resilience with the hope that, over time, the efforts and contribution of Māori to our society will be nationally recognised.

Kingi Smiler

CHAIRMAN (2007-2021)
Ahuwhenua Trophy Management Committee

Grazing sheep, Gisborne, 2009
(Ahuwhenua Archives, Pākārae Whangara B5 Field Day7–25)

Introduction
─────────

The Ahuwhenua Trophy

'We've Got Unfinished Business'

When the Te Awahohonu Forest Trust of Te Hāroto was contemplating entering the Ahuwhenua Trophy competition in 2013, it had a lot on its collective mind.

The Trust had entered the competition once before, in 2004. However, that year, severe flooding in the lower North Island and the South Island had led to significant rural disruptions; other prospective entrants had withdrawn and field days had been cancelled.

The Ahuwhenua Trophy Management Committee had therefore decided in fairness not to award a national trophy for 2004. Instead, two regional finalists were each presented with a regional award: Kuratau Trust of Turangi and the Te Awahohonu Forest Trust.

By 2013, the Te Awahohonu Forest Trust had moved a long way from its 2004 position. The Trust now owned 2623 hectares of rolling hill country, with 16,000 ewes and 5000 replacements on site. A Highlander herd of 1000 breeding cows plus heifer replacements with finishing steers and bulls also grazed on the Trust's expansive holdings.

A further 10,000 hectares of mature native forest owned by the Trust had been set aside as a forest reserve; it was now home to protected bird species such as the kiwi and the whio (blue-billed duck).

However, because a significant part of the Trust's land adjoined production forests and undeveloped lands, possums and other vector pests and tuberculosis in cattle had become major issues. Annual cattle testing, managed by the national Animal Health Board, was now seen as essential.

Despite this, the Te Awahohonu Forest Trust valued opportunities to benchmark itself against peers by participating in farming competitions. But would it try again, in 2013, for the Ahuwhenua Trophy? Yes, it would. As one kaumātua explained, 'we've got unfinished business.'[1]

This decision proved to be a good one. At the awards dinner held at the Pettigrew Green Arena, Taradale, on Friday, 7 June 2013, attended by over 850 people, the Te Awahohonu Forest Trust was announced as the winner of the 2013 Bank of New Zealand Māori Excellence in Farming Ahuwhenua Trophy.

Early History of the Competition

The Ahuwhenua competition was established by Governor-General Lord Bledisloe and Minister of Native Affairs Sir Apirana Ngata in 1933. The competition was launched at Ōhinemutu in Rotorua to 'high enthusiasm', as Lord Bledisloe reported. However, it then immediately went into a two-year recess. This disturbed Lord Bledisloe, who wrote to the prime minister in 1935 to express his fears that the competition might not survive for very long. 'It is now at least eighteen months,' he wrote, 'since I formally at Ohinemutu made the first presentation of the Trophy and Medals to the winning Maori Tribe and to individual competitors in the presence of representatives of several tribes … I am particularly anxious that this competition should not be allowed to lapse, especially in view of the initial enthusiasm of the Natives regarding it.'

The Ahuwhenua Trophy
(John Cowpland – Alphapix)

Pōwhiri at Ahuwhenua Awards dinner, Taradale, 2013
(John Cowpland – Alphapix-0706130836)

Noticing the large number of unplaced competitors also present, who were 'of almost equal merit to the winners,' Lord Bledisloe had contributed additional endowment monies so that extra bronze medals could be awarded. He was now concerned that he had heard nothing of any subsequent competition, or, if such competition had taken place, the names of any winners.

Fortunately for Lord Bledisloe's sensibilities, the Ahuwhenua Trophy did not lapse. The full competition returned in 1936. Although it had originally been confined to the Waiariki and Bay of Plenty Native Districts, the competition was extended in 1936 to all tribes. This extension had always been intended by Lord Bledisloe, with the trophy first mooted following his tour of the new land development schemes in those native districts in 1931. Thereafter, with the exception of 1937 when fire destroyed the actual cup, the trophy was contested annually for the next forty-two years, from 1938 to 1979.

A Competition for Iwi?

In 1939, Lord Bledisloe, prolific letter writer that he was, again wrote to the prime minister pointing out that his original desire for the trophy had been that it focus, not on individual Māori farmers, but on tribal winners, suggesting that this emphasis should be maintained: 'I specifically intended the trophy to be competed for among the tribes (and to carry the names of the winning tribes on the trophy); an inter-tribal and not an individual competition.' The inaugural 'Bledisloe Cup' had been presented to one of the East Coast tribes.

For individual encouragement, in addition to the trophy, Lord Bledisloe had also gifted 'a beautifully designed die and adequate endowment for the yearly provision of one large silver medal and two large bronze medals'. These were to be presented to the 'individual protagonists or champions' who contributed towards their tribe's trophy win. 'I hope these medals have been continued; I see no reference to them in the judge's reports,' he told the prime minister.[2]

However, by 1939, when Lord Bledisloe wrote his letter, sixteen top-four placings had already been awarded to individual Māori farmers, including two women, Mrs Tatai Hall of Te Teko and Mrs Huinga Nepia of Tikitiki.

Native Minister Apirana Ngata's private secretary, H R N Balneavis, responded to Lord Bledisloe's letter, advising his minister that he had looked carefully at the conditions in the original Ahuwhenua Deed and had determined that the position was 'other than as stated by Lord Bledisloe'.

Ahuwhenua Medal, showing both sides
(Ross Setford Photographer)

Experience had demonstrated that all of the competitors belonged to two or more different tribes; it was considered inappropriate 'to inscribe the name of one of [a competitor's] tribes only and too expensive to inscribe the names of all the tribes of which he was a member'. In addition, despite the absence of references to tribes in judges' reports (judges would not necessarily know to which tribes a competitor belonged), the tribes of trophy winners had always evinced a great pride in such prestigious successes of their members.[3]

Accordingly, the competition remained open to Māori farming individuals.

In 1954, Lord Bledisloe applauded the continuance of the trophy by granting a further sum for the creation of a second trophy, enabling the competition to be split into two sections: sheep and cattle, and dairy. Thereafter, the competition gathered further momentum and ran until 1979.

Between 1980 and 2002, the Ahuwhenua Trophy was staged only three times before being reintroduced with the support of Meat & Wool New Zealand in 2002.

This reintroduction was instigated by two Māori board directors, Mr Wayne Walden and Ms Gina Rudland, who urged Meat & Wool New Zealand to support Māori farming in this positive way. In seeking to increase the representation, economic potential and profile of Māori farmers, Meat & Wool New Zealand agreed, acknowledging the trophy as an opportunity to 'recognise excellence in Maori farming'.

Ahuwhenua Farmers over the Years

When Te Awahohonu Forest Trust chairman Tamihana Nuku stepped up to accept the Ahuwhenua Trophy for 2013, the Trust became the 213th individual Māori farmer, couple or corporate collective to win an Ahuwhenua top-four placing since 1933.

Over the years, Māori farmers came forward from the remotest parts of the country to participate in the competition. In the earliest decades, the hardships experienced by some of the farmers were immeasurable; some lived on earthen floors, dealing with uncontrollable infestations of bracken and ragwort. Others travelled for days across broken country on packhorses to collect supplies.

Māori women ran their own farms or worked their properties with their husbands, often entering the competition in their own name where succession rights had accrued only to them. School-age children were a common sight working on farms. Sickness and 'hard living' were not uncommon; nor were disabilities and injuries. One tragic example – a Bay of Plenty farmer withdrew from the 1969 competition, at the judge's suggestion, because of the 'sad loss of his daughter in a tractor accident earlier that year'. This young woman had been a great help on the farm, and this tragedy had 'disorganised' the farmer's work, the judge noted.[4]

Two generations of Māori ex-servicemen later joined those who had farmed the land developments from the very beginning; one veteran, Tikirau Callaghan, was fittingly presented with his trophy in 1948 by former Lieutenant General (now Governor-General) Sir Bernard Freyberg VC, who had earlier commanded the 2nd New Zealand Expeditionary Force with distinction in the Mediterranean.

Witness to all these developments were the Ahuwhenua judges who were in the main senior advisory officers appointed by the Department of Agriculture for their expertise, experience and tact. Normally appointed for three-year terms, the judges visited Māori farms and reported on what they saw, often in language that made clear their admiration

Celebrating members of Te Awahohonu Forest Trust, winners of the Sheep and Beef Competition 2013, including Deputy Prime Minister Hon Bill English (centre behind cup); Tamihana Nuku (second from right), Chairman Te Awahohonu Forest Trust; and Hon Pita Sharples (second left) Minister of Māori Affairs
(John Cowpland – Alphapix-0706132665)

for and empathy with these farmers. Their detailed analyses of farming operations were checked by the economics section of the Agriculture Department and sent on to the Minister of Māori Affairs, along with recommendations for winners and placings.

These early reports and other papers related to the early competitions present an invaluable insight into the lived realities of generations of Māori farming families, as difficult and yet as gratifying as they were. As one judge commented in 1948, when reflecting on the quality of entrants that year; such competitors could truly be termed 'sons of the soil'.

In later decades, Māori farmers became more prosperous, and more expert, keeping well abreast of the rapid changes occurring in the agricultural sector. Emphasis in the competition shifted from old land development districts like the Waiariki to new dairy hubs like Taranaki, the Waikato and Northland. By 2002, the changing face of Māori farming saw the increasing importance of Māori incorporations and trusts, which found new expression in the rejuvenated competition.

Looking Back, Looking Forward

By 2013, the Ahuwhenua Trophy BNZ Award for Excellence in Māori farming had come a long way.

Eighty years earlier, the trophy had been launched, making it 'by far the oldest (if not) the most prestigious agricultural competition in New Zealand', according to the current chairman of the Ahuwhenua Trophy Management Committee, Mr Kingi Smiler.

However, the competition is now more than a showcase for excellence in Māori farming: 'It's a barometer of Māori business acumen and success'. At its heart is a commitment to retaining the whenua as a sustainable resource for present and future generations, Smiler says.[5]

The competition also demonstrates best practice in environmental and economically sustainable farming. Finalists are widely regarded as being among the best farmers in New Zealand, invariably running large-scale operations. While the award is open to all Māori sheep and beef or dairy farmers, the majority of competitors are now whānau trusts or incorporations operating on behalf of shareholders or owners.

Māori farmers came to the Ahuwhenua competition from most parts of New Zealand: initially from established development areas like Waiariki and the East Coast, but increasingly from Northland, the Waikato, King Country and Taranaki, and later from Southland, as these areas developed.

Māori group at a farm in Winiata, 1895; photograph by Edward George Child (Alexander Turnbull Library, 1/2-032309-G)

Today's competition is still about working the land most efficiently and economically, as it was when first inaugurated by Sir Apirana Ngata and Lord Bledisloe in 1933. However, today, Māori hoping to win the coveted trophy are also required to demonstrate their commitment to tikanga, to their environment and to the wider communities that support them.

Leaders of whānau trusts and incorporations today are demonstrating impressive leadership skills in long-term environment and agribusiness management, while increasingly exploring links with international markets.

The contemporary Ahuwhenua Trophy competition highlights successful collaborations between Māori farmers and the wider business, banking and farming communities: 'The longevity of the Ahuwhenua Trophy competition reflects the commitment Maori farmers have as kaitiaki of their lands for future generations to their whenua and whanau,' says Mr Smiler; 'and from where I'm standing, the future for Maori farmers, and their children, looks bright.'[6]

Mr Dana Blackburn of Ātihau-Whanganui Incorporation has agreed with this assessment. The Ahuwhenua Trophy was attracting the support and interest of Māori farmers at the top of their game, he said when appointed as the competition's new chief judge of sheep and beef in 2010, succeeding former chair of the Meat Board, John Acland. Mr Acland had undertaken the role of chief judge in sheep and beef in 2003, and had since then significantly enhanced the standards of judging and reporting back, where business operational strengths and the potential for improvements were concerned.

Entrants now needed to be in top form, said Mr Blackburn. 'Through the competition, we take a great deal of pride in the knowledge that like most Maori farmers we look to the land to ensure that future generations can continue to gain a livelihood from it.' Some years ahead would be tougher than others,

Pōwhiri, Pākarae Whāngārā B5 field day, 14 May 2009
(Ahuwhenua Archive, Field Day-20)

but if you were committed to the long haul, making well-informed and intelligent business decisions, 'things would always even out'.⁷

Interest in the competition was increasing every year, observed Mr Bob Cottrell, a former chairman of the Ahuwhenua Trophy Management Committee, who had spent many years in the commercial sector working in finance, farm management and real estate. Mr Cottrell had also farmed for twenty-two years and had had significant involvement in large-scale Māori commercial forestry and farm interests. An awareness of the Ahuwhenua Trophy was definitely growing, he said. It was now fair to say that Māori farmers were becoming increasingly aware of the many benefits that came from taking part in the competition.⁸

Such interest in the competition continued to be heartening, said Mr Doug Leeder, the chief judge of the Ahuwhenua dairying competition from 2006 to 2012. Mr Leeder ran a dairy farming business of 500 cows in the eastern Bay of Plenty and had been involved in industry governance since 1987, having been a director and chairman of both Bay Milk Products and New Zealand Dairy Group and at the forefront of the major changes in the industry that had led to the formation of Fonterra.

The Ahuwhenua competition's fundamental purpose, said Mr Leeder, was to highlight governance and management features that generated profit and growth while not forgetting critical elements like environmental factors. 'The farming industry as a whole was now beginning to realise that there was a lot to be learned from Maori traditional attitudes to the sustainability and stewardship of the land', he said. When one looked at what individual Māori farmers and trusts and incorporations had achieved, it was undoubtedly substantial.⁹

Long-term Investment

Long before the 1930s, wrote the Minister of Māori Affairs, Hon Peter Sharples, in 2013, Māori had already been leading some of New Zealand's earliest trade missions overseas, and had been helping to establish the country's sheep industry. Māori had also been exporting produce direct to Sydney on their own ships.

Māori enterprises 'like our farms', said Dr Sharples, remained some of New Zealand's oldest existing businesses. 'The establishment of the Ahuwhenua Trophy by Lord Bledisloe and Sir Apirana Ngata secured the future of best practice in agribusiness for Maori owned farms.'

Māori governance was second to none in New Zealand, said Dr Sharples. 'Our Māori economy represented generations of Māori people whose endeavours had been passed on to those alive today.'

Judges of the Ahuwhenua Trophy BNZ Māori Excellence in Farming Award – Sheep and Beef Category 2013: Sam Johnson, Rob Davison, Abe Seymour (Kaumātua), Dana Blackburn (Chief Judge) and Dr Gavin Sheath
(John Cowpland – Alphapix-0905131250)

Eighty years of Māori farmers and farming families had contributed hugely to this burgeoning economy since 1933. The Ahuwhenua Trophy competition had played a role 'beyond comprehension' in ensuring that the Māori powerhouse economy could continue into the future.

The efforts and hard work of the old people on broken land development schemes prepared the way for the prosperity of future generations of Māori, including those yet unborn. This, said Dr Sharples, was what long-term investment looked like to Māori.[10]

(John Cowpland – Alphapix)

Chapter One

Retaining the Land — to 1933

'A Path Our Tūpuna Used to Walk'

When Pah Hill Station of Ohakune, run by the Ātihau-Whanganui Incorporation, was picked from sixteen Māori-owned sheep and cattle farms as the winner of the Ahuwhenua Trophy in 2007, the victory was described by the Member of Parliament for Te Tai Hauāuru, Mrs Tariana Tūria, as 'an enormous honour' for the station's management and shareholders alike. Pah Hill station had been a formidable entry in the competition; it ran 48,000 sheep on 3950 hectares of isolated back country while leasing land to and supporting adjoining farmers under vested lands regimes or through standard commercial arrangements. Pah Hill also managed its own forests by arrangement with Winstone Pulp, including 240 hectares of native bush set aside as a conservation area (through the government's Nga Whenua Rahui fund). Power companies were also knocking on Pah Hill's 'whare door', looking to develop power generation opportunities.[11]

The station also occupied part of an ancient Māori walking track that ran from Rānana on the Whanganui River to Karioi, about 40 kilometres away to the north-east. It was in the vicinity of marae at Ngāmoki and Tirorangi and ran beneath the historic Te Ranga a Kauika Pā site, which was known to house a number of urupā. This pā was the principal historic settlement of Ngāti Rangi, which was the main tribe of the area, and the one to which most Whanganui tribes connected.

Mrs Turia said that the track was 'used by our tūpuna to walk from Rānana to Karioi, so I can't help thinking that Ātihau-Whanganui Incorporation is truly walking with its ancestors.' The incorporation had been able to energise and develop land that had originally been vested with the Aotea Māori Land Council by Whanganui Māori as far back as 1903, a contentious move at the time because vesting land in this way risked its alienation. Now, said Mrs Turia, it was 'great to see Ātihau farming Ātihau land and doing so well'.[12]

Pah Hill Station
(John Cowpland – Alphapix)

Māori acquired generations of customary experience in planting, harvesting and storing crops (and maintaining a supply of seeds) through the winter.

Man seated before a kūmara pit built with rough planks dug into the hill at Waikanae, c. 1847; drawing by William Swainson (Alexander Turnbull Library, A-023-014)

The 'New Maori Economy'

The Ātihau-Whanganui Incorporation represented the new type of large-scale Māori corporate agribusiness that had by that time become integral to the 'new Maori economy now dominated by economic, environmental, social and cultural frames of development'.[13] Pah Hill Station had once been marked out as a potential hydro lake. Its impressive 2007 structure had evolved from an initial 1900-hectare sheep and cattle farm by way of modern governance and on-farm/business management plans, physical and financial key performance indicators, environmental protection strategies and social, community and ngā tikanga Māori policies.

The Work of Generations

Pah Hill Station represented a Māori economic aspiration inherited from Apirana Ngata and his generation of Māori farmers, as expressed in its vision statement: 'taonga tuku iho; toitū whenua, toitū tangata – sustainable wealth creation from pastoral farming'.[14] In 1991, when the control of Pah Hill Station returned to its Atihau owners from lessees, a huge amount of work and money had been necessary to bring the sheep and beef unit up to standard, reflecting the labours of earlier generations of Māori farmers who had turned their hands to the difficult task of clearing the land for the raising of sheep and cattle.

Many of the early Māori sheep and cattle farmers came from Hawke's Bay and Wairarapa, where, for a time, the industry was heavily focused. Sheep farming had originally been concentrated in the South Island; in 1867, there were 6,683,000 sheep in the South Island and only 1,788,000 in the North. From the 1870s, however, the rate of growth was faster in the North Island; by 1915, there were 13,314,000 and 11,585,000 sheep in the North and South Islands respectively.

For a time, Napier was New Zealand's most important wool port and the hub of the North Island sheep industry, providing facilities for a lucrative trade in wool. This all changed on 15 February 1882 when the *Dunedin* set sail from Port Chalmers for London, carrying 5000 sheep carcasses and a small amount of butter. The *Dunedin* arrived safely three months later with its cargo intact. A new export market in frozen lamb quickly developed. The first shipment of lamb to depart from Napier left for London in 1884, heralding the beginning of a local frozen lamb industry that led to the construction of freezing works at Tōmoana, Napier and Whakatū.[15]

Māori were not major sheep or cattle run holders at this time; their rates of participation in the Hawke's Bay rural economy were low. In 1891, Māori owned a total of 17,006 sheep in Wairoa County, 28,200 in Hawke's Bay, 3400 in Waipawa and 9500 in Pātangata County. The combined total of sheep owned by Māori amounted to less than those run by any number of Pākehā pastoralists on their own.

As Richard Boast has argued, however, some Māori did have reasonable flocks. Ani Kingi, who was married to a Pākehā pastoralist, ran 6000 sheep in her own right in 1886 at Moawhango. Henare Kepa had 5500 sheep, also at Moawhango. Renata Kawepo, who lost his right eye pursuing Te Kooti Arikirangi at the Battle of Te Pōrere in June 1869, also ran 7000 sheep near Hastings.

Further north, the most significant area for Māori sheep farming was Waiapu where Ngāti Porou had managed to retain most of their land. By one estimate, in 1873, there were 20,000 sheep being farmed by Māori in this area. Ngāti Porou leaders like Rapata Wahawaha and Mokena Kohere owned large flocks. Ngāti Porou developed good working relationships with local Pākehā farmers, who trained their young people in farming skills and were prepared to advance capital for land developments. Apirana Ngata made efforts to improve and modernise Māori stock raising and was a director of the Tikitiki station.[16]

Māori Customary Arable Farming

Throughout the early and mid-nineteenth century, Māori arable farming and the marketing of crops had developed in some areas into a highly lucrative income earner. Some whānau and hapū worked sizable plantations. In Waitara in the late 1850s, for example, Wiremu Kingi Te Rangitake's Te Ātiawa people possessed thirty-five ploughs, twenty pairs of harrows, forty carts, 300 cattle, 150 horses and a small flotilla of sailing boats. Their produce contributed at least £6000 a year to the Taranaki Provincial Treasury, even though settler vessels from New Plymouth were banned from calling into Waitara to load Māori produce.[17]

However, as G V Butterworth has argued, with the growth of large Pākehā market gardens throughout New Zealand, located around most North Island towns after 1870, Māori arable farming enterprises like the one at Waitara lost their competitive edge. The Māori economic collapse that followed was assisted in no small measure by the devastations inflicted on whānau and hapū by the Land Wars. In some areas, like Taranaki, these devastations exacerbated the dramatic slump in Māori rural economies after the 1870s.

In these circumstances, some Māori abandoned arable farming, turning instead to stock rearing. This was especially true of Māori from the East Coast, Hawke's Bay and Rangitīkei, where by the mid-1870s some chiefs were running sizable flocks. These 'experiments' in raising stock, however, tended to fail, Butterworth suggests, because Māori had no tradition of animal husbandry, and did not appreciate the need for essentials such as fencing, culling of unproductive stock or the treatment of grievous stock ailments like scab, which devastated many of the Ngāti Porou flocks in particular. Scab was easily transmitted and caused rapid stock degeneration and even death. In 1849, New Zealand's first port inspectors were appointed in an effort to control the spread of scab. Despite sheep branding becoming compulsory in the 1860s, the country was not entirely free of scab until 1893.

Had Māori remained focused on arable farming, argues Butterworth, the fate of the Māori agricultural economy after the 1870s may well have been different.[18]

Arable Farming before Colonisation

Māori were of course highly skilled and productive arable farmers before the arrival of Pākehā in New Zealand. 'When the ancestors of the Maori arrived from Eastern Polynesia and settled on these shores,' wrote Elsdon Best in 1925, 'they found that the conditions of life in these isles differed widely from those of tropical Polynesia.' Māori were compelled to become agriculturalists; to make drastic cultural changes, devoting more time to the planting, growing and harvesting of food than had been necessary prior to their settling in Aotearoa.

Cultivations generally failed in the colder southern regions, where the hunting of birds and the gathering of shellfish, fern root and berries soon replaced crops. What was grown in the high-lying areas of the South Island was produced for cultural and ceremonial purposes only. In Te Urewera and through the National Park region in the North Island, Tūhoe, Te Arawa and Ngāti Tūwharetoa were similarly compelled to rely upon products of the forest. Small quantities of kūmara were grown, but the crop never properly matured and would not keep.[19]

In more favoured areas, like the Auckland isthmus and certain alluvial valleys, Māori learned to produce crops sufficient to provide for their main food supply. But few parts of Aotearoa were so favoured. Soils in some of the coastal regions, like those around Wellington, were sterile and unsuited to the cultivation of the kūmara.

When James Cook first observed the papakāinga of Queen Charlotte Sound, he noted that, to the north, there were plantations of yams, sweet potatoes and cocoas, but little was seen to the south. He concluded that the inhabitants of the Marlborough area 'must subsist wholly upon fern root and fish, except the scanty and accidental resource which they may find in sea fowl and dogs'. However, during Cook's third voyage, his men were informed by local Māori that, at certain times of the year, they did in fact migrate away to other parts of the country to work their cultivations.

Farming the Soil

Despite these problems, the kūmara and other root crops were more widely cultivated than was first thought. The Nelson district once possessed extensive kūmara gardens, as did Blenheim and the Wairau Valley to the south, though greater care was needed in tending crops down the exposed southern coast. Māori living at Kaiapoi were observed 'devoting much of their time to the cultivation of the kūmara and to the preparation of the kāuru, or cabbage tree stems,' which were gathered, transported and used as barter with Māori in other parts of the South Island. Frost greatly affected kūmara grown this far south. Māori made attempts to regulate the temperature of the soil and to secure adequate drainage by covering the planted ground with fine gravel.

According to John Puahau Rakiraki of Teroro Port Molyneux, some attempts were made much later to cultivate taro in carefully selected and protected areas, like underneath papakāinga outbuildings. The crops were carefully tended but never amounted to anything fit to eat.

Ancient Agriculturalists

When Raka Hautu came to the South Island forty-two generations before 1925, tradition states that he carried his kō or wooden spade with him. That he had carried this all the way from Patu Nui A Aio in his vessel Uruao suggests that his people were intuitive agriculturalists who had every intention of cultivating the new land.

When traversing the countryside, especially the fertile north, early Pākehā travellers noted earthwork defences of huge villages with storage pits for crops and forests cleared for planting. It was evident that vast tracts of country that had lain wild since time out of mind had once been fully cultivated. Ditches for drainage could still be seen alongside food storage pits. Who such sites had belonged to, even Māori could no longer say.

The first plantations James Cook saw were situated between Poverty Bay and Te Māhia. Cook saw Māori 'assembling in great numbers' there; he was able to distinguish cultivated ground that had recently been turned up, with furrows and new plants clearly visible. The first plantation Cook's men actually visited was located just north of Tolaga Bay. In evidence were kūmara, taro and a plant 'of the cucumber kind', which was just appearing above the ground.

The Spread of Plantations

In pre-contact times, plantations varied in size from 1 acre to 10 acres. In one of the East Coast Bays, Cook's men saw 150–200 acres of cultivated ground. Each distinct patch was fenced in, generally with reeds placed close by one another 'so that a mouse could scarcely creep through'. Cook's men observed that the tillage was excellent, presumably owing to the necessity Māori were under to cultivate or run the risk of starving.[20]

Early Pākehā also observed the insecurity of papakāinga from sudden attack: 'I consider it a great proof of the insecurity in which these people live that their grounds are rarely cultivated to any extent in the immediate vicinity of those places where they reside in congregated bodies.' Plantations were generally located at some considerable distance from villages, which were invariably constructed either on the summit or at the foot of some high and almost inaccessible hill.

Feudal enmity among tribes all too often meant that gardens were not sited where the best soils were. Instead, Māori were obliged to choose the best concealed or protected planting places. In 1772, Jules Crozet observed that Māori cultivators had 'no idea of concentrating their industry', instead maintaining any number of disparate yet easily defendable gardens. If only the tribes could be 'brought

In 1772, Jules Crozet observed huge areas set aside for growing crops near papakāinga. He described Māori as expert in 'the art of agriculture'.

Te Ariki Pā, Kāwhia, with an extensive area (left) cleared for arable plantation, before the Tarawera eruption in 1886 that destroyed this village; photograph by the Burton Brothers (Alexander Turnbull Library, PA1-q-153-04)

to live in amity with each other and build their villages on fertile grounds', Crozet wrote, their respective districts would in a short time assume a much more economically vibrant and 'civilised' appearance.

'A Start in the Art of Agriculture'

Jules Crozet visited New Zealand with French explorer Marc-Joseph Marion du Fresne in 1772, observing that Māori had already made a 'start in the art of agriculture'. Crozet noted Māori planting techniques in some detail, also recording a range of cultural practices that accompanied planting and harvesting. All persons took part in the tasks of clearing and preparing the ground for planting and tending and gathering the crop, he wrote: 'chiefs, warriors, commoners, and women – old and young – all assisted in some way'. Only chiefs and freemen, however, were allowed take part in the labours affected by the laws of tapu. It was assuredly a fact, wrote Crozet, that the art of agriculture was one held in high respect by the Māori.

Crozet did observe some limitations; for example, Māori confined their 'whole agriculture to two or three objects'. They had no knowledge of any sort of grain, and 'I saw nothing which might be taken for an orchard and I did not even meet with the least fruit either wild or cultivated'. However, Māori took great trouble to protect crops from strong winds, erecting temporary fences or screens and expending considerable labour in conveying gravel to shore up (and heat) their staple and most important crop, the kūmara. As Elsdon Best later concluded, the amount of patient care and selection required in raising new kūmara varieties ('it was not generally known that more than 50 varieties of the kumara were being cultivated') exemplified the fact that 'Maori were patient, careful and expert agriculturalists.'[21]

Adapting to Changing Agricultural Technologies

The old world of the Māori agriculturalist changed very quickly after large-scale Pākehā settlement in the early 1800s.

As Hazel Petrie has argued, Māori were not slow to adapt to the new technologies introduced by settlers. Once acquired by one tribe or hapū group, the new technologies – methods, ideas, implements, seeds – were quickly disseminated through tribal trading networks, often faster than the movement of Pākehā themselves. Ngāti Awa obtained the Whakatāne district's first potatoes, guns and steel axes from Ngā Puhi in Northland, not from Pākehā. Inland Tūhoe received theirs from Ngāti Awa. Similarly, the people of Gisborne traded imported goods such as blankets, muskets and powder with Te Wairoa who were renowned for their canoes, sails and carvings.[22]

That Māori would quickly adopt European methods of cultivation and agriculture was seen as inevitable by the early missionaries, especially Samuel Marsden. With the arrival of the first ploughs, axes and bullocks, wrote Marsden, came a 'new era of progress' for Māori. But this came at a cost.

During the 1830s and 1840s, as new plantations were developed to meet the settler demand for food, Māori worked them effectively, breaking the land, planting and taking up the crop. Enough of their pre-Pākehā expertise and instincts survived to ensure plentiful crops of wheat, maize and potatoes.

However, Māori agricultural outputs in real terms (that is, in light of an expanding Pākehā industry) soon went into decline. Māori were unable to manage demand for an ever-increasing supply because of ever-increasing costs, the need for new seeds to improve a deteriorating product, a lack of suitable implements to break new ground plus difficulties in transporting produce to market. According to Petrie, in some areas, Māori became disheartened and cultivated less and less every year, leaving large areas of fertile land undeveloped. At this time, many native districts also faced a chronic shortage of provisions, and many faced starvation.

Compounding this situation was the fact that historically, for Māori, food had fulfilled a critical cultural function including the social obligations of hosting. Food represented social status and economic power. As Ann Parsonson has argued, Māori were often compelled by such cultural pressures to alienate their land in the 'pursuit of mana' as well as economic security.[23]

Peace-loving Agriculturalists
Eventually, as Pākehā settlements spread and Pākehā influence increased, sharply divergent expectations of Māori agriculture developed. Settler ideas of 'desirable' agricultural changes appealed to some Māori, but not to others. At the heart of these innovations lay the thorny issue of the status of the land; specifically, how could the land be best farmed and rendered economic, and who should do this – Māori or settler?

Where the land was concerned, in the minds of government, settlers and missionaries alike, property rights and capitalism – economic development based on individual ownership – were closely linked, as indeed were colonisation and Christianity. As Petrie has argued, many settlers shared the vision of English philosophers like John Locke who insisted that land ownership rights should be based on the land being cultivated, or in some way used economically.

Biblical references to planting and ploughs motivated missionaries into believing that training Māori to become good farmers was an essential part of the missionary endeavour. Such training would also secure Māori their lands from a Lockean perspective.

According to Petrie, the missionaries envisaged an ideal Māori society of 'peace loving god-fearing agriculturalists'.

Such aspirations were part of a complex missionary aspiration that also included Māori desisting from taking slaves as cheap labour to work their plantations, much less treating them with wanton brutality. The value of reliability, regularity and hard work, especially working the land, were also in the mix, as indeed were expectations that Māori would desist from waging war. Implicit within the missionaries' vision were Māori rights to their own lands, confirmed by cultivation and farming.[24]

Young Māori men were encouraged to enrol in the mission schools to acquire the full range of agricultural skills such as planting, harvesting and animal husbandry. An early example of a school set up for this purpose was the mission station farm established at Waimate in 1831 by Henry Williams, believed to have been New Zealand's first farm. (Williams' Mission House, which still survives, is the second oldest building still standing in New Zealand.)

An Experimental Māori Farm
Sheep and cattle rearing and breeding had first arrived in New Zealand in 1814 when Samuel Marsden established a mission station some distance east of Waimate, at Rangihoua. At the time, Marsden owned extensive sheep holdings in Sydney, and he was able to ship some of his stock across the Tasman Sea.

In 1831, Henry Williams trekked inland from Paihia and established the mission station and farm at Waimate on land gifted by local Māori. When first built by a local Māori work force using local materials, the

Waimate Mission Station and farm 1845 – 'New Zealand's first farm', where young Māori were trained as 'peace-loving agriculturalists'; drawing by Lieutenant Colonel Cyprian Bridge
(Alexander Turnbull Library, PUBL-0144-1-330)

station comprised three mission houses, a chapel and a school. Waimate was regarded as an 'experimental farm' and was visited by many prominent people, including Charles Darwin, who would write that he had come across 'an English farm house and its well dressed fields, placed there as if by an enchanter's wand'. The station also included dwellings for Māori pupils who worked on the farm, receiving instruction in new farming and agricultural techniques. By such labours, Māori were able to contribute to the economy of the mission stations, which enabled those stations to feed their pupils and in many cases stave off heavy debts.

Differing Māori and Pākehā Perspectives

The training of young Māori as farmers and agriculturalists received the support of New Zealand's early Crown Colony Government, which paid generous subsidies to most mission stations that took training on. But other colonists saw the mission societies as obstructing the aspirations of settlers.

Colonial politicians like William Fox resented the influence of missionaries over Māori, claiming that Māori agricultural endeavours were better free of missionary interference. In centres like Nelson, Fox noted, Māori rates of cultivation had been high

New Zealand's First Commercial Dairy Farmer

New Zealand's first commercial dairy farmer was Rawiri Taiwhanga. Originally one of Hongi Hika's warriors, Taiwhanga began working on the farm of missionary John Butler at the Kerikeri Mission Station in 1821, gaining some experience in the ways of European farmers. He then travelled to Sydney and stayed with Samuel Marsden for eighteen months, learning more about the farming trade before returning to Paihia as a sawyer. In Paihia, he developed a garden producing corn, potatoes, cucumber and pumpkins. He also had an acre of wheat, making himself a plough and purchasing cattle and sheep. Taiwhanga is thought to have been New Zealand's first commercial dairy farmer because he was selling butter to sailors in 1838 at the rate of 8 lbs a week.

Source: Petrie, *Chiefs of Industry*, p. 6.

– 770 acres per 615 persons – when compared with those of Pākehā – 3465 acres per 2867 people. Māori agriculture had benefitted because Māori were able to enter into trade and commercial relationships with settlers, unhindered by missionaries. Fox was critical of the missionaries for encouraging Māori to hold onto their lands and render their lands economically viable by planting cultivations or learning the new techniques of raising stock.

Fox asserted that it was not true that colonists coveted the lands of Māori and were determined to take it by any means possible. He maintained that every acre currently being farmed by Europeans had in fact been purchased at prices 'equal to any value the land would have had' if colonists had not come to New Zealand, and improved the land 'by our capital and our own labour'. According to Hazel Petrie, Fox's view reconciled the philosophies of Christianity, political economy, systematic emigration and the virtues of intensive land use, all within 'an ideology that endorsed the righteousness of all four'.[25]

Māori were endeavouring to establish a viable agriculture in the wake of large-scale settlement and were facing multiple challenges from Pākehā. Christian beliefs about the sanctity of Māori working the soil went against prevailing colonist dreams (as promoted by Fox) of an efficient farming economy led by Pākehā. Against this background, immigration and land values were rapidly escalating.

To most Pākehā, Fox was making a valid point – the country could not economically prosper unless settlers acquired Māori land and farmed it themselves. Britain had insisted that the New Zealand colony was to be self-funding, but its financial viability was in question. It was expected that by 1843 the proceeds of land sales to new settlers by the government would be high enough to comprise the overwhelming portion of colony revenues. However, the reverse occurred; income from land sales after 1842 dropped very quickly, forcing Governor William Hobson to 'bequeath an insolvent ministry' to Governor Robert FitzRoy, his successor.

In 1845, new Governor George Grey strongly supported efforts to render Māori agriculture economic and profitable. Grey looked closely at the colony's slumping export figures and discerned a growing Māori preference for the much expanded internal market. Wheat and farm produce comprised the most profitable venture, and this was to be encouraged, thought Grey. Farming at this time was expanding rapidly, especially in Hawke's Bay, Wairarapa and the South Island. Sheep and cattle were being imported in significant numbers from Australia, though government regulations and restrictions soon became necessary because of the inadvertent introduction of stock diseases like foot rot and catarrh.

Grey noted that Māori were developing considerable commercial interests and would soon be substantial contributors to colonial revenues. Prudent expenditure on roads and infrastructure, he said, would ensure that agriculture expanded and that the colony could become self-sustaining. Māori were cultivating huge amounts of wheat, maize, flax, potatoes, peaches and other produce. But because of the condition of the roads, most Māori were hindered in conveying their produce to market.

Land Loss and Māori Economic Decline

Grey's support for Māori agricultural development in the mid-1840s was genuine enough; drawing Māori into the mainstream as substantial economic contributors made good financial sense. However, somewhat paradoxically, during the late 1840s, Grey and his chief land purchase commissioner, Donald McLean, also pursued an aggressive land purchase programme.

In Hawke's Bay at this time, colonists were leasing extensive blocks of land from Māori for sheep grazing. Such leases, however, were illegal, having been prohibited in 1846 by Grey's Native Land Purchase Ordinance. Grey eventually turned his attention to such aggregating leases, determined to purchase them outright before their true value could be realised by owner chiefs. Huge blocks of such lands were subsequently acquired in Hawke's Bay and Wairarapa.

In 1850, land purchases were also negotiated in Taranaki, Whanganui and the Rangitīkei. Almost 30 million acres of the South Island were also purchased during Grey's tenure for a total sum of about £8000, extending from the sale of Ōtākou in 1844 to the sale of Murihiku in 1853. Towards the end of his governorship, Grey also exerted intense pressure on Ngāti Kahungunu to sell extensive parcels of land in south Wairarapa. Fragments of these lands that remained in Māori hands were eventually consolidated under new incorporations like Parininihi ki Waitōtara (Taranaki), Ātihau-Whanganui Incorporation (Whanganui) and Wairarapa Moana Incorporation (south Wairarapa).

Such land losses inflicted significant damage upon the Māori economy. Continuing losses after Grey's departure in 1853 exacerbated the rapid erosion of any

In 1855, flour, potato and maize prices fell quickly, exacerbating issues already facing Māori arable farmers like problem soils, inappropriate climate, deficient seeds and crop disease.

Young Māori girl at Te Ariki Pā in front of whare and small family vegetable garden, c. 1880s; photograph by the Burton Brothers (Alexander Turnbull Library, 1/2-004619-F)

chance Māori may have had of attaining a strong if not dominant economic position in arable agriculture.[26]

The Decline in Māori Agriculture

Other factors besides land loss contributed to the rapid decline in Māori agriculture. Throughout the 1830s and 1840s, at the behest of the missionaries, many Māori had concentrated their economic efforts into a few limited areas, especially wheat-growing, which led to a narrowing of the Māori economic base.

By the mid-1850s, a number of negative factors were impacting on wheat-growing, such as the advent of steam shipping. Given the amounts of capital required, this soon led to the growth of private steamship companies. The substantial and speedy steamships quickly rendered small Māori sailing craft for transporting produce obsolete. By the 1870s, steamships were carrying as much as 80 percent of New Zealand's coastal tonnage, posing difficulties for Māori requiring long-distance haulage to markets. A sharp drop in demand for Māori agricultural products followed.

In 1855, there was a sudden collapse in the wheat and flour market, caused by the end of the Victorian gold era – strong competition suddenly emerged from Tasmanian and Chilean flour producers. A collapse in potato and maize prices followed. Market forces were now revealing to Māori agriculturalists the dangers of concentrating economic activity in too few areas, especially as long-term factors like problem soils and climate began to exert their influence. At the same time issues were arising as to the quality of Māori wheat with some Māori producers unable to eliminate 'humpback wheat' or fungal disease from their product. Consumers started to see Māori seeds as deficient – wheat harvests were increasingly demonstrating the effects of soil exhaustion and over-cropping.[27]

One other factor was of course the advent of war, with Māori pitted against the government during the

1840s and the 1860s. In some areas, the destruction wrought by the fighting was immense.

The invasion of Rangiaowhia on 21 February 1864 was but one example. The British Army approached the village during the night and attacked at dawn, galloping past 'a scene of peace and beauty ... fields of wheat, maize and potatoes extended over long gentle slopes and peach groves shading clusters of thatched houses'.

The British Army approached the village through the Pekapeka Rau Valley and along the southern rim of the Rangiaowhia Basin, where the lagoon that supplied the power for local flour mills was situated. After sacking the village, the soldiers put the 'breadbasket of the Waikato' to the torch. Few native districts were left unaffected by the wars, and some areas, like Taranaki and the Waikato, faced a near-total economic collapse.[28]

The attack on Rangiaowhia in 1864 constituted the penultimate British Army action against the forces of the Māori King Tawhiao, who had long since taken refuge deep in the King Country. Four years earlier, the Māori King had asked Wi Tako Ngatata of Wellington to investigate the reasons behind the outbreak of the war at Waitara in 1860. As Wi Tako reported in a memorandum, which also went to Te Wetini of Hangatiki, Te Heu Heu of Taupō, Wiremu Tamihana of Tamahere and Rewi Maniapoto of Ngāruawāhia, 'you have asked me to investigate and send you truth which is this; Friends, Listen to me! The cause of this war is the land!'[29]

Post-war Developments

By the mid-1860s, most of the South Island had been acquired by the Crown along with significant land blocks in the lower North Island and Northland. The government was making moves to standardise agriculture; for example, it introduced sheep and cattle regulations in the 1860s in order to monitor the quality of stock being imported from Australia. In 1861, the Provincial Government of Canterbury banned the importation of cattle from areas with a history of disease. Otago did not follow suit and suffered huge losses as a consequence.

Fencing regulations were also standardised in the 1860s having first been introduced by Ordinance in 1847. A national transport infrastructure began to take shape, mainly in the form of small but important provincial railway connections between important primary producing centres. A large-scale plan for a national railways grid came a decade later, funded by massive government loans raised on the London market. Because of the war in the Waikato, the main trunk railway line was not commenced in Auckland until 1873, and it was not until April 1885 that Ngāti Maniapoto gave their consent for the railway to extend south of Te Awamutu. By 1908, the 681-kilometre railway finally joined Auckland to Wellington.

The Land Conundrum

The Treaty of Waitangi represented one of the earliest steps taken by government to acquire Māori land.

According to Evelyn Stokes, Māori were never fully informed in 1840 as to the radical nature of the land tenure reform facing them if they signed the Treaty. What was never adequately explained, Stokes writes, was the fact that 'the Crown by the Treaty assumed title over all of the lands of the country'. This assertion of Crown ownership represented a 'tenurial revolution' that was not fully understood by Māori who continued to live under the Māori customary law system, never suspecting the enormity of what they had signed away to the Crown. 'In this legal fiction,' Stokes argues, 'all land titles were vested in the Crown, though subject to the rights of Maori to alienate (or not) as guaranteed under Section Two of the Treaty.'[30]

The Crown retained the sole right to negotiate alienations with Māori in order to protect its own interests, not those of Māori, and most significantly to extinguish native title to land. By 1856, when Parliament passed into the hands of the settlers, politicians were determined to enact these extinguishing provisions in the face of Māori resistance to land sales. Conflicts between tribes and settlers over disputed titles were increasing.

The solution from the government's point of view was simple – Māori land titles needed to be given a status recognisable under English law. The government attempted to pass legislation to make this happen: in 1858, it introduced the Native Territorial Rights Bill, which set out to convert Māori land titles. However, because the proposed conversion process also involved the extinguishing of Māori land titles as they then existed, the British government refused to sanction the measure.

In 1862, with the war at Waitara now over, it was time to try again. At that time, the country still effectively lay in Māori hands. Māori still possessed almost 80 percent of the country – 23.2 million acres – and were consistently refusing to sell.

Parliament now had the political will to push through legislation that would extinguish Māori land titles and expedite land sales to settlers. The answer was to establish a new court to enforce these new legislative provisions.

'The Land-taking Court'

In 1862, the Native Land Court was established to 'promote the peaceful settlement of the country' by determining Māori ownership of specific lands and then 'assimilating those rights into British law'.

The actual court structure was not decided until 1865, however, when a whole new legal framework was introduced that made the extinguishing of customary tenure possible. The extinguishing process would begin with an 'investigation of title hearing' before a Pākehā judge and Māori assessor. All evidence deemed relevant by the Court would be introduced, and a decision made as to which of the Māori applicants possessed which rights to the land in question. The Court imposed the idea of individual ownership and insisted that, among a group of Māori owners, all owners had equal rights – thus undermining centuries of functioning political and social structures.

The impact of the Native Land Court on Māori agricultural development can barely be measured. In 1867, the Court was labelled 'Te Kooti tango whenua – the land-taking court' by a Crown official named Reginald Biggs who was anxious to encourage East Coast Māori to voluntarily vest large blocks of land in the Court. In reality, as Biggs was aware, the Court already possessed the powers to strip huge parcels of land from unsuspecting Māori owners.[31]

Customary Māori ownership of land was thus disregarded by the Court. Māori who had long since moved away from holdings were now assigned 'absentee' rights, and migrant Māori communities were given rights to stay where they were, even if they were squatting on the lands of others. The adversarial court system invited contention between parties, which produced a 'morass in which Maori floundered for decades, frittering away their estates in ruinous expenses for often negligible reward'.

An investigatory commission into the state of Native Land Laws in 1891, of which rising Māori politician James Carroll was a member, strongly criticised the 'pernicious consequences of Native Land Legislation' that had led to decades of disputes and litigation and thousands of petitions to the government. By 1891, the legislation had become so complex as to be unworkable; seventeen statutes regulating the ownership and use of Māori land had been passed in the previous two years alone. According to the commission's chairman, William Rees, there was 'confusion in law and practice', creating a state of near anarchy.[32]

The Fate of the Tenurial Revolution

After the 1870s, Māori participation in the now rapidly expanding farming economy of New Zealand continued its sharp decline. Māori agricultural economies were still trying to navigate the shift from the quasi-subsistence days of the early 1800s to the modern market economies of the 1870s and the radically reformed patterns of land ownership imposed by new laws.

Māori access to resources to underpin development was poor, as the Crown acquired the best and most productive land. A major lack of finance also hampered Māori farmers; the proceeds of selling and leasing land fell well short of the funding levels required to provide for capital reinvestment in agricultural development. High surveying and legal costs also added to Māori difficulties, and access to government funding schemes was severely limited. Though they had been granted a limited franchise in 1867, Māori were also effectively losing political power. Stokes suggests that Māori land was never successfully and completely 'commodified' – instead it was subject to partial commodification, imposing on Māori the 'worst of all possible worlds, stranding them somewhere between true customary tenure and true private property rights'.

For Māori, land became valueless as a security and, although freely alienable, it was seldom able to reach its full market value because it was artificially depressed by legal complexities. Māori freedom to sell their interests in a free market was illusory; the government always dominated, regulating and manipulating the land market to coerce sales and drive down prices. Māori land sales to the government bore no relation to the free market concept of sale. Māori were deprived of the market value of their land as security by the legal changes made to their land titles. They were losing their land base and only fractionally integrated into the capitalist economy.[33]

The most successful Māori sheep farms after the 1860s were in the Waiapu area, where Ngāti Porou were farming 20,000 sheep in 1873.

Māori man with a bullock team pulling a wagon of wool bales, 1900s; photograph by William Williams (Alexander Turnbull Library, 1/4-055489-G)

Financing Māori Agriculture

The most rapidly developing part of the New Zealand economy after 1870 was dairying, participation in which required access to credit, considerable technical and management expertise and a close interconnectedness with vets, dairy factories, stock and station agents and banks. Māori largely lacked access to these resources, and the government made little effort to provide them.

The Government Advances to Settlers Act of 1894 created a system of cheap credit available to all farmers. It established a special office empowered to advance monies as a first mortgage. By 1901, the government had loaned £2,679,520 to settlers under this legislation. In the same period, it paid £1,010,140 – under half that amount – in order to acquire 3.2 million acres of Māori land. According to Richard Boast, settlers therefore had access to more money by way of loans than Māori did by way of sales.

Māori were effectively excluded from the Advances to Settlers scheme because of the categories of land available for assistance, as defined by section 25 of the Act. To be eligible, Māori had to first obtain a Land Transfer Act certificate of title to their land, which was always a daunting prospect. Multiply owned lands were ineligible for advances because the land was not seen as good security. Moreover, where a block had hundreds of owners, applying for loan finance was logistically difficult.

Wiremu Pere and Māori Land Reform

Wi Pere (1837–1915), prior to his departure to England in 1889; photograph by P J Gordon
(Alexander Turnbull Library, 1/2-034936-F)

One of the first Māori leaders to grapple with the difficult issue of consolidating and incorporating Māori land titles to promote economic development was Wiremu Pere.

Pere was of Te Aitanga-a-Māhaki, Te Whānau a Kai and Rongowhakaata descent and was born in Gisborne in 1837. He grew up with a strong sense of economic land use.

In the 1870s, Pere supported the Ngāti Kauhanganui Repudiation Movement founded by Henare Matua and Karaitiana Takamoana, which sought a reversal of the fraudulent purchases of huge Hawke's Bay pastoral estates, made possible through iniquitous Native Lands legislation. Pere also opposed the Native Land Court's practice of granting land titles to individuals, believing that customary lands should remain in communal possession.

One of Pere's early reform initiatives was his attempt to simplify Native Lands laws, then mired in paralytic complexity, especially where Māori multiple ownership was concerned. From 1880, he worked with liberal lawyer W L Rees to persuade East Coast Māori to consolidate their lands into a protective trust. However, the scheme failed because of economic depression and hostility from politicians.

Wi Pere remained concerned that Māori keep their lands and farm it themselves. He was elected to Parliament as MP for Eastern Maori in 1884, speaking strongly against the Native Land Court and its practice of dealing with individual Māori rather than tribal groups, instead advocating that Māori communities maintain control over their lands through elected 'block committees'.

In 1887, Pere was defeated by James Carroll but re-entered Parliament in 1894, once again criticising the government's Māori land policies. He called for a boycott of the Native Land Court and an end to land sales and leasing, and continued to press for greater Māori control over land. Pere lost his seat to the young Apirana Ngata in 1905. In 1907, he was appointed to the Legislative Council, where he remained until 1912. He died in 1915.

Sources: Walker, *He Tipua*, pp. 106–108; Ministry for Culture and Heritage, 'Wiremu Pere', www.nzhistory.net.nz

Leasing Māori Land

The urgent need for government finance to fund Māori development work was gradually realised. The 1900 Māori Land Administration Act did make some monies available for Māori land development, empowering the Minister of Lands to lend up to £10,000 per year to any Maori Land Council established under the Act on the same terms as those set by the 1894 Government Advances to Settlers Act. However, instead of providing finance or public credit for Māori land development, remedying the dire shortage of capital, the councils were only allowed to on-lend money to Māori for title-related expenses incurred in the six preceding years.

Māori development finance never really became available until the government established the Native Trust Office in 1920. The Trust Office's primary function was to aggregate all Māori monies held by the government into a single development fund. Most of those funds would not be unlocked until new Māori lands legislation was passed in 1929. In the 1900s, land purchasing was considered much more important than enabling Māori to retain and develop their land.[34]

Throughout the early 1900s, a great deal of Māori land was leased by Pākehā farmers. Unable to finance the agricultural development of their lands themselves, many Māori chose to lease their lands instead. However, politicians and the press lamented the high acreage of Māori land under lease; it was often claimed that Māori were idly getting rich on the backs of hard-working farmers. Carroll's advice to Māori at this time, that they desist from selling their lands, was subject to bitter criticism from Pākehā politicians who accused the government of 'locking up Maori land, holding back European settlement and thereby jeopardising the prosperity of the colony'.[35]

But the returns Māori received from such rentals were meagre. The fragmented nature of Māori land titles meant that those with shares in multiple blocks could generally only expect small payments scattered across an indeterminate period. In this way, Māori

The Mangatū Incorporation – 'The First and Still Surviving'

One of the earliest incorporations established to protect Māori land from 'predatory Crown purchasing' was the Mangatū Incorporation, located near Gisborne. Its origin was a decision taken in 1881 by Wiremu Pere to file a claim for right of ownership of the land with the Native Land Court in the names of the Wahia and Ngāriki hapū. The original Mangatū lands consisted of 65,026 hectares partitioned into six blocks. In 1881, the Native Land Court granted ownership of the Mangatū lands to twelve individuals who were to hold the land in trust.

Despite the 'years of difficulty' that followed, the Mangatū owners managed the land 'with devotion and diligence', seeking to turn tribal fragmentation and leasehold management into legal consolidation and economic prosperity. The Mangatū Incorporation was established to represent those beneficially entitled to the block and to resist pressures to sell.

In 1917, as a result of financial difficulties, the Mangatū lands were placed under the control of the East Coast Commissioner. A year later, the Whānau a Taupara people were added to the list of owners. In 1947, after twenty-nine years of commissioner control, the lands were handed back to the Mangatū Blocks Committee of Management, which became the legal administrator of the incorporation.

Today the Mangatū Incorporation has 5000 owners and manages resources in the important agribusiness, forestry and viticulture economic sectors. Under its subsidiary company, Integrated Foods Ltd (IFL), Mangatū has created a vertically integrated sheep and lamb business selling its products domestically and exporting to international markets.

Mangatū is today 'committed to the sustainable management of assets in order to become a key player in the economic outcomes of the Te Tairāwhiti region, contributing to the prosperity of our people without compromising our roots and identity'.

Source: Walker, He Tipua, pp. 106–108; www.mangatu.co.nz

had no chance of accumulating capital for their own agricultural developments.

Some Māori did farm their own blocks, particularly in the Wairarapa, Hawke's Bay and the East Coast. But renting was often the best option because of the capital costs involved in setting up farms with stock, plant, equipment, housing, grass seed, fertiliser and tools. Sheep farming required a substantial amount of capital with dairying and cattle farming requiring even more. Farming was only possible for Māori if they saved and invested their rents and additional monies obtained by labouring, bush-felling or shearing.[36]

Māori Involvement in Dairying

Māori involvement in dairying occurred much later than their participation in the sheep and cattle industry; large-scale confiscations of primary dairying land in Taranaki and the Waikato were a major reason for this.

The New Zealand dairy industry itself had been a late starter; the earliest dairy factory was not opened until 1882 in Te Awamutu. A second factory was established in Lepperton the following year. Dairy farming thereafter struggled to establish a dominant economic foothold; lands suitable for dairying often needed to be burned out and broken in, a long and painstaking process. In 1890, the South Island Dairy Association was established; the North Island equivalent was set up in Hāwera in 1894.

The percentage of national export income from dairying rose steeply, from about 13 percent in 1900 to 20 percent by 1914, with dairy farming continuing to earn good returns into the 1920s. A unique feature of dairying was the establishment of farmer-owned co-operatives, most of which owned and operated a small factory. Dairying was more complex than the wool industry; it required expertise in management, marketing, transport, shipping of perishable products and the local manufacture of butter and cheese.

Dairying also required ancillary specialists like advisers, agents and suppliers. Dairying could spread its wealth more broadly than sheep farming; the large number of small towns over the Waikato and Taranaki wool-growing areas did not spawn such an infrastructure.

Māori participation in dairying was initially limited. In addition to the factors already discussed, dairying's demand for intensive round-the-clock labour may not have appealed to Māori families unwilling to sever wider and time-consuming kinship responsibilities. Māori participation in dairy co-operatives was also marginal, at least until the 1930s. Māori landowners had to resist continuing government attempts to acquire their land for dairying for Pākehā farmers, especially after 1911 when the Reform Government of William Massey, supported by the Farmers Union, set about aggressively purchasing lands.[37]

Some Māori families did surmount these difficulties and became successful dairy farmers before 1932. Kurupō Tareha owned a dairy farm near Taradale in 1908. In 1907, the Opouriao Dairy Company established a cheese factory in Rūātoki, which was substantially a Māori village. The Opouriao dairy factory is thought to have been the first to have the bulk of its milk supplied by Māori farmers. In 1920, a butter factory was established on the southern shore of Lake Taupō; the suppliers were mainly Māori. In 1925, two dairy factories were established in the Māori farming areas of Te Kaha and Ruatōria. The Ruatōria factory in particular was an economic success, renowned for the quality of its butter.

The Ruatōria dairy farms were established following an initiative by Apirana Ngata, who persuaded local Māori to switch from sheep to dairy, financing the scheme with a loan from the Māori Trustee in 1924. One other successful scheme was established in the far north. It was known as the Te Aupōuri scheme and was established with the assistance of Judge Acheson of the Native Land Court. The scheme involved title improvements as well as the establishment of dairy farms and a cream truck service between Te Kao and Kaitāia. Māori also provided butterfat to the dairy factory at Te Kūiti.

These dairy factories were all successful, by and large, but it is significant that they did not develop until the 1920s and needed finance from outside the districts to get under way.

A profitable and innovative industry allowing for rapid capital formation therefore passed Māori by until a comparatively late stage in the industry's development, despite high concentrations of Māori in rural Taranaki and Waikato. Had the land confiscations of the 1860s not occurred, Māori in those affected regions might have been better placed to take advantage of the dairying boom. As it was, Māori dairy farming did not really take off until the time of Ngata's land development schemes after 1929.[38]

As with so many aspects of the development of Māori agriculture and farming, Ngata's land development schemes were the key.

Sir Apirana Ngata and the Land

Sir Apirana Turupa Ngata (1874–1950) while he was the Member for Eastern Māori, 1914; photograph by Stanley Andrew
(Alexander Turnbull Library, 1/1-014489-G)

Māori land reform, protection of economic assets and agricultural development were the hallmarks of Māoridom's most significant political figure of the early twentieth century, Sir Apirana Ngata.

Apirana Turupa Ngata was born at Te Araroa on 3 July 1874 with affiliations to Te Whānau-a-Te Ao, Ngāti Rangi, Te Whānau-a-Karuai and Ngāti Rākairoa of Ngāti Porou.

Ngata's mother was Katerina Naki, the daughter of a Scots trader named Abel Knox. His father was Paratene Ngata, a storekeeper who also managed the Waiomatatini sheep station. Paratene was trained in agriculture by Pākehā farmers and went on to provide prodigious leadership to Ngāti Porou in their efforts to establish economic security through self-managed farming. Paratene also advocated land reform, favouring autonomous tribal committees unfettered by the Native Land Court. His achievements in agriculture exerted an undoubted influence on the young Apirana.

Ngata was educated at the Waiomatatini Native School and Te Aute College, receiving instruction in mathematics, English literature and classics. In the early 1890s, Ngata and colleagues travelled the length of the East Coast, visiting papakāinga and urging Māori to adopt a range of sanitary practices like boiling their water and washing their blankets. After eight years at Te Aute, Ngata was awarded a Te Makarini Scholarship to study arts at Canterbury College; he later transferred to Auckland to study law.

From an early stage, Ngata was determined to effect Māori land development and title reform as a response to a nineteenth century characterised by severe land loss and communal disruption. Fortunately, Ngāti Porou still possessed most of their tribal lands, which were now being used for running sheep. In 1916, Ngāti Porou had 156 flocks and a total of 180,919 sheep, and they were investing heavily in pasture improvement, buildings, equipment and mechanical shearing machines. Ngata advocated a system of incorporations that kept land titles in tribal ownership while allowing properties to be developed as viable units.

Ngata also advocated the consolidation of individual tribal fragments of land into aggregated blocks, beginning with the Waipiro block in 1911. In 1912, he founded the Waiapu Farmers' Co-operative Company, which was owned and managed by Ngāti Porou farmers. Ngata also assisted Native Minister

James Carroll with the drafting of the Maori Lands Administration Act 1900, which enabled Māori to establish land boards to administer the sale or lease of their land.

In 1905, Ngata won the Eastern Māori seat, which he held until 1943. He was 'a superb parliamentarian', quickly becoming Carroll's indispensible protégé. In 1907, he was appointed to the Stout-Ngata Commission, which conducted a detailed enquiry into the economic value of remaining Māori lands 'not currently being developed'. Tribes with ample lands were encouraged to sell or lease lands for further development; Ngata's advice to them was effectively 'use it or lose it'. Ngata also assisted in drafting the Native Land Act 1909, which was a massive consolidating measure.

After the war of 1914–1918, Ngata agitated for land to be made available to returned Māori servicemen in recognition of the Māori who had died overseas. Because of his close friendship with Native Minister Gordon Coates after 1921, Ngata was able to establish bodies like the 1924 Maori Purposes Fund Control Board, which collected and administered funds from unclaimed Māori land rentals. Ngata also promoted Māori arts, music, composition and sport.

At Waiomatatini, Ngata arranged a subdivision of tribal holdings in the Waiapu Valley, enabling Ngāti Porou to move into dairying. Progress was impressive, following the introduction of graded cows, new milking machines and a co-operative dairy factory at Ruatōria.

In December 1928, following his appointment as Native Minister, Ngata pressed ahead with his land development schemes. Māori land was cleared of bush or scrub, ploughed, grassed, fenced and stocked, then subdivided into individual farms. In 1933, Ngata co-launched the Ahuwhenua Trophy competition, to reward Māori famers working the difficult land blocks.

Ngata's vision for land reform and development emphasised the need to work through traditional chiefly organisations, turning old tribal suspicions into friendly rivalries focused on land and farming developments. Ngata saw his land development schemes not as an end in themselves but as a catalyst for economic and social regeneration. Land reforms also turned the emphasis away from a succession of individuals – a legacy of the Native Land Court – to the development of strong and influential collectives that might take a lead in the fast-growing Māori economic sector.

Influenced by his father's enormous contributions to the expansion of agriculture on the East Coast, Ngata saw government funding as the key to Māori recovering and developing land blocks in order to re-establish a critical economic base that had been, like the land itself, stripped away during the nineteenth century. Getting Māori themselves involved in developing the land was the key – not only on the land as farmers but also in governance and management of agricultural assets driving a strong economic sector. These principles remain important imperatives of Māori farming today.

Sources: Walker, *He Tipua*; Sorrenson, 'Ngata, Apirana Turupa', www.teara.govt.nz

Ngata's Reforms

Throughout the 1920s, the Crown's acquisition of Māori land continued apace. Extra holdings were needed for the rehabilitation of ex-servicemen, Māori and Pākehā, after 1918. Farm balloting was instituted after the war, but was not applied equally to Māori and Pākehā ex-soldiers. Accordingly, the Maori Patriotic Committee was set up in 1917 by Māori leaders who did not trust government intentions. However, it had insufficient funds to help Māori returnees to find farms; instead, a number of stations were purchased, the incomes of which could be used for relief purposes.[39]

In 1918, Māori still retained about 1 million acres of their lands, having lost 62 million acres to Pākehā over the previous century. A fear emerged that if land purchasing were to continue, Māori would face total landlessness. After the war, the government conceded that Māori should be actively encouraged to improve productivity on their remaining lands, though land title reform would be necessary if this were to happen. Most New Zealanders had been impressed with the tremendous sacrifice of the 2227 Māori who had served overseas in the Pioneer Battalion; 336 of their number had been killed and 734 wounded.

Apirana Ngata urged the farming families of Ngāti Porou to consolidate their fragmented land holdings; this was a painstaking process that signalled Māori determination to develop their 'uneconomic lands' themselves.

Māori woolshed at Tokomaru, East Coast, early 1900s; photograph by the Oates Brothers
(Alexander Turnbull Library, 1/2-001243-G)

In the early 1920s, Ngata argued for the government to finance Māori sufficiently to develop their land blocks to commence large-scale farming. Ngata also insisted that Māori be empowered to retain communal ownership where the organisation of their land was concerned.

Ngāti Porou Forerunners

Since his election to Parliament in 1905, Ngata had prevailed upon Ngāti Porou farmers to establish incorporations to be tasked with developing large areas of multiply owned land. Such incorporations would essentially act as private companies, bringing together the fragments of Māori-owned land scattered over a wider regional landscape. Incorporations would be managed by committees elected by owners.

Incorporations would represent their Māori owners in all farming matters, including the selling and leasing of land blocks. However, Ngata was insistent that Māori should not sell their land; they could lease it, but only those areas that they could not farm themselves.

In 1911, Ngāti Porou ventured into the consolidating of individual shares, painstakingly amalgamating various fragmented blocks. Such lands were consolidated as a single economic unit, once again allowing agricultural development to proceed. Pioneering land tenure reform like this helped revive the economy of the East Coast – Ngāti Porou were seen as the most successful farming iwi. Such consolidations of shares also signalled a greater Māori willingness to develop their uneconomic lands themselves, thus easing government concern over lands lying idle.

The government supported Ngata's consolidations though it was slow to recognise his determination that Māori should be able to farm the consolidated blocks themselves. It did concede that the schemes would contribute to the national economy if managed well and that Māori could be thereby brought into the

In 1921, the new Native Trust Office was established for the purpose of investing Māori monies in land developments and Māori farming. Maori Land Boards could now provide mortgages and underwrite land development work.

Young Maori woman with plough, and man spreading seeds, with family, c. 1900; photograph by William Henry Thomas Partington (Alexander Turnbull Library, 1/1-003133)

modern farming economy. In fact, the Ngāti Porou experiments had encouraged the government to sponsor large-scale consolidations in other areas, at first in the Urewera, and thereafter elsewhere.

In 1922, Ngata issued a series of 'consolidation and incorporation guidelines' that eventually became the 1923 Native Land Amendment and Native Land Claims Act. This important measure aggregated existing legislation dealing with land consolidations and authorised the Minister of Native Affairs to approve projected schemes. In 1924, land consolidations became a part of government policy.

From 1926, consolidation schemes extended all the way to Northland. Some Māori opposed the consolidation policy, arguing that centralised consolidations 'trampled on the right of individuals and collectives'. Others supported the policy, believing that the new family-based farming units would foster productivity gains and capital formation.[40]

State-assisted Land Developments

Not all Māori land problems were solved by the consolidations. Fragmentation of Māori land would continue, it was argued, unless the damaging 'succession policy' of the Native Land Court was addressed. Also, the consolidations did not guarantee Māori access to much-needed development finance. Lending institutions were still shy of Māori as clients when their holdings remained communally owned and highly fragmented.[41]

For his part, Ngata argued for Māori access to development capital, finance and expertise. In their absence, Māori had turned to lending agencies like the Public Trustee, which acted as sole administrator of Māori reserve lands on behalf of beneficial owners. Such lands had been leased out to Pākehā farmers, in some cases in perpetuity. Māori began to resent the Public Trustee's exercise of this power over their own lands, arguing, as Ngata did, that they should now be allowed to administer these lands themselves.

In 1920, the government decided that Māori assets should be managed by a dedicated Native Trustee. In 1921, a Native Trust Office was established as part of the Native Department, though it operated autonomously. The Native Trustee would manage the Māori monies, ensuring that they were responsibly invested. Accordingly, the first Native Trustee, Judge W E Rawson, approved the investment of loan monies to Maori Land Boards engaged in land development and farming operations. From 1922, Maori Land Boards themselves were allowed to provide mortgages to Māori or their lessees. From 1926, Land Boards could also underwrite land development work with Ngata's approval.[42]

Land Development Schemes

Throughout the 1920s, Ngata attempted to persuade the government to finance large-scale Māori land developments. He argued that such development schemes would greatly benefit the national economic good. He also pointed out that, once their land was developed and rendered economically viable, Māori could make good on decades of unpaid rates and other local authority levies – an argument that local authorities strongly supported.

In 1927, Apirana Ngata was knighted. At that time, he was also the chairman of the Native Affairs Committee of Parliament, and presided over the Native Lands Consolidation Commission, which sought to persuade Māori that the state's involvement would not be excessive and state-financed developments and consolidations were worth pursuing.

In December 1928, when the government changed hands, Ngata was appointed Native Minister and also received a high Cabinet ranking. Up until that time, the pace of Māori land development had been frustratingly slow, undermined by an uncertain economy. Ngata sought to fast-track Māori land developments on a national scale and was able to forge ahead despite the onset of the depression in 1929, largely because of Pākehā fears that Māori might become destitute and end up an enormous burden on the state.

In 1929, Ngata introduced the Native Land Amendment and Native Land Claims Adjustment Act, which offered extensive state assistance for Māori farming developments. Under the scheme of the Act, committees of management would be formed from within a consolidation, representing the owners.

Application would be made for financial assistance on their behalf to develop a particular block within the consolidation. The Minister of Native Affairs would then consult with Crown Advisory Committees and make a decision on funding.[43]

Once a consolidation had been approved for finance, and the land had been gazetted for a scheme, the land could not be leased or sold without the permission of the Minister of Native Affairs – Ngata himself. Further extensions of Ngata's powers came in 1930 and 1931. One key aspect of his ministerial intervention was his call to tribal leaders to become hands-on in the management of the schemes as foremen or supervisors.

The earliest of the land development schemes fared well. By 1931, there were forty-one schemes operating with Ngata at the helm. Finance could now be approved for schemes that had not yet been consolidated. Ngata had wanted development to follow consolidation but this had proved to be too slow. Massive state managerial control was the price of a speedier approach.

Land destined for development would be placed in the hands of the Native Department and officials; owners would temporarily surrender proprietary rights. Scheme supervisors were required to deal directly with the Native Minister who also appointed members of the advisory committees. While owners could nominate the farmers who would move onto the blocks, the Department performed the final selection of 'competent' Māori to farm the development lands in order to protect its investment.

As the schemes proceeded, farmers who moved onto them were able to gain a degree of protected legal status, ultimately at the expense of the collective owners, contrary to initial Crown assurances that, once debts had been repaid, land and stock would revert fully to the Māori owners' collective control.[44]

Te Ahuwhenua Competition

From the beginning, as we will see in the next chapters, the Ahuwhenua competition focused on small farmers working lands situated within one of Apirana Ngata's land development schemes (specifically, those lands that came under section 522 of the Native Land Act 1931).

Over the years, the Ahuwhenua focus on small farmers remained in place despite the growth of whānau trusts and incorporations as contributors to the Māori agricultural economy. This focus on small farms reflected departmental policies of facilitating

Māori economic security based on land development properties and farms. However, this focus turned away from the land development blocks to include Māori farmers at large. Instructions emanating from the Under Secretary of Native Affairs, or (after 1949) the Secretary of Māori Affairs, continued to focus the attention of supervisory staff on small farmers. From official memoranda, and from the judges' reports, this focus was not intended to exclude whānau trusts or incorporations; rather, the focus on small farms was seen as an affirmative policy designed to encourage Māori farmers to persist in their difficult farming endeavours.

In the early 1930s, when the Ahuwhenua Trophy competition was launched, most of the land development schemes were just getting under way, posing enormous challenges for collective owners and individual farmers. With most of the country's prime real estate now in Pākehā hands, Māori were presented with a prodigious set of obstacles – as well as opportunities – in 'bringing in' some of the most inhospitable landscapes possible.

However, guided by Apirana Ngata's singular determination that Māori re-establish a strong economic base centred on the land, and by Lord Bledisloe's desire that Māori farmers be supported in these endeavours, the Ahuwhenua Trophy presented an important benchmark for Māori farmers to strive for in clearing, burning, ploughing, planting, watering, fertilising, cultivating, subdividing, fencing and stocking those lands in pursuit of economic and social sustainability.

(John Cowpland – Alphapix)

Chapter Two

The Early Years of the Ahuwhenua Trophy – 1933–1939

1933

A 'Spirit of Friendly Rivalry'

In early 1931, Governor-General Lord Bledisloe visited several of the Māori land blocks being farmed in the Bay of Plenty and Rotorua under a number of land development schemes being organised by the Department of Native Affairs. He was impressed and determined that the prodigious efforts of Māori farmers, as he had witnessed, 'might be even more effective if a spirit of friendly rivalry' could be introduced to raise the standard.[45]

One of the blocks inspected by Lord Bledisloe was the Horohoro Block, which comprised 21,000 acres located along the Rotorua–Atiamuri Road. One of the first blocks to be 'opened up' under Apirana Ngata's land development schemes, the Horohoro Block belonged to the Ngāti Tūhourangi people of Whakarewarewa. Development had only been possible after Ngata consulted with the aged tribal chief Mita Taupopoki in 1929.[46]

In order to commemorate his tour of the Waiariki land developments, Lord Bledisloe presented Māori with a silver cup to be known as the Ahuwhenua (Sons of the soil) Trophy. He also provided an endowment fund to provide for annual prizes. Ngata responded warmly to Lord Bledisloe's gesture, saying that 'the sympathetic interest shown by his Excellency in the efforts of the Maori to develop their land and so assist the whole of the Dominion' was widely appreciated by the native race. Ngata was certain that this 'further token of his Excellency's interest' would stimulate and encourage Māori landholders to become more efficient farmers.[47]

Ngata and the Governor-General drew up conditions to govern the trophy competition. Although the trophy would ultimately be nationally contested, the inaugural 1932/1933 competition would be limited to the areas that Lord Bledisloe had toured in 1931. Māori farmers from Waiariki who were occupying lands currently being managed under a native land development scheme qualified for entry. The criteria for judging would include good husbandry, economy of management with a view to commercial profit, and the 'cleanliness and neatness' of the farming unit.

Another important criterion was the progress a farmer had made in developing his or her block since initial occupation. Account would be taken of certain 'advantages' that some competitors may have had over others, such as access to finance and prior supply of implements, seeds and fertilisers. The original state of the development block prior to occupation would also be considered, since the nature and terrain of farming units throughout Waiariki and the Bay of Plenty varied considerably.

The competition conditions stipulated that the trophy would be held by the winner for one year and the winner would not be allowed to enter the competition again for three years. The Governor-General's endowment would provide a large silver medal for the first-prize winner and a large bronze medal for the farmer placed second. Certificates of merit could also be awarded to all place-getters, including fourth place.

Lord Bledisloe proposed a trust board in 1933 to oversee the running of the annual competition; trustees were the Minister of Native Affairs, the Minister of Agriculture and the Under Secretary of Native Affairs. The board would take over the management of Lord Bledisloe's endowment of £151.16s, which had since been converted into sixty-three Bank of New Zealand shares. In November 1933, Lord Bledisloe and Native Minister Ngata signed a formal trust deed setting out the terms and conditions of the competition

On 13 December 1933, the Ahuwhenua Trust Board was formally established. Thereafter, the Under Secretary of Native Affairs would provide the administration services to run the competition.[48]

Lord Bledisloe and Agriculture

Governor-General Lord Bledisloe supported Ngata's work in promoting Māori land developments. As an experienced agriculturalist himself, he empathised with Māori farmers working to bring difficult land blocks into production.

Lord Bledisloe – Charles Bathurst – was born in London on 21 September 1867. He trained as a barrister before attending the Royal Agricultural College at Cirencester, obtaining a Diploma in Agriculture. He thereafter practised law and was in 1910 elected to Parliament where he was allocated responsibilities for food production and supply. He was knighted in 1917 and elevated to the peerage. He served as Parliamentary Secretary at the Ministry of Agriculture and Fisheries between 1924 and 1928.

Bledisloe resigned from Parliament in 1928 and became chairman of the Imperial Grasslands Association, an appropriate appointment for one whose passion was agriculture; he was well known for advocating a scientific approach to farming.

Formal portrait of Sir Charles Bathurst Bledisloe and Lady Alina Kate Elaine Bledisloe, taken between 1930 and 1935

Photograph by Herman John Schmidt
(Alexander Turnbull Library, 1/1-001242-F)

In 1929, Bledisloe was appointed Governor-General of New Zealand. In 1930, he took up office in Wellington, where his interest in agriculture was soon apparent.

The Bledisloes are probably best remembered for their purchase of the Treaty grounds in Waitangi, which included Resident Agent James Busby's original house and 1000 acres of adjacent land, all of which the Bledisloe family presented to the nation in May 1932 as a national memorial. Bledisloe also donated £500 towards restoration of the Treaty House and was present at the first ceremony at Waitangi to mark the signing of the Treaty on 6 February 1934.

During his stewardship as Governor-General, Bledisloe promoted a number of causes for which he donated trophies, including landscape painting, gardens of native plants and rugby played between New Zealand and Australia.

Upon his return to England, Lord Bledisloe was created Viscount Bledisloe of Lydney and thereafter was awarded numerous honorary titles. Bledisloe devoted his last years to rearing cattle, pigs and dairy cows on his estate, as well as growing potatoes and grain and maintaining a huge orchard. He died aged ninety at Lydney on 3 July 1958.

Source: Marshall, 'Bledisloe, Charles Bathurst', www.teara.govt.nz

The Inaugural Competitors

The inaugural Ahuwhenua competition generated a huge amount of interest among Māori farming families of the Waiariki and Bay of Plenty districts.

Seventy-two people entered the competition. Their names were divided into seven groups, denoting areas within Waiariki and the Bay of Plenty where land development schemes were located and, in the view of Native Department officials, had been developed to a sufficient standard: Te Kaha, Tōrere, Whakatōhea, Rūātoki, Horohoro, Mourea and Maketū.

Twenty-six competitors came from Rūātoki alone; the region also produced the eventual second, third and fourth place-getters. Te Kaha accounted for fourteen competitors; twenty entrants came from Horohoro. Only one entrant came from Mourea and two from Maketū.[49]

Assessment

The competition was judged by Mr W Dempster, a dairy instructor and grader with the Department of Agriculture based in Hamilton. This inaugural judging was time-consuming and required extensive travelling. By March 1933, Mr Dempster had visited seventy-one of the farms recommended to him and another ten 'as a matter of courtesy and policy'.

The visited farms extended from the Horohoro Block, Rotorua, out to the Maketū Block, Te Puke, and then down the coast to Cape Runaway. Mr Dempster had three criteria in mind as he inspected the farms: the 'wise expenditure of money', the personal application of labour and the prospects of 'a high standard quality of milk or cream'.

Mr Dempster concluded that the units on the Horohoro Block had not yet managed to fully prepare for the competition because of the demands of conserving supplies of winter feed, which continued to be organised on a communal basis. Neither did he consider the Maketū and Ōpape Block farms when awarding prizes although they were in a 'high state of productivity' and recent developments showed value for the money being expended. Were a prize to be awarded to one of these farms, said Mr Dempster, it might detract from 'that spirit of emulation' that the award and the development schemes themselves were calculated to create. The Ahuwhenua prize, it was implied, constituted a standard to be aspired to.

Judge Dempster observed that the largest acreage of development work in the area was to be found on the newest blocks, which comprised both flat and undulating surfaces. The flat land in particular was of a very good quality. He noted that, during the two years that the Maketū-Ōpape development scheme had been in operation, drains had been laid and new farms had been created, and new roads also built. He had heard that a ten-vat cheese factory was being planned in the Maketū-Ōpape area.[50]

Dempster noted that, from Ōpōtiki down the coast, blocks of 'idle land' were now being 'brought in', and Māori settlers were being established on new farms. Cream from as far away as Maraenui, over 20 miles away, was being transported to the Ōpōtiki factory. The milk supply from the new farms would go to the Te Kaha factory where a new water scheme had been developed to make land more suitable for dairying. With an increase in dairying, Te Kaha's future was looking more assured.

New lands had also been developed near Cape Runaway and would be available for dairying over the next few years. The dairy cattle supplied to the settlers were high-grade jersey stock with a pedigree bull at the head of each herd. However, in many of the herds, Dempster observed that 'scrub cows and scrub bulls' were still evident, to the detriment of the quality of the herds.

Dempster noted that the standard of the buildings and sheds he had inspected throughout the area was consistently high: concrete floors and drains were common. The cleanliness of equipment and utensils was also without a single exception the highest he had seen in dairy sheds; in this respect he singled out the Horohoro Block. As a whole, Dempster believed that the 'most adverse critic would be silenced' by making a personal inspection of the Waiariki area and various surrounding localities.

Inaugural Winners

Apirana Ngata announced the results of the Ahuwhenua Trophy competition on 27 March 1933, as reported that evening in Wellington's *Evening Post*.

First place went to William Swinton, of Whānau a Apanui descent, of Raukokore. Second place went to Jack Black of Rūātoki; Tawera Kopae, also of Rūātoki, came third. Three other competitors received a special mention – P Mihaere of Horohoro, Te Katene of Tikitere and Charles Oneroa of Horohoro.

In his report, which had little to say about individual winners, Mr Dempster singled out James Swinton for special praise. He was complimented for his 'judicious top-dressing and surface growing

William Swinton, first winner of the Ahuwhenua Trophy, in 1933, pictured with daughter Roka and the trophy
(Photo courtesy of Parehau Richards)

of grass, and for converting ratstail paddocks into good grass'. Dempster noted that he had built a two-bail cowshed out of materials considered worthless by other settlers.

Although Swinton had no concrete floor in his milking shed, it was nonetheless perfectly sanitary, with all utensils in perfect condition. He was running fourteen cows on 20 acres and had ample feed. All he needed, wrote Mr Dempster, was more land to show his abilities; with thirteen children, he possessed 'ample material to dissipate his knowledge'. For his efforts, James Swinton was 'highly commended' and presented with a cream can.

On 31 May 1933, Lord and Lady Bledisloe and Apirana Ngata attended a special presentation hui at Ōhinemutu. They met with the competitors and their families, taking special pride in presenting the inaugural 'Bledisloe Cup' to William Swinton.[51]

The Trophy's Design

Lord Bledisloe's Ahuwhenua Trophy had the following inscriptions:

> *Te Tohu Ahuwhenua*
>
> *Na Rore Peraro*
>
> *Tihema 1932*

> *The Son of the Soil Trophy*
>
> *Presented by His Excellency Lord Bledisloe*
>
> *December 1932*

Copy.

Department of Agriculture,
Rotorua, March 7th., 1933.

The Director of the Dairy Division,
WELLINGTON.

Under the guidance of Mr. Scott, Rotorua and Mr. Royal, Ruatoki, I have just completed the visiting of 71 of the farms on the list sent by you. In addition 10 additional farms were visited as a matter of courtesy and policy. The farm of Hirina Waititi, Cape Runaway was not visited. Mr. Royal communicated with him by phone and was informed that he was not ready but that he hoped to win first prize next year. The farms visited extended from the Horohoro Block, Rotorua, out to the Maketu Block, Te Puke, and down the coast to Cape Runaway. In judging the farms besides taking into consideration the wise expenditure of money and the personal application of labour I also took into consideration the possibility of delivering a high standard quality of milk or cream.

After visiting all the farms I came to the conclusion that the settlers on the Horohoro Block have not had sufficient time to display their initiative, as the conservation of winter feed has been done on a communal basis (an ample supply is available). I have therefore selected three of the farmers on the Horohoro Block which have shown the most initiative and placed them in order of merit.

The Maketu and Opape Block farm I have not taken into consideration in awarding the prizes, although they are both in a high state of productivity, and show value for money expended; were a prize awarded to either of these farms, I believe it would tend to destroy that spirit of emulation which the awarding of the prizes is calculated to create. As you are aware a number of Natives have been supplying milk at Ruatoki for some years, and have been working on indifferent lines. Under the development scheme a spirit of emulation has been created among the older suppliers around the factory and there are abundant signs of a forward movement. The bulk of the Development scheme is on new country on the West bank of the river and includes both flat and undulating country. Some two years ago, I viewed this country from the eastern bank of the river but did not realise the extent of country. During the two years which the Development Scheme has been in operation, drains have been made and new farms created, while at the present time new roads and more drains are being made. The flat land is very good quality and will compare favourably with the best land in the Opouriao district. The undulating land compares favourably with similar country in the Waikato and King Country. This block of land is eminently suited for the establishment of a central cheese factory. I do not think I am unduly optimistic in stating that sufficient milk will be produced to warrant the building of a ten-vat factory. Plenty of river shingle is available so that good roads could be made at low cost.

From Opotiki down the Coast blocks of idle land are being brought in and settlers are established on the farms. The cream from down the coast as far as Marainui (28 miles) is taken to the Opotiki factory. Beyond the Motu river new country has been brought in and will be settled this year. This supply will go to the Te Kaha factory. Around Te Kaha assistance has been given to a number of settlers and a comprehensive water scheme has been introduced which will make more land suitable for dairying. The settlement scheme will save the situation at Te Kaha as with the increased supply the future of the Te Kaha Dairy Company is assured.

Beyond Te Kaha as far as Cape Runaway new country has been brought in and will be available for dairying next season.

The first Ahuwhenua judge's report, prepared in 1933 by Judge W Dempster

2.

The quality of the dairy cattle supplied to the settlers is high standard grade Jersey and at the head of each herd is a pedigree bull. In many of the herds of the old established natives the scrub cow and scrub bull is much in evidence. The milking sheds and yards are well finished and are kept in good order. This does not apply to sheds built previous to the development scheme.

The standard of cleanliness of the concrete floors, drains and all utensils without a single exception on the Horohoro block, was the highest I have ever seen in dairy sheds.

Taking the Development Scheme, as a whole, I believe the most adverse critic would be silenced by making a personal inspection of the various localities. I would award the prizes as follows:-

William Swinton	Raukokore	First
Jack Black	Ruatoki	Second
Tarewa Kopae	"	Third
James Swinton	Raukokore	Highly commended.

Special Mention.

P. Mihaere	Horohoro	First on Block.
Te Katene	Tikitere	
Chas. Oneroa	Horohoro	Second on Block.

I would specially commend the efforts of Jas. Swinton as, First: He has by judicious top-dressing and surface sowing of grass, brought ratstail paddocks into good grass. Second: He has used material to build a two bail cow shed which would be considered valueless by either Maori or Pakeha. Third: Although he has no concrete floor in shed he is milking under most sanitary conditions; his dairy utensils being in perfect condition. He is running 14 cows on 20 acres and he has ample winter feed. If this man could get more land he could display to greater advantage his undoubted ability as a farmer. He has 13 children and has ample material to disseminate his knowledge.

(Sgd). W. Dempster,

Dairy Instructor.

According to H McMillan Bull, who judged the Ahuwhenua competition from 1962 to 1964, no trouble was spared at the time to ensure that the trophy fulfilled every wish of Lord Bledisloe, literally and figuratively. The designer of the cup was G F Goldie, the well-known painter of Māori; a Mr Wright, a sculptor at the Auckland School of Art, did the modelling. Mr Tai Mitchell of Rotorua was responsible for the trophy's 'historic accuracy'.

Safeguarding the Trophy

After the 1933 competition, Tai Mitchell, on behalf of the Te Arawa iwi, suggested that the Ahuwhenua Trophy itself be lodged in the Auckland Museum because of a fear that the thermal atmosphere of Rotorua would have a 'tarnishing effect upon [its] silverwork'.

Mitchell suggested that the 'Onewa Patu', which had been recently presented by the Federation of Ngā Puhi Tribes to Te Arawa to mark 'the extinguishing of the slumbering embers of enmity' caused by past inter-tribal wars, also be displayed with the cup. The patu had been presented to Sir Apirana Ngata during his visit to the North Auckland district the previous May by Henare Kingi, chief of Ngāti Kahu. Ngata had then conveyed the patu 'with peace overtures' to a large gathering of Te Arawa hapū attending the presentation of the Ahuwhenua Trophy at Ōhinemutu. Mitchell noted: 'we feel that the remarkable interest taken by [the Bledisloes] in all matters pertaining to the welfare of the Maori and their history was undoubtedly inspired to a very large degree by the Nga Puhi gesture which produced a patu as a symbol of that inspiration'.[52] He added that a valuable greenstone mere and a Māori head modelled life-size in kauri gum had recently been acquired in London and that these 'Te Arawa heirlooms' should also be lodged in the Auckland Museum. Accordingly, all these trophies and heirlooms were taken to the museum.

Ahuwhenua competition judges paid great attention to the 'neatness and cleanliness' of the living quarters of Māori farmers. Fears still persisted that rural Māori living conditions had not improved since the 1900s, when authorities considered papakāinga in some parts of the country (but not all) to be largely uninhabitable.

Group of Māori standing in front of a whare and storehouse, c. 1909; photograph by William A Price (Alexander Turnbull Library, 1/2-001938-G)

1936

Ahuwhenua for All Regions

After 1933, the competition went into an immediate recess until 1935, when entries were sought once again. The competition for 1935/1936 was extended to the whole of the country, in accordance with a pledge to that effect made by Governor Bledisloe in 1932.

After the initial 1933 competition, nominated farms were to be limited to thirteen, given that the 'cost of asking the Award Judge to inspect the farms of every Maori farmer who may desire to enter the competition would be prohibitive'. Three nominations were allowed from the North Auckland District, two from the South Auckland District, three each from Rotorua and Gisborne, and one each from Whanganui and Wellington/South Island.[53]

By the beginning of May 1936, eight farms had been nominated. They were located near Rotorua, in the Bay of Plenty, at Rūātoki, at Cape Runaway and at Tikitiki, near the East Cape. As before, Māori applicants needed to qualify as farmers occupying land under development, subject to the provisions of section 522 of the Native Land Act 1931.[54]

On 19 May 1936, Native Minister Michael Joseph Savage (who had not long been elected prime minister) asked the Minister of Agriculture to appoint a member of his department to inspect and judge the farms. The judge appointed was Mr C Walker, instructor in agriculture from Tauranga, who was to judge the competition for the next four years. It is to be noted that at least six later judges would similarly serve over multiple years in this way, investing a much-needed level of continuity of criteria and assessment

Mr Henry Dewes' headstone, which displays his Ahuwhenua medals. Standing by the gate are Mr Dewes' grandson Campbell Dewes and Henry's daughter Mereheni Dewes.

(Photo courtesy of Lyn Harrison)

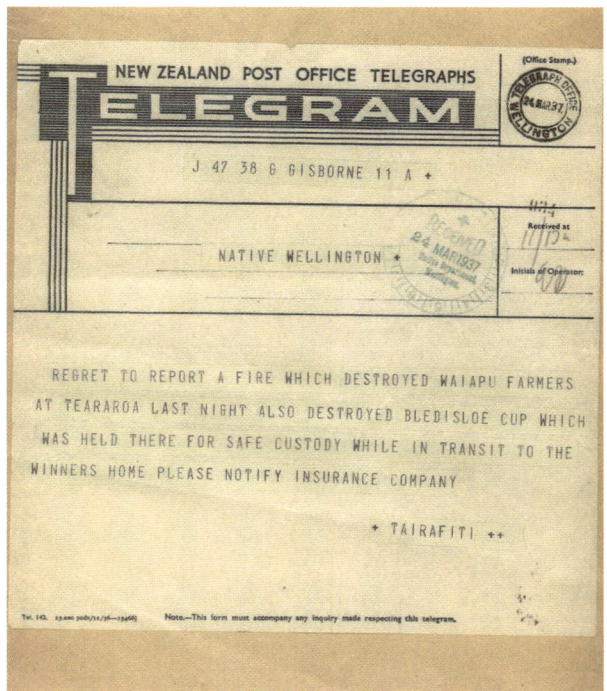

Telegram of the Bledisloe Cup loss in the Waiapu Farmers Store fire

Replica cup presented to 1936 winner H Dewes
(Photo courtesy of Lyn Harrison)

standards, as well as admiration and empathy, into the competition.

Inspection and Assessment

Mr Walker inspected farms in July 1936, allocating points for good husbandry, economy of management, neatness and cleanliness, overall progress with development, quality of stock and quality and quantity of farm produce. Walker ranked competitors in order of merit.

The winning competitor for 1936 was Mr Henry J Dewes, a sheep farmer of Tikitiki, who was awarded a score of 80 percent. It had been a close competition: the scores of the other competitors ranged from 61 percent to 79 percent. Second place went to Robert Clarke of Rotorua, and a third-equal pairing was awarded to Jack Black and Pat Raharuhi, also of Rotorua.

The overall standard of the competition, said Mr Walker, was 'highly satisfactory and reflected credit on the competitors as well as those responsible for their supervision'. Walker also generally commended standards of management. He considered that pasture management, especially, was being approached intelligently, as was the management of flocks and herds. General farm maintenance and upkeep of fences were also very satisfactory in Walker's eyes.[55]

However, Walker found that other aspects of the farming operations he inspected were not performing so well. The provision of winter feed was poor in some areas; it was noted that in the colder areas a good supply of winter feed was essential. Equally, the management of pig stocks and run conditions varied greatly, to the detriment of productivity.

A major problem facing Māori farmers was a lack of consistent supply. Walker noted that, considering the limited nature of essential materials, Māori farmers were making the best use of those provisions that they could. Conversely, on some farms, there was a tendency to overstock, and Walker observed a consequent lack of condition in those herds. He advised some culling of herds if a high quality of stock was to be aspired to. On another note, on some farms, Walker saw that the manuring of inferior pastures was scarce; some farmers seemed reluctant to engage in this useful practice.[56]

Many competitors lost points for the quality of their stock, a situation that Walker noted would be vastly improved by some form of herd testing. He found the quality of pig stocks in particular disappointing. In many cases, butter fat returns were low – as much a result of overstocking as it was of poor quality stock. More planning was needed to ensure that stock could be wintered in good condition.[57]

'The standard of farming management amongst competitors was particularly creditable.'
(Judge C A Walker, Report on Ahuwhenua Trophy Competition, 1936)

Four pupils from Māori Agricultural College, Hastings, 1930; photograph by Henry Whitehead
(Alexander Turnbull Library, 1/1-004724-G)

In general, Walker noted, most Māori farmers settling on new development blocks had made 'very commendable progress since taking over their respective farms'. A good deal of assistance had fortunately been available to them; this was taken into account. Pasture improvement was noticeable, as was tree planting on all farms. Also, on most of the farms, drainage, suppression of weeds and secondary growth, subdivision and fencing were directions in which good progress had been made.

When notifying the Registrars of Native Land Court Districts of the confirmation of winners, the Under Secretary of the Native Department suggested that the judge's overall appraisal of the competitors' farms should be brought to the attention of Native Land Court land supervisors and, ultimately, the competitors themselves. This would ensure that Māori farmers could 'take steps to remedy those defects which make the farms of the settlers generally fall short of the desired standard'.[58]

'Neatness and Cleanliness'

One important and revealing aspect of Walker's judging criteria was his focus on the cleanliness and neatness of farm sheds, buildings, yards and homesteads. Many of the Māori farmers of the 1930s were new farmers occupying virgin development lands, engaged in the difficult task of 'bringing those lands in' while also seeking to render them as economically profitable. To do this, of course, they needed satisfactory living quarters and clean places in which to work.

A few generations earlier, Native Affairs officers had reported that Māori rural living conditions were unsatisfactory and required urgent attention. One government programme of the 1900s sanctioned the enforced burning down of whare deemed to be

In 1937, the Waiapu Farmers Store in Te Araroa was burned to the ground, destroying the original 'Bledisloe Cup' (Ahuwhenua Trophy) while it was on display.

Man on a horse on the main street of Te Araroa, 1943; photograph by John Dobree Pascoe (Alexander Turnbull Library, 1/4-000661-F)

no longer habitable. In 1906, for example, Health Inspector Riapo Puhipi from Mangonui had informed officials that many of the Māori houses in his area had been condemned and would be destroyed, as did Hori Pukehika of Putiki when reporting on a number of papakāinga up the Whanganui River.

Haimona Apete reported from Arapawa that a substantial new building programme was needed for Arapawa, including thirty-nine new 'outhouses.' Māori were however opposed to the building of such facilities, wrote Elsdon Best in 1906 'and that is the truth', he said, adding that he would strongly oppose moves being made by Maui Pomare to take legal action against papakāinga unwilling to see their 'inhabitable whare' destroyed.[59]

By the 1930s, most of the sanitary and rebuilding measures introduced by Māori leaders like the young Apirana Ngata had reached the most isolated of the native districts. But the government was still clearly wary of Māori rural living conditions, particularly on the development blocks, many of which were often quite isolated. The emphasis of the Ahuwhenua Trophy judging criteria on 'neatness and cleanliness' was a consequence of this.

In 1936, Judge Walker rated 'neatness and cleanliness' as third in his list of five essential criteria, after 'good husbandry' and 'economic management' but ahead of 'development progress' and 'quality of stock and quanity of produce'. He found that, in practically every case, sheds and buildings were 'models of cleanliness', and in no case was a single 'dirty shed' even observed. He found that holdings and homesteads were also neatly kept.

Disaster after Presentation

After the presentation of awards, some controversy emerged over possession of the trophy itself. Although Mr Dewes believed the trophy was his to look after, Tairāwhiti Native Land Court Register Mr R J Thompson insisted that it should remain stored in the strong room of the Waiapu Farmers Store. Meanwhile, against Thompson's instructions, Mr Omundsen, the Native Affairs Supervisor, allowed the store to display the trophy. Two weeks after its presentation, a fire broke out, burning the store to the ground and destroying the cup in the process. The manager of the store, Mr Pepere, made 'repeated efforts to enter the burning building to look for the valuable trophy, the loss of which has cast a deep gloom over the Ngati Porou natives'.[60]

Portions of the metal recovered from the ashes after the fire were later confirmed by the jewellers Good & Co, who had cleaned and engraved the trophy, as having been part of it. Registrar Thompson reported the holding of these fragments at the Native Land Court in Gisborne pending further instructions from the Under Secretary of Native Affairs.[61]

1938

A New Trophy

As a result of the destruction of the trophy, the competition was not staged in 1937; there were some concerns as to whether a replacement trophy could be arranged. However, in July 1937 the Finance Minister approved the placing of an additional £150 into 'Vote: Native' for the purposes of replacing the trophy,[62] and Wellington jewellers Walker & Hall took on the job, at a price of £72.10s.[63]

Thus the competition was able to resume in 1938. Again, the Under Secretary of Native Affairs asked Native Land Court registrars to nominate worthy Māori farmers from their districts. Perhaps because of the recess, interest in the competition had increased considerably: a near full complement of twelve nominations was received. Mr Walker returned as judge.

Judging

Judge Walker inspected the twelve competing farms under an expanded set of criteria that included 'quality of farm produce and winter feed supplies'. Eventually, he decided upon the unusual step of awarding the trophy to two winners: Whareparoa Rewharewha of Tōrere and Jack Black of Rotorua. Mr Black's success was especially satisfying; he had competed for the trophy since its inauguration in 1933, having been placed second that year and third equal in 1936. Johnny Edwards of Rotorua was placed third, and Mrs Huinga Nepia of Tikitiki (wife of All Black great George Nepia) was fourth. Highly commended were Joe Wharekura and Ruhi Vercoe, both of Rotorua.

Both farms were described by the judge as being 'of outstanding merit'; the hard work of both farmers would have done credit 'to many leading European farmers'. Both men showed an excellent knowledge of farm management, and their pastures were being 'intelligently handled'. The farm stock on both units was of a high quality and showed the benefits of good management. Subdivisions were also planned well and fences well cared for. Additional crops like lucerne and maize were being grown, and pedigree pigs employed to improve the quality of pig production.

The trophy was presented to the joint winners at Wairoa in June 1938.

Assessment

All other farmers in this year's competition, wrote Judge Walker, had attained 'a fairly high standard'; their farms fell short of the standards set by the two winners only by 'quite small margins'.

Overall, Walker noted, the stock quality of competitors was fairly good; some herds reached 'quite a high standard'. Some system of herd testing and culling still seemed necessary; though Native Affairs seemed to be aware of the problem, no assistance with this was presently being provided. Pig husbandry had also improved significantly: in some cases, pedigree boars and sows or good cross-bred sows had been introduced. Butterfat returns were 'somewhat low' when compared with those of local European farmers, but were showing 'an upward trend'. This would no doubt improve with herd testing and better winter feed supplies.[64]

Walker commented that the pastures of competitors were generally of a high standard and were being well managed. Manuring was a regular feature of fertilising, as was chain harrowing and topping. He also noted that, for all the farms, more shelter needed to be provided, especially in cold and wet areas. The general standard of cleanliness and neatness was outstanding in all but two of the competing farms. Attention was also being paid to homestead grounds: in some cases, Walker saw wonderful displays of flower beds, flowering shrubs and neatly kept lawns.

Included in Walker's report were comments on the farm of Robert Clarke, who came seventh, but

The Māori Economy 1930–1950

Ngata's land development programmes of the 1930s provided a significant impetus to the Māori farming economy. By 1937, 750,000 acres were being developed under the 1929 Native Land Amendment and Native Land Claims Adjustment Act, and 200,000 acres of lands were being cultivated on 1500 Māori farms: mostly small dairy units. In the 1930s, a typical Māori farm had twenty cows on a 40-acre property.

Māori farms continued to develop in the 1940s; £4 million of government money had been invested in Māori farming by 1950.

By 1950, according to the Census of Agriculture, nearly 10,000 Māori farmers were farming 1.1 million acres of sheep, cattle or dairy farms. These comprised 700,000 sheep, 100,000 dairy cattle and 200,000 beef cattle. While Māori farms were smaller, less livestock-intensive and less productive than most Pākehā farms, the land development programme was raising Māori incomes at a time when Māori were still significantly poor and rural.

The urban drift of the Māori workforce affected the Māori economy as the century progressed. Mainly male Māori took work in the expanding manufacturing and construction sectors: the Māori labour force increased from 9 percent in 1936 to 37 percent in 1951. Over this time, however, the largest group of Māori still worked in forestry and agriculture. The census of 1945 reveals that the most commonly held occupations of Māori men at the time were farmer or farm worker, forestry worker, freezing works employee, dairy factory assistant, road grader, road construction labourer, lorry driver and labourer. The most common occupations of Māori women were farm worker, school teacher, cook, housemaid, domestic servant and waitress.

Source: Coleman, Dixon and Maré, 'Māori Economic Development – Glimpses From Statistical Sources', pp. 16–18.

The joint trophy winners in 1938 were farmers of 'outstanding merit' whose hard work and efforts would have 'done credit to many leading European farmers' (Judge C A Walker).

Māori farmers harvesting corn in Christchurch, 1910; photograph by Steffano Webb (Alexander Turnbull Library, 1/1-019450-G)

depended a great deal on hired labour that had been difficult to obtain, and whose farm had therefore fallen into neglect in some key farm operations, especially in drainage, pasture quality and infectious weeds control. Walker also commented on some particular entrants from North Auckland and Kāwhia: 'While not wishing to discourage them, I do not consider that their farms have yet reached a sufficiently high standard for this competition.' Walker doubted whether they would in time be competent and permanent farmers.

Judge Walker's Overarching Comments

Judge Walker recorded some perhaps subtle observations as to the nature of Māori farms and the readiness of competitors for a trophy competition of this nature. For the benefit of those who might be selecting farms for this competition in the future, he wrote in his final report, 'I would stress the fact that neatness and cleanliness must be features of all entries.' He also emphasised the need for clear indicators that nominated farms could soon be profitable if they were not already so. This had not always been apparent on some farms located on difficult landscapes, where, for example, the clearing of 'stone and troublesome weeds' would have to be tackled at the very least 'before the native farmer could expect any return'.

One striking feature of the competition, in Walker's eyes, was the keenness and enthusiasm shown by the competitors. There was no doubt that the entrants were much appreciative of the opportunities offered them to become farmers in their own right, he said. All entrants were working hard towards this goal – Walker noted that it must have been gratifying for supervisors to witness such keenness.

1939

End of the First Decade

By 1939, the Ahuwhenua Trophy competition had generated a significant amount of positive attention among Māori farmers, and interest in the competition was high. The judge appointed for 1939 was once again Mr C Walker, who had by now accumulated much experience and knowledge of Māori farming and farmers, especially those attuned to the performance expectations incumbent in the prestigious Ahuwhenua Trophy.

By late February 1939, Judge Walker had inspected all of the nominated farms. He placed the top seven competitors in order of merit, nominating his 1939 winner as Johnny Edwards, who farmed along the Rotorua–Atiamuri Road. Second place was awarded to Mrs Tatai Hall of Te Teko and third place to Ngarangi Kohere of Rangitukia. A fourth place was not always awarded, though it was within the judge's discretion. This year, Walker awarded a fourth-place prize to Charles Sergeant of Te Awamutu.[65] The trophy was presented to Johnny Edwards by Lord Galway in April 1939 at Horohoro.

Walker again praised the general standard reached by all competitors, noting that the two top place-getters were outstanding examples of native farmers who had applied themselves 'diligently and intelligently' to the task of developing their holdings. Much progress had been made by both farmers in the breaking up of new country and the building up of pastures, and a general improvement of stock quality was the result. In their provision of winter feed, both Mr Edwards and Mrs Hall had demonstrated 'excellent foresight in building up reserves in the form of hay, silage and farm crops'.

Looking to the future, Mr Walker saw good prospects for Māori farming; strong economic gains were to be earned from 'consistent efforts and continuing application'.

Recurring Issues: Shelter, Stock Quality and Winter Feed

In his comments for this year, Walker considered that a number of recurring issues were affecting Māori farming at large.

One was the issue of shelter. Although many farmers had made commendable efforts in this regard, Walker strongly recommended that Māori farmers be encouraged to plant more trees, especially eucalypts and macrocarpa, which had considerable value in later years as farm timber. He noted that some Māori farmers had wisely preferred to save native trees as shelter, especially pūriri and karaka.

Another issue was herd testing. Walker again emphasised the need for Māori farmers to have urgent access to a herd-testing system. Testing should be extended to all districts; if associated with rigorous culling, this could bring marked improvement in yields. Farms also needed access to bulls of 'good butter fat backing'. Walker noted that some competitors had managed to purchase pedigree bulls in an effort to raise the standard of their herds.

Walker observed that considerable progress had been made in regard to pig quality, and it was becoming rare to see 'scrub pigs', which were once such a feature of Māori farms. But he said it was still

Ahuwhenua judges often reminded Māori farmers of the need to provide sufficient crops, hay and silage for their stock during winter, otherwise, 'prospects were bleak'.

Elderly Māori man sorting kūmara, c. 1920; photograph by Arthur James Northwood (Alexander Turnbull Library, 1/1-006227-G)

important to utilise pedigree sires and even pedigree sows, to continue improvement of pig production.

More work needed to be done to improve the housing of pigs, he said. Farmers needed to appreciate that pigs thrived better in warm, dry and sanitary conditions. Pig housing needed to be sunny and dry with no draughts; the practice of providing a totally enclosed pig house needed to be strongly discouraged. Yards used for pigs to run about in needed to be drained and open to the sunlight. Too much shade led to damp conditions, leading to ailments like pleurisy and pneumonia.

Another recurring issue Walker identified was that of winter feed. Some farmers had made great efforts to ensure access to sufficient crops, hay and silage to see them through harsh winters. However, many others had still not realised the necessity for ample and adequate winter feed. Some of the farms Walker inspected had no hay or crops. Unless provision was made in the form of oats and barley, he said, prospects for the winter seemed to be bleak indeed.

Conclusions

Mr Walker once again commented favourably on the general cleanliness of buildings, sheds, homesteads and farms, and in particular, on care taken to control weeds: even in districts where ragwort, blackberry and gorse were causing so many problems, most of the farms Walker visited were almost totally free of such noxious infestations.

The judge's concluding comments were that the general standard of entries for 1939 was well up on the previous two years, a fact that could be gratifying to both competitors and supervisors.[66]

(John Cowpland – Alphapix)

Chapter Three

The War Years – 1940–1945

1940

Māori and the War

On 28 August 1939, on the eve of the Second World War, Registrars of the Native Land Court were advised that the Ahuwhenua competition for 1940 would be proceeding as normal, given the high level of interest.

With the advent of war in Europe, many Māori men enlisted for the 2nd New Zealand Expeditionary Force, which had left New Zealand by January 1940. Thereafter, Māori men enlisted throughout the other services. Māori women also served abroad, with the New Zealand Army Nursing Service or as voluntary aids with the Women's Auxiliary Army Corps. However, the major contribution of Māori to the war was through the 28th (Māori) Battalion (though enlistment in the Battalion was low among Māori from native districts that had suffered British Army invasions and land confiscations during the 1860s).

Many Māori also served at home in territorial units and in the Home Guard. With the threat of a Japanese invasion in 1942, all Māori who were registered for overseas service were called up and formed into special units stationed near their homes.

Māori on the Home Front

Māori also participated in the huge fundraising and production drive that swept across the country in 1940. In 1942, the Māori War Effort Organisation was formed. It was responsible to a special minister in charge of the Māori war effort, Paraire Paikea from Rātana. Paraire oversaw the organisation of native districts into twenty-one administrative zones and 315 tribal committees for this purpose.[67]

Many Māori at this time moved to the cities to work for munitions and other industries. This, according to some historians, constituted the beginnings of the Māori migration to urban centres, a trend that would continue and greatly increase after the war. However, in some areas, Māori migrations had already started well before the war, as Apirana Ngata was acutely aware. These earliest migrations occurred as a consequence of continuing land losses and a Māori rural economy ensnared by underdevelopment, overpopulation, economic sluggishness and high debt.

In 1940, Ngata viewed the Ahuwhenua scheme as a significant tool in assisting Māori in the recovery of their rural economic base while they remained resident on tribal lands. The Under Secretary of Native Affairs therefore urged Native Land Court registrars to actively recruit quality entries into the 1940 competition.

The 1940 Competition: Honouring Women Farmers

Mr Walker returned as judge of the competition. He proclaimed the winner of the 1940 competition to be Mrs Tatai Hall of Te Teko, who had been placed second in 1939. She became the first woman to win the trophy. Another woman entrant, Mrs Ngarangi Kohere of Rangitukia, was placed third; she had also been placed third the previous year. Mrs Huinga Nepia had been the only previous woman place-getter, having been placed fourth in 1938. Such accolades were significant achievements for Māori women dedicated to making a success of their land development and farming opportunities.

Second place in 1940 was awarded to Mr Fred Amoamo of Ōmarumutu, near Ōpātiki. Mrs Hall was awarded the trophy by Apirana Ngata in September 1940.

Judging Standards

As he travelled about, Mr Walker was once again impressed with the high standards of Māori farming, which bore favourable comparison to those of 1940; it was clear to him that Māori farmers were taking full advantage of what the competition had to offer.

Walker said that there was little to separate this year's first four place-getters: 'anyone of them could have been awarded the top prize'. They had all worked diligently and had adopted advanced ideas with marked advantage to their holdings. Below the top four competitors, other entrants who had not been as highly placed were only separated 'by the slightest of margins'. In order to assist those who did not rise to the standards of the winner, Judge Walker suggested a range of issues that might be addressed.

The Judge's Observations of the 1940 Competition

Once again, the judge drew attention to the issue of winter feed and also to herd testing, a significant determinant in ensuring a good herd. It was an old axiom of farming that 'bad cows got the same attention as good cows', he said, to the detriment of improving the prospects of higher graded animals.

Judge Walker noted improvements in the care of pigs; especially the introduction of pedigree stock of both genders. He noted that some competitors had acted as the custodians of local pig clubs; others had purchased stock from pig clubs in order to improve their own stock.

Generally, Walker reported favourably again on the cleanliness of sheds and homesteads and the care of grounds and property. He said that one noticeable aspect of the competition was the involvement of women; he had taken note of the planting of flowers and shrubs, kitchen gardens and small orchards; this compared very favourably 'with the best of the European standards'.[68]

1941

The War Years Continue

The Under Secretary of Native Affairs encouraged Native Land Court registrars to support the Ahuwhenua Trophy for 1941 by seeking out quality entries from Māori farmers who, throughout the war years, had remained on their land blocks at the behest of the government in order to contribute to the national wartime economy.

During the Second World War, Māori women were recruited to farm service through the Maori War Effort Organisation. Māori women also served with the New Zealand Army Nursing Service and as voluntary aids with the Women's Auxiliary Army Corps.

Woman in overalls, Rotorua, 1943; photograph by John Pascoe (Alexander Turnbull Library, 1/4-000544-F)

Once again, a limit of thirteen entrants was imposed because of the 'prohibitive costs' of assessing any more. Interestingly, the judge's expenses papers filed for the 1941 competition show that the overall cost of judging the 1940/1941 competition was £23.16s.6d. This included the cost of 497 car miles at 3.25 pence per mile, a daily allowance covering thirteen days and fifteen hours, train fares and sundry expenses like reservation costs and luggage.[69]

A New Judge

Soon after filing his competition report for 1940, the previous year's judge, Mr Walker, informed the Under Secretary of Native Affairs that he would not be available to judge the competition in 1941, thus bringing to a close a renowned tenure. Mr Walker had attended to his judging responsibilities with proficiency and clarity of purpose. Although brief and often lacking any contextual commentary (for example upon the wider significance of his choosing a female winner in 1940), his reports had provided details, expressed with some empathy, of a Māori farming industry attempting to modernise itself and to attain economic viability.

Replacing Mr Walker was Mr C R Taylor, an instructor with the Department of Agriculture in Hamilton. Taylor immediately set a new tenor in assessing the competition; or, at least, he certainly picked up on many of the concerns Judge Walker had expressed as to the many issues facing Māori land development and farming.

Judge Taylor's 1941 competition report shed light on such issues as how Māori farmers were to be assessed and even whether the Ahuwhenua competition was working to their favour.

The central issue raised by Judge Taylor concerned the enormous disparity between farmers on land conducive to farming (which he labelled 'A Division') and those farmers who were essentially working hardscrabble on difficult land and coming away disillusioned by the experience. Given such disparities, which were an unavoidable fact of the current land development system, he argued that arriving at a 'just' decision as to the winning entry was difficult.

Judging the 1941 Competitors

After giving 'all of the issues his fullest consideration', Judge Taylor nominated Fred Amoamo of Ōpōtiki as the winner of the Ahuwhenua Trophy. This success was especially pleasing to Mr Amoamo, who had been placed second in the previous year's competition.

After naming the winner, Judge Taylor went on to discuss issues pertaining to Māori farming. If his report seemed at times to be critical, he said, this should not detract from his 'long and lively' interest in assisting with the development of Māori people through the activities of the Native Department.

Judge Taylor's Notes on the Competition

Taylor considered that all the farms he visited were developing well under the various land development schemes being administered by the Native Department. He noted that a number of competitors in 1941 came from farms where troublesome lands were being cleared, planted, developed and stocked; others were based on land that had been entirely improved before their occupation. This led to an inequitable judging situation that made his decision challenging.

Taylor raised some concern about the objectives of the competition itself. He was compelled to conclude that, in some areas, the objectives of the competition were being insufficiently achieved; in fact, in some districts he thought the competition might have been doing some harm, both to the individual farmer and to the land development scheme itself. 'On more than one occasion,' Taylor reported, 'I sensed a definite feeling that it was quite useless to compete in the competition owing to the natural difficulties with broken and undeveloped land, because such landscapes invariably precluded success.'

Part of this problem lay in the generally disappointing quality of Māori land: 'One effect of this outlook could quite easily be in the direction of awakening in the Maori mind a realisation that his land was poor in comparison with that of his race in other parts and thus he might lose heart and give in.'

Taylor acknowledged that some might argue that a mere competition could not have such an effect, 'but I would suggest that there was a significant history in the trophy already to warrant such a response'. Judge Taylor suggested that entrants in future be classified into particular groups – for example, an A division to a D division – 'in much the same way as Military Honours are ranked'. Such a categorisation would assist judges and ensure the more equitable treatment of competitors. Re-categorisation could also lead to greater interest in the competition and in development schemes themselves.[70] Taylor's vision was that the poorest farmer on the poorest land could have a chance at winning against his more favoured brethren who were farming lands already broken

In 1940, Mrs Tatai Hall of Te Teko became the first woman to win the Ahuwhenua Trophy, having been placed second in 1939. Earlier, Mrs Huinga Nepia was the first woman to gain a top placing, earning fourth place in 1938. In 1940, third place was also awarded to a woman: Mrs Ngarangi Kohere of Rangitukia.
Young Māori women tending forestry seedlings, c. 1940s (Alexander Turnbull Library, PAColl-8983-64)

in when they were occupied, and who were always more likely to win the cup.

Taylor also suggested that judges should visit farms later so that farmers could be observed going about their haymaking: 'that most important area of winter feed'.

The Winning Entry

Taylor observed that winner Fred Amoamo had 'a very nice farm' and was managing it well, though it did include some pumice. Good pasture, with fencing well maintained, complemented satisfactory subdivisions arranged 'along modern lines'. Winter feed was well catered for through hay, silage and root crops, and shelter and water were well provided. The cowshed and plant were adequate, but no satisfactory method of housing equipment was evident, which was unfortunate given the expensive equipment involved.

The dairy herd was in good condition; good butterfat backing was used in the herd. Pigs were well cared for, with provision given to pedigree breeding stock. Dwellings and grounds were well looked after, providing a home that was modern and satisfying. The competitor's wife could be clearly seen keeping the home well cared for 'and also in the flower gardens outside'.

Other Place-getters

Taylor noted that the second-placed competitor, J D Jones, farmed a property quite unlike that of winner Fred Amoamo, in that it was at least 50 percent undulating, with the balance in a poor state requiring an inordinate amount of work to 'bring it in'. Fencing was in good order and well planned. The provision of shelter and water was 'not quite up to the standards required', though Taylor noted that plans 'for an attack on these problems' were well in hand. Mr Jones also had a plantation where fine trees were growing well. Taylor commended him for his efforts to provide for winter feed.

Taylor reported that the third place-getter, Eruera Hoera, had also made great strides in developing what had been from the outset a particularly difficult piece of land with little in the way of good pastures. Much had been done to bring the land into permanent grass, but with land as undulating as this, it was a continuing and difficult problem. The herd – mainly shorthorn stock – was in a splendid condition and was producing well given the type of country concerned.

The young stock and other dry cattle were well grown and thriving, 'which is a good index as to a person's ability as a farmer'. The same could not be said, however, of the pigs and the way that they were housed: seemingly, this competitor had little time for pigs, and gave them little of his time or attention.

The homestead was a very old one, but was kept clean and tidy. Surrounding the home was a 'nice garden and orchard', indicating an appreciation of the fact that 'a house alone can never make a home'.[71]

Other Competitors

Mr Taylor then ranked remaining competitors from fourth to ninth. As they had been for Judge Walker, herd testing, shelter, water and winter feed were recurring issues, as was the keeping of pigs. Taylor commented variously on the type of land being farmed, which was often difficult and challenging. For example, one competitor's farm was on alluvial soil, located alongside a river. Its area of pasture was low, in spite of extensive ongoing clearing. Another had a low-lying farm constantly being flooded, despite which he was managing to maintain a modern system of rotational grazing. A further competitor's farm was barely ploughable at all, and another possessed 'the most difficult' of all: pipe clay.[72]

1942

Ahuwhenua Unchanged

Despite Mr Taylor's suggestions of the previous year, the 1942 competition got under way in the same manner as it always had. The competitors were still limited to thirteen: three from North Auckland, two from South Auckland, three from Rotorua, three from Gisborne, one from Whanganui and one from Wellington or the South Island. Registrars were asked to 'arrange for your supervisors to select the most promising competitors and, if necessary, confer amongst themselves with a view to reducing the number of competitors'. The suggestion here was that, if a district had too many potential nominees, perhaps the competitors themselves could choose who among them would compete. If a supervisor thought that there were one or two outstanding competitors above all others, he should 'submit their names, together with a short explanation of the position'.[73]

The Glanville Points System

The competition for 1942 attracted nine nominations. The judge was E B Glanville, an instructor in agriculture from Auckland.

It was the practice in the Ahuwhenua competition for competitors to provide judges with information about their farming operation prior to the judge's tour of inspection. Details included the condition of the property when development began; crop production figures; details of stock being carried; butter fat figures for the previous two seasons showing 'percentage of finest, first and second grade' and showing the average production per cow for each season; wool and sheep returns from the previous two years; returns from other sources of production (such as pigs or crops); financial advances and repayment details; and details of any special handicaps the competitor may have overcome.[74]

In determining the winners and place-getters for 1942, Glanville commented that he had made every allowance for developments and improvements made since holdings had been occupied. He also paid some focus to recent increases in production, alongside financial assistance that had been given and rates of repayment, irrespective of the type of farm under consideration.

With this information to hand, Judge Glanville then devised a points system whereby points would be allocated to each farmer under a range of categories, including condition of pastures, water supply, shelter, condition and type of stock, financial assistance and 'general initiative'. Some of the points would be added prior to the farm inspections, such as those related to financial matters. This system would be used by judges following Glanville for at least the next decade.

Assessment

In his competition report, the judge reported that the standard of the seven farms he visited overall was very good; this suggested that competitors were taking advantage of advice and financial assistance

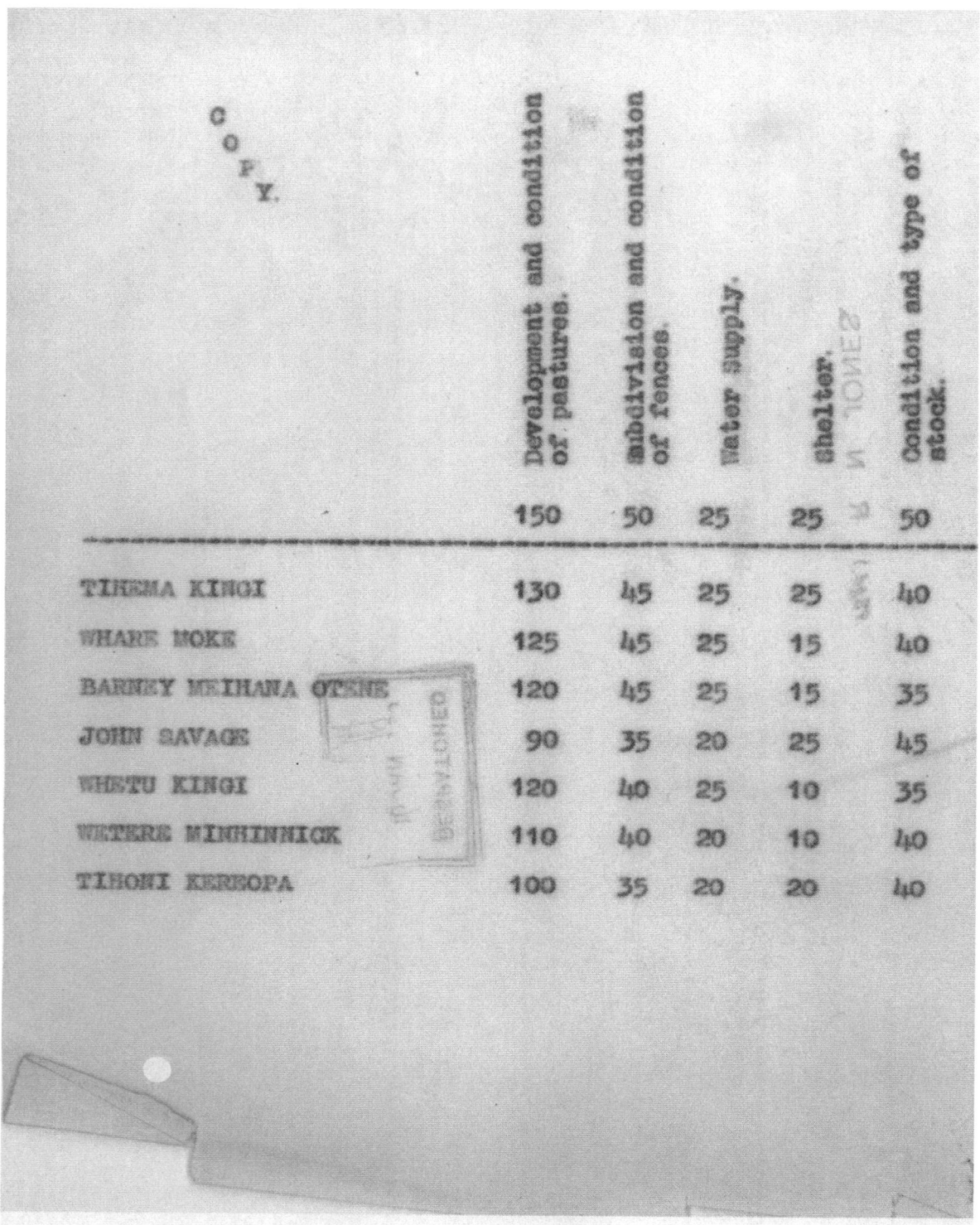

The judge's points table, 1941
(Ahuwhenua archives)

Farm buildings, piggory, yards etc.	Production	Supplementary feed.	Homestead and gardens.	General cleanliness - shed, utensils etc.	Financial assistance	General initiative.	Total points.	Place in competition.
50	50	50	50	50	50	50	650	
50	45	50	35	50	45	50	590	1st.
45	45	45	40	45	45	50	565	2nd.
45	45	50	35	50	50	45	560	3rd.
35	40	50	40	45	45	45	515	4th.
40	40	45	30	40	40	45	510	5th.
45	40	50	20	35	50	45	505	6th.
35	40	40	30	45	40	50	495	7th.

Mr Tame Pukunui of Te Kūiti, winner of the 1942 Ahuwhenua Trophy
(Photo courtesy of Barney Anderson)

now readily available to them. Glanville nominated King Country farmer Mr Tame Pukunui from Te Kūiti as first-place winner. Second place was awarded to H and T Paraone of Clevedon and third place to Tihema Kingi of Rotorua. A fourth place was also awarded, to Charles Wells of Whakatāne. Mr Pukunui was awarded the Ahuwhenua Trophy by Hon Mr H G R Mason in Te Kūiti in June 1942.

Mr Glanville reported that the winning nominee had an 'exceptionally well kept farm' that showed the effects of high-level management skills, especially in terms of increasing returns. The farm was the 'best by a clear margin'. Glanville said that little separated the second and third place-getters: Mr and Mrs Paraone and Mr Kingi had clearly worked diligently, putting original ideas into practice while using modern methods in their development programmes. The fourth place-getter, Charles Wells, was worthy of commendation, Glanville said, for the work he had done on his farm in the short time that he had been in occupation.

Glanville also congratulated the remaining five competitors. From the units he had inspected, it was noticeable to him that 'native farmers' were performing well and generally making good use of the assistance they were receiving.

Glanville's Observations on Māori Farming

Judge Glanville noted that, in general, pig production was in a 'fairly good' state – most farmers were aware of the breeding potential of pigs. Pig housing and running conditions, however, continued to be poor. Only one competitor had a modern lay-out with good housing and concrete feeding platforms; one other had commenced constructing a very efficient layout with good housing.

Glanville generally commented favourably on the perennial issues of winter feed, herd testing, shelter and the cleanliness of sheds, utensils and homesteads.

1943

The Competition to Date

By 1943, the Ahuwhenua Trophy had been contested seven times; twenty-eight competitors had attained a top-four placing, of which five were women. At least eighteen of the twenty-eight had come from the Waiariki District (noting that, in 1933, the competition had been limited to that area), confirming Waiariki as a district with a prodigious Māori farming sector. Four judges had presided over the competition.

Evaluation of Place-getters in 1943

The 1943 competition attracted seven entries from as far afield as Kāwhia, Hastings and Ōpoutere. The judge was Mr K M Montgomery, a fields instructor from Auckland.

In reaching a decision as to the ranking of the seven competitors, Judge Montgomery indicated that he had used Glanville's points system of the previous year. Montgomery announced that his nominee for winner of the 1943 competition was Tihema Kingi of Rotorua, who had been awarded 590 points out of a total 650. Second place was awarded to Mr Whare Moke of Kāwhia, and third place to Barney (Bunny) Meihana Otene of Hastings. Fourth place went to Mr John Savage of Ōpoutere, who was only five points behind Mr Otene. This year, a relatively large gap distinguished place-getters from remaining entrants.

Mr Kingi had been placed fourth in 1941 and third in 1942. His purchasing of sheep on his own behalf;

his clearing of weeds, scrub and ragwort; his building of a new and modern piggery; and his growing of root crops like turnips and carrots for winter feed, supplementing his existing stocks, all pointed to initiative and intelligence in his farm management, Montgomery considered. Mr Kingi was presented with the Ahuwhenua Trophy at Horohoro in September 1943.

The second-placed competitor, Mr Moke, had successfully replaced heavy mānuka with fine pasture, showing good soil husbandry. He had also built an 'excellent woolshed with attached lean-to implement shed', showing commendable use of available materials. His herd showed the benefits of herd testing and the careful selection of bulls. Montgomery saw further room for improvement, however, in the condition of his piggery.[75]

From the judge's report, it seems that Mr Bunny Otene was awarded third place in his absence: 'this competitor is very fortunate in having such a capable wife to carry on whilst he is serving in the Armed Forces, and Mrs Otene deserves great credit for the exceedingly neat appearance of the homestead lawn and flower gardens.' Mrs and Mr Otene had a 'compact little farm' near Hastings, with an operation that consisted of fat lamb-raising and grass seed production. Some of their fields carried good ryegrass sward, Montgomery noted, but were very weak in clovers. If pasture was to be productive throughout the year, it needed to be balanced with clovers. White clover in particular was required to supply the nitrogen so essential for the growth of grasses. It was not considered good practice to sow ryegrass and rely on volunteer white clover becoming sufficiently dense to furnish the nitrogen supply.

Overall Assessment

Mr Montgomery reported that drought conditions had seriously affected all regions of New Zealand

Māori farmers in 1943 were commended for their efforts to maintain farming operations in the face of great difficulties, especially the devastating drought of the 1942/1943 summer.

Scene at Maraenui showing a Māori dairy farm and farmhouse by the coast, 1944; photograph by John Dobree Pascoe (Alexander Turnbull Library, 1/4-001098-F)

in early 1943. As a consequence, pastures were not up to their expected standard by late summer, when judging took place. Despite the lack of rain, Montgomery found that farm stock was generally of a high standard, and most farms had managed to put aside good supplies of winter feed.

As previous judges had done, Montgomery commented favourably on cleanliness and not so favourably on the state of piggeries. He noted that a number of competitors had lost points because of a lack of proper care of their calves: too many had been fed solely on skim milk and were undersized. Montgomery strongly recommended the use of a suitable calf meal.

The judge congratulated all competitors on their sterling efforts in the face of numerous difficulties, not least of which was the drought. He said that, in a few years, all of the farms that did not win a place in this year's competition could be serious contenders.

Montgomery noted that the competitor placed last had a particularly difficult farm to manage, with very little land suitable for ploughing or crops. However, his method of spelling hill pasture in the autumn to produce rank grass growth to supplement his mangold and small hay stocks demonstrated that he had a very sound knowledge of stock husbandry. Further subdivision, top dressing and over-sowing the hill pastures with legumes would, in a very short order, make a marked difference on this otherwise challenging property.[76]

1944

Judge Shepherd

The judge for the 1944 competition was Mr J F Shepherd, a field instructor from Hamilton. Judge Shepherd visited nine nominated farms, assisted by Native Department supervisors. He used the now customary point categories, appending his points allocations chart to his report and indicating that he hoped it would help competitors to ascertain where improvements could be made.

The Successful Competitors

The winner of the 1943 Ahuwhenua Trophy was Mr Bunny Otene from Hastings, who had come third the previous year. He was presented with the trophy by Hon H G R Mason at Hastings in September 1944. Also present were Mr T Omana MP, Mr E L Cullen MP, the Bishop of Aotearoa and senior members of the judiciary.

Judge Shepherd reported that Mr Otene had a small farm where 'everything was in its place'. He ran a highly productive unit raising fat lambs and producing grass seed and crops for canning. Mr Otene had just been released from the army; 'gratitude was due to his wife for carrying on successfully during his absence'.

Henare Paraone of Clevedon, who was placed second, had a difficult farm to manage, there being very little flat land. 'Brown-top' had originally been sown all over the farm, but with good management this undesirable type of grass was generally being superseded by ryegrass and white clover.

Joe Wharekura of Rotorua, who was placed third, also had a difficult farm to manage, most of it being undulating or hilly. Being on the central plateau, it was subject to long and severe winters, and maximum crops of roots and hay were necessary to carry stock through the winter. Ragwort on an adjoining farm was a menace to this property; as a consequence, 'no little attempt was being spared' to eradicate or defoliate the pest.

Shepherd was pleased to note that all competitors had shown a fairly sound knowledge of stock husbandry, demonstrating that the assistance and advice of Native Affairs supervisors was being taken advantage of.[77]

Overall Assessment

Once again, the country had suffered from a prolonged drought through the summer months. However, except for Northland, all parts of the country had recovered sufficiently to present good pastures. Shepherd found that competitors were paying good attention to supplies of winter feed; excellent root crops, hay and silage were much in evidence. Stock conditions were generally good, and the use of purebred sires was marked in some herds.

With one or two exceptions, Shepherd noted, this year's calves were an excellent group. Milking sheds were generally well kept and beautifully clean, but there was room for improvement in some yards.

In terms of pigs, breeding sows and boars were generally good, but housing conditions 'left a lot to be desired' in nearly every farm Shepherd visited. Shepherd was glad to note that some farmers were erecting up-to-date layouts in their piggeries. He suggested that the services of local Pig Council supervisors should be made use of in remodelling some of the existing layouts.

The standard of Māori farms as assessed in the competition in 1942 was 'very high'; Judge J R Taylor suggested that Māori farmers were clearly 'taking advantage of every advice and assistance available to them'.

Two Māori women crutching sheep, 1940s (Alexander Turnbull Library, PAColl-6348-42)

Ahuwhenua officials were advised to select the most promising farmers for the trophy competition; nominations were initially restricted to a maximum of thirteen nationwide because of costs. Native Department field supervisors were advised to consult with local interested Māori farmers before making their nominations.

Māori family group with maize crop, Christchurch, 1920s; photograph by Steffano Webb (Alexander Turnbull Library, 1/1-019452-G)

(John Cowpland – Alphapix)

Chapter Four

Māori 'Economic Advancement' and Farming – 1945–1961

1945

The End of the War

On 8 May 1945, Germany formally surrendered, bringing the war in Europe to a close. On 2 September, Japan surrendered.

About 15,750 Māori men and women had volunteered for war service (3600 of them serving with the 28th Māori Battalion), from a Māori population base estimated to be fewer than 100,000. The Māori Battalion suffered 2628 casualties, comprising 649 dead, 1712 wounded and 267 taken as prisoners. Prior to the posting of Māori volunteers abroad in 1940, Apirana Ngata had conceded that there would be Māori losses: 'We will lose some of the most promising of our young leaders, and we have lost a few already. But we will gain the respect of our Pakeha brothers and the future of our race as a component and respected part of the New Zealand people will be less precarious.' After the war, in January 1946, the Māori Battalion returned home to a huge welcome in Wellington, which included a pōwhiri and a grand hākari (banquet).[78]

The Māori Post-war Economy

After the war, the government acknowledged the considerable contribution that Māori had made to the national economy during the war. The Maori War Effort Organisation, established on 3 June 1942, had been particularly effective in mobilising Māori into essential industries like munitions. Tribal committees established around the country, comprising in the main elders with good local knowledge, had also given advice on such things as appropriate land use and food production. Committees had also handled weightier tasks like employer–employee relations and the welfare of young women sent to work in the cities.

Its functions at an end, the War Effort Organisation was closed down on 31 January 1944 to the disappointment of many Māori; the disbandment was hastened along by the Native Department. As a compromise for those Māori who had wanted the organisation retained, the Maori Social and Economic Advancement Act 1945 was enacted shortly afterwards, setting the tone for Māori development over the next decade and beyond. This legislation, which few Māori supported, represented the government's expectation that Māori committees and papakāinga would continue to develop and economically expand under the firm control of Native Affairs.

Many Māori saw the Ahuwhenua Trophy as an essential means by which this Māori economic drive could be facilitated.

The 1945 Ahuwhenua Competition

Mr A D Mercer, a field instructor from Auckland, was appointed to preside over the 1945 competition. Five Māori farmers were also nominated: the lowest number of entrants thus far.

During early 1945, Judge Mercer visited the farms of the entrants, accompanied by supervisors from the revamped Native Department. In his report, he commented that the development work being carried out by Māori farmers on their isolated farms was a 'great credit to all concerned', as was the standard of farming and the volumes of production. If the same degrees of effort could be applied, 'as it could be', to the enormous tracts of North Island land currently lying empty, production would increase many times over, Mercer said. All competitors had shown what was possible with the application of 'energy, determination and intelligence', assisted where possible by the Native Department. The efforts of such Māori farmers as he had seen 'should serve as an encouragement and example to others'.

The Winning Entry

The winning nominee for 1945 was Mr Joe Wharekura, who was farming near Rotorua on the Rongomaipapa Development Scheme. Mr Wharekura, who managed a farm of 125 acres with fifty-five milking cows, had been awarded third place in the previous year's competition. Fifteen years earlier, along with another thirty Ngāti Kahungunu famers, Mr Wharekura had moved from his home in Wairoa in order to assist in the development of the pumice country near Rotorua. The 'gratifying results' of their labours were now much in evidence, wrote Judge Mercer, in the excellent farms to be found in the Rongomaipapa Valley today.

Mr Wharekura was a returned soldier of the First World War. He had been confined to army camp for two years between 1942 and 1944, not being permitted to proceed overseas on account of his age. Reports of officers recorded his excellent qualities of leadership and general ability. During her husband's absence, Mrs Wharekura had carried on managing the farm with the help of a young girl. The hard work of all parties had now ensured that excellent standards were maintained, as the farm's production returns ably demonstrated.

Judge Mercer found that Mr Wharekura's property was well laid out and carefully managed, especially with regard to ragwort control, considering the lack of sheep until this year and the state of the surrounding country. The layout and condition of the farm's piggery was also excellent, and was obviously the product of 'much careful thought'. The pigs were of a 'very good bacon type'. The provision of shelter trees was a feature of Mr Wharekura's farm that merited comment, as was the ample supply of winter feed for all stock, which was outstanding. Haystacks were well built, covered and fenced and implements were cared for and kept under cover. Pastures were also renewed each year, and rotational ploughing was well planned and executed. Overall, the farm was instructive as to what could be achieved in difficult pumice country.

Mr Wharekura was presented with a photograph of the Ahuwhenua Trophy at Horohoro on 22 September 1945 by Sir Cyril Newell, Governor-

'My koro won the Ahuwhenua Trophy on the 22 September 1945. It was a huge affair for the district. The Governor-General came all the way from Wellington to present my koro the prize of the Ahuwhenua Trophy, all the way to Horohoro. When the Governor-General came to present the prize of winning the Ahuwhenua Trophy, he actually presented a photo of the trophy to my koro. It's a beautiful trophy, but in 1944 it was misplaced during a rail journey; in 1945 my koro won it; and in 1946 it reappeared. So it seems as if that was the only time my koro was unlucky enough ... fortunate enough to be involved in the occasion that he won it, but unlucky enough at the same time because it was the one year the trophy was misplaced.'

Left to right: Sir Cyril Newell, Governor-General, Taiti Wharekura and Joe Wharekura

(Photo provided by Philip Wharekura)

General. Invitees to the presentation function included Sir Apirana Ngata, Mr T Anaru, Mr R T Carroll, members of the Te Arawa Trust Board, the Mayor and councillors of the Rotorua Country Council and the Rev. Panapa from Ōhinemutu.

Other Competitors

Second place was awarded to Whare Mill of Tikitiki on the East Coast. Mr Mill was from Ngāti Porou, of the Whānau a Pākai hapū. At the time of the competition, he was farming 55.5 acres and milking thirty cows. His butterfat production was recorded as up 559 lbs on the previous year. Mercer observed that some of the most noteworthy features of this farm were the fine management and condition of its pastures and the farm's well-designed layout from a central race. The homestead and the garden were also very well kept. However, the judge noted that the farm overall was handicapped by the lack of a good water supply. Further shelter was also required, especially given the severity of winter around Tikitiki, which was also prone to flooding.

The farmer awarded third place was Eurera Hoera of Takahue, about 10 kilometres south-east of Kaitāia. Mr Hoera was especially commended for the transformation of his land from rough bush to pasture, more so because of the lack of finance that had been available to him, as well as his acute problem of accessibility and isolation. Mercer noted that a flock of sheep on land such as this was essential to maintain pasture control, particularly because of the paspalum. Because summers in this area were dry and severely felt, a reserve of silage would be invaluable.

The fourth-placed competitor was Robert Tapa of Rānana, north-east of Whanganui. Judge Mercer said that the contrast between Mr Tapa's farm and the surrounding countryside was 'ample evidence of the industry of this entrant'. Mr Tapa farmed 50 acres, and was carrying a good type of stock, though winter crops for the coming winter had failed. A projected piggery would prove to be of great benefit to this farm.

The fifth-placed farmer, Wiremu Karaka, was making good use of a very small area: production figures and the condition of his herd reflected good management practices. Mercer commented that his decision to move his piggery to a new site would in time greatly improve his unit, and that what was now required was some intensive cropping for winter pig feed.

'Great Opportunities for Enterprising Farmers'

Overall, wrote Judge Mercer, the quality of the stock and standards of care observed throughout the competition was very high. However, he noted a tendency, not confined to Māori farmers, in districts favoured with a mild winter climate to rely too much on roughage and winter grass growth for feed. Adequate winter feed was essential because it laid the foundation for the following seasons' production, while contributing to the disease-resisting capacity of the stock. Mercer said: 'There is here a great opportunity for the enterprising Maori farmer with a well-developed property to lead the way in his own district.' Judge Mercer ended by commending the Native Department for the 'excellent work being done in the development of backward, remote and difficult areas'. Encouragement of more entries for the competition in future would be of great benefit to all concerned, he concluded.[79]

1946

A Post-war Competition

In 1946, New Zealand servicemen who had gone abroad during the war continued to return home for rehabilitation and reintegration into the new post-war society. Training and employment opportunities were greatly assisted by a buoyant economy based on high export prices.

In early February 1946, Mr T E Rodda, field instructor of Hamilton, was appointed to preside over the judging of the ten Māori farmers nominated for this year's Ahuwhenua competition. In his report, Mr Rodda complimented all of the competitors before nominating Henare Paraone from Clevedon as the winner of the 1946 trophy.

The Judge's Travels

The itinerary followed by Mr Rodda in his 1946 judging was typical of those followed by all Ahuwhenua judges and reveals how intensive and engaging these inspection tours were.

Judge Rodda's visits began on 8 March 1946, with a visit to eighth-placed Mr Tiaki Tamaki of Pirongia in the Waikato. Mr Rodda then travelled to Waiotemarama, in Northland, for a 14 March inspection. The next day, he travelled to Tautoro and Takahue, also in Northland, to visit two farms on the same day. These were ultimately placed seventh and tenth, suggesting that the visits may not have lasted overly long. On

Whānau and dignitaries celebrate the Ahuwhenua Trophy award, 1946. Centre is Te Puea Hērangi, a frequent visitor to Clevedon, with Viscount Bledisloe and Lady Bledisloe.
(Photo courtesy of Karena Paraone)

Henare Paraone, winner of the Ahuwhenua Trophy 1946, with his whānau
(Photo courtesy of Karena Paraone)

19 March, the judge was in the north, at Clevedon, and then, two days later, he returned south to Rānana, in Upper Whanganui, on 21 March. He then travelled to Hastings, visiting the ninth place-getter at Ōmaranui. Then he journeyed up the East Coast to Ruatōria to visit Kopua Waihi, eventually judged fifth. Finally, the judge visited Tōrere, near Ōpōtiki, and then Tikitere, near Rotorua, before returning to Hamilton, bringing to a close a hectic twenty days of travel.[80]

Bringing Poor Country into Profit

Mr Rodda commended Mr Paraone for the manner in which he had brought 'such a quantity of poor and hard country' into profitable production. Owing to the shortage of phosphatic fertiliser, Mr Paraone had ceased breaking in further acreages for the time being and was instead concentrating on maintaining his established pastures with fertiliser that had been allocated to him. He was 'good at all round farming', and clearly had a good knowledge of dairying, fat lamb rearing and bacon and beef production. He also did his own buying and selling of livestock. Rodda said he had shown 'great wisdom' in making a large amount of hay and silage during the previous year, when grass was so plentiful. Although he had only made one-third of his usual quantum of winter feed, he would have more than sufficient to get through the coming autumn and winter. Hay had been stacked in good time to prevent deterioration, and the most recent hay had been baled and stacked securely. His farm machinery was well protected and well maintained.

Rodda observed that Mr Paraone's home and buildings were well kept, 'and the furnishings in the house were exceptionally good'. He had recommended to Mr Paraone that he use a greater quantity of subterranean clover seed per acre and that he use all of the high country for sheep grazing.[81] Mr Paraone was presented with the Ahuwhenua Trophy by Viscount Bledisloe at Clevedon on 25 March 1947.

Judicious Use of Cattle and Sheep

Second place was awarded to Robert Tanginoa Tapa of Rānana, who had been placed fourth in the previous year's competition. Mr Tapa was commended by Judge Rodda for his highly effective breaking in of rough country that had once been infected with blackberries, gorse and ragwort, bringing small areas into production while gradually exterminating remaining weeds by judicious use of cattle and sheep.

Mr Tapa's pastures were becoming well established; 'eventually this will be a very good farm'. Rodda noted that good use had been made of pines

During the 1940s, Māori farmers were obtaining good financial returns from their farms; pastures were steadily improving, with liberal applications of phosphatic top dressing.

Māori woman and child harvesting potatoes at Whakatāne, c. 1930s (Alexander Turnbull Library, PAColl-8841)

for killing gorse and blackberry on hills too steep for grazing. He also observed that Mr Tapa had taken great pride in the home and surrounding gardens. However, he noted that more shelter was needed.

Bringing Ragwort under Control

Third place was awarded to Wiremu Matene Naera from Waiotemarama, just south of Ōpononi on Northland's East Coast. Mr Naera was described as a 'thorough and economical farmer' who would in time become very successful. Mr Naera was very keen to expand his knowledge of farming, it was noted.

Fourth place was awarded to Heemi Lawson of Tōrere, who Judge Rodda described as a keen and intelligent man who had done well in developing an area that was once badly infected with ragwort. By bringing the ragwort under control with the use of sheep, and also by top dressing, Mr Lawson had been able to obtain good financial returns from both cows and sheep. His pastures were steadily improving, helped along with liberal applications of phosphatic top dressing. Judge Rodda commended Mr Lawson in particular for the way in which he looked after his equipment and machinery.

Remaining Entrants

The farm of Kopua Waihi of Ruatōria, judged fifth in this year's competition, had unfortunately not shown to the best advantage owing to the impacts of a prolonged drought, though allowances were made for this. Mr Waihi was a man of ability 'somewhat above the average Maori'. He had made a good job of subdividing his farm, and his fences were the best of any examined during this year's entire tour. Judge Rodda recommended the planting of some extra shelter and a system of tile drainage.

Katene Huriwai of Tikitere, near Rotorua, was placed sixth. On the day of Judge Rodda's inspection, Mr Huriwai was 'laid up on account of a strained back caused by attempting to do a job beyond his strength'. Judge Rodda commended him on his progress thus far in the face of numerous difficulties. His pastures were in generally good condition, and would improve with further top dressing.

Tupu Erueti of Tautoro, just south of Kaikohe, was placed seventh. Judge Rodda described Mr Erueti's property as 'an exceptionally hard place to farm': it was always likely to be extremely hard and unproductive during the dry weather. Shortly before Judge Rodda's visit, Mr Erueti's water supply had failed: his livestock had to be driven to water every day. All the land suitable for grass was in fair pasture, but the remaining 30 acres was not at all ideal for dairying and could possibly be better off farmed with sheep – provided it could be managed in conjunction with an outlying area. Mr Erueti was a hard-working and reliable person: 'it must be admitted that he has done very well under the conditions he has been up against'.

Tiaka Tāmaki of Pirongia, who was placed eighth, had made good progress on his farm, and much greater improvements were to be expected during the next two years. Judge Rodda noted that pastures were also likely to improve with fertiliser. Under further supervision, Mr Tāmaki would become 'a reasonably good farmer'.

Ninth-placed James Waitaringa Mapu of Hastings had 49 acres of pasture near Pākōwhai in good condition. Grass in this area of Ōmaranui was notoriously weedy and required constant renewal. Mr Mapu's farm was nicely situated, and good work was being done to eradicate tall fescue.

Eruera Hoera was in last place. A vast amount of work had been done on this farm over the years to bring it up to its present admirable condition. However, the farm's overall production levels were not as high as one might have expected, largely because of the broken and difficult nature of the country. Judge Rodda recommended a more productive type of cow: possibly a Jersey-shorthorn cross, which would greatly improve Mr Hoera's butter fat production. Rodda noted that 'Great pride is taken by this family in their home and garden, which are exceptionally well kept.'

The Trophy is Missing and Then Found

A *New Zealand Herald* report of the results of the competition of this year reveals the curious fact that the Ahuwhenua Trophy itself (that is, the cup that had been created anew after the fire of 1937 destroyed the original) had been missing for the past three years and had only now been found. The *Herald* noted: 'The cup itself does not pass into the hands of the winners, who instead receive medallions and certificates.' It had gone missing in 1943 while in transit between Rotorua and Wellington. It was at first thought that it may even have been sent overseas by a mistake involving military stores. Inexplicably, it was eventually traced to Hamilton, in a storehouse where it was mixed in with some personal luggage. Meanwhile, insurance on the trophy had been collected and a new trophy commissioned. It was noted that this new 'Bledisloe Cup' was one of the largest trophies in New Zealand, standing two feet tall.[82]

1947

A Judge from the North

A new judge was once again appointed for the 1947 competition: Mr E H Arnold, an instructor in agriculture from Whāngarei. Judge Arnold continued to use Mr Glanville's points system to evaluate competitors. It is to be noted that the detailed points charts that resulted from this system were used by Native Affairs supervisors when conducting follow-up visits with all competitors, including the winners. In fact, the Under Secretary of Native Affairs frequently wrote to his field supervisors, reminding them that follow-up visits with farmers were important and constituted an obligation under the original rules drawn up to administer the Ahuwhenua competition in 1932.

Although Judge Arnold awarded near-identical points in most categories, there were notable differences in his assessment of provision of water supplies, provision of shelter, condition of 'homestead and gardens' and the cleanliness of equipment and sheds.[83]

'An Efficient Producing Combination'

There were five entrants in the 1947 competition. Wiremu Matene Naera of Waiotemarama was selected as the winner; he had been placed third the previous year.

Although Mr Naera's property was separated by a road, and contained a fair amount of poor hill country, the results he had achieved showed his ability, energy and initiative. The systematic improvement of his pastures on growing crops and sowing seeds down to good pastures was to be commended. The maintenance and cleanliness of his sheds and utensils was of a very high standard 'and was not equalled on many Pakeha farms'. The farmer and the farm together formed 'an efficient producing combination'.[84] Mr Naera was presented with the Ahuwhenua Trophy by Hon C F Skinner on 19 July 1948 at Waiotemarama.

Good Farmers Working Efficiently

Second place was awarded to John Savage of Ōpoutere. Mr Savage farmed in a very isolated valley of a standard 'seldom met with in dairy farming', said Judge Arnold. Despite his numerous difficulties, Mr Savage was performing very well. His homestead, surroundings, orchards and gardens were particularly commended. Production from improved pastures was at a high level, and his young herd was being raised excellently. The judge noted that, when fertiliser was again available and unrationed, the area of pasture on the poor sandy country needed to be top dressed with super phosphate and oversown with 4 lbs of subterranean clover seed per acre.

Third place was awarded to Mrs Mihi Stevens of Rangiāhua, whose farm was described as 'quite a good farm unit worked in quite an efficient manner'. The property was well laid out in the circumstances but was handicapped by its limited access across an unbridged river. Stock was of a good type, but would be improved by herd testing. Extra lines of shelter would also be of great benefit to this unit. Top dressing and the use of lime would greatly increase further production.

Among remaining competitors, the unit of Tiaka Tamaki in Pirongia was a farm founded on 'rather poorer soil than the others'. It, too, was awaiting the availability of unrationed fertiliser. Fencing was in a good condition, and shelter belts were well planned and protected. The outstanding feature of Euera Hoera's property was a reserve fund that allowed for further improvements to be made when materials were more freely available. Mr Hoera was a good and efficient farmer, said Judge Arnold. However, his production would be greatly improved if his herd could be gradually changed over from shorthorn to Jerseys. This would best be achieved by acquiring at least ten good Jersey in-calf heifers and a good Jersey bull to cross with the shorthorns. In five or six years, the herd would be predominantly Jersey. Mr Hoera was also commended for the quality and neatness of his homestead and garden, which were nestled into a clump of native trees.

1948

Changing Names

The Ahuwhenua Trophy competition in 1948 attracted twelve competitors, judged by Mr G A Blake from Matamata, who reported that he had visited all of the farms in early 1948, accompanied by either a 'Maori Affairs Supervisor or the Welfare Officer'. This statement reflects changes introduced by the Maori Social and Economic Advancement Act passed in 1945. One was a name change from the 'Native Affairs Department' to the 'Department of Maori Affairs'. For some time, the word 'native' had been receding in use under the weight of its mostly negative nineteenth century connotations. The name change took effect on 17 December 1947, after which it was gradually introduced throughout the public service; in 1954, for example, the Native Land Court became the Maori Land Court.[85] A second important change was the creation of the senior positions of chief welfare

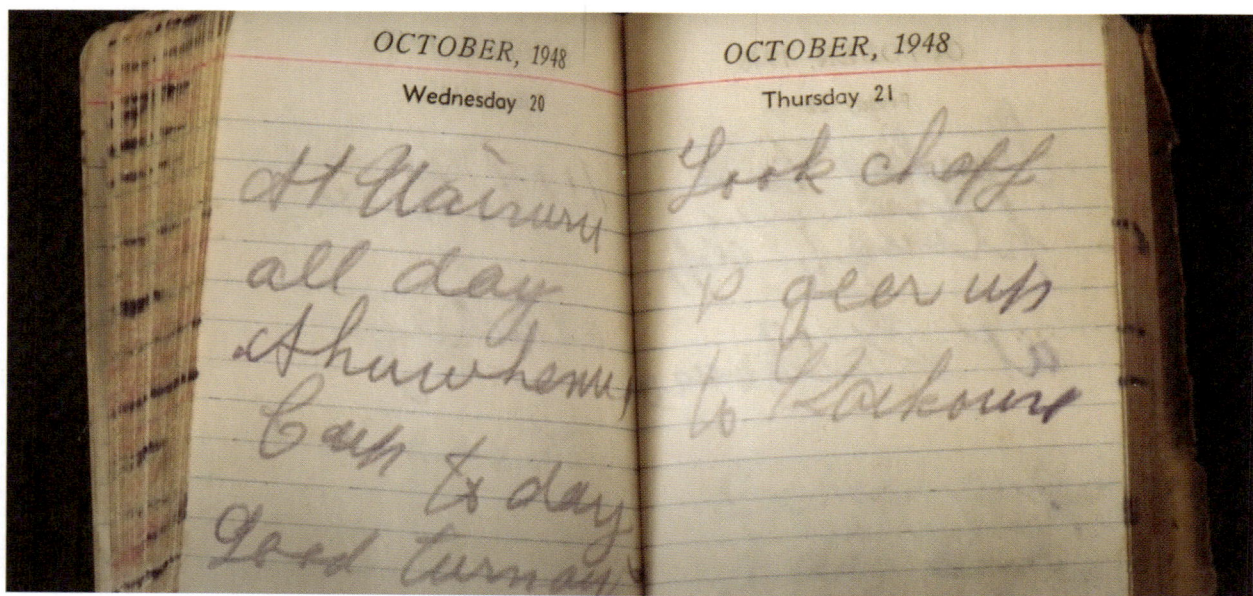

Top: Tikirau and Urukakengarangi Callaghan
Bottom: Diary entry from Tikirau Callaghan the day he was presented with the trophy. The entry reads:
'At Wairuru all day Ahuwhenua Cup today good turnout.'
(Images courtesy of the Callaghan whānau)

officer and district welfare officers. These were in the main Māori men and women charged with ensuring that Māori welfare issues were attended to in the rapidly changing post-war New Zealand society. Most welfare officers would not be appointed until the 1960s to deal with issues arising from Māori urbanisation. However, it is interesting to note that, from 1948, welfare officers often accompanied Ahuwhenua judges as they toured often remote rural areas visiting isolated Māori families, many of whom, as the judges were reporting, were 'living rough'.

A Winner from Raukokore

The winner of the 1948 competition was Tikirau Callaghan of Raukokore in the Bay of Plenty. Mr Callaghan was of Whānau a Apanui descent, from the Whānau a Pararaki hapu. His farm, located at Ōrete, near Waihau Bay, was 45 acres in size, mostly flat, and covered by low-lying creeping blackberry broken by swampy gullies without fences or buildings.

After service during the Second World War, Mr Callaghan purchased land at Ōrete using his war gratuity payments. He lived 'for some years in an old whare with an earth floor' before managing to build a house with a Native Affairs subsidy, which he also used to improve his property's drainage. Judge Blake described Mr Callaghan as a 'silent lone hand worker full of initiative, preferring to push ahead in this way'. He was an 'exceptionally good dairy farmer' whose management style stood out as an example to Pākehā farmers – he had hewn out a model farm from a blackberry- and weed-infested land holding.

Following years of careful management, Mr Callaghan was now in financial credit. His current stock comprised thirty-four milking cows, two dry cows, five yearling heifers, six calves, one bull, two pigs (sows) and four horses. His butterfat yield for 1945/1946 had been 8666 lbs, compared with 9739 lbs for the current year.[86] Judge Blake said that his pasture control, his piggeries and his production per acre were outstanding.[87]

No Fences and No Access

Second place was awarded to Patuwahine Albert (Arapeta) of Ōmaio, near Ōpōtiki in the Bay of Plenty. Mr Albert was Whānau a Apanui of the Ngāti Horowai hapū. His farm was currently stocked with twenty-two milking cows, ten yearling heifers, eleven calves, one pedigree bull, one pig and five working horses. Mr Albert's farm comprised 30 flat and 51 hill acres; Judge Blake described blackberry and ragwort as the 'worst of his enemies'. The Ōmaio area had originally been infested with heavy scrub, with a small creek that had now been harnessed for water. Originally, there had been no fences or buildings on the property, and there was once no road access.

Mr Albert was described as an excellent worker and exceedingly cooperative with Maori Affairs supervisors. The outstanding feature of his farm was its remarkable conversion from bush to quality pasture. Mr Albert's pasture management was of the highest order. Judge Blake suggested that Mrs Albert must also have been a major contributor to the success of the farm; 'this old couple have done remarkably well and now have a well divided farm with clumps of native bush reserved for shelter.'

Other Competitors

Mr Bunny Otene of Hastings, who was awarded third place, had achieved similarly in the 1943 competition. Judge Blake noted that the most outstanding feature of his farm was the homestead and garden, 'with its beautifully furnished home which was kept spotlessly clean'. Mr Otene was a good farmer engaged mainly in agricultural work. The type of stock he ran, however, left much to be desired. His main need

By 1945, Māori welfare officers often accompanied Ahuwhenua judges when visiting Māori farmers because many farmers were still observed to be living rough.
Māori family in front of a small house on a rural property, 1937 (Alexander Turnbull Library, MS-Papers-0270-027-01)

In 1949, Judge Allo noted that Māori farmers had 'significant natural difficulties to deal with in rough topography, poor drainage, swampy and weedy ground'.

Two Māori work on a felled tree in native bush, c. 1930s
(Alexander Turnbull Library, PAColl-6585-31)

at the moment was a suitable shed to house the valuable machinery that he had recently acquired.

Fourth place was awarded to Mrs E F Clements of Tauranga, who 'deserved special praise' for the manner in which she had made a farm of such a gorse-infested area and for the shelter she provided for her stock. At the time of the inspection, however, pastures were poor, and Indian doab was in evidence. The practice of cutting the same field each year for hay had caused the quality of the grass to deteriorate.

Judge Blake commented variously on the remaining competitors. Fifth-ranked John Savage had brought his farm to a high state of production from difficult scrub, but his pastures were paspalum-dominant and needed more drastic harrowing. Mr Heemi Lawson, who was placed sixth, had developed a good farm, though his location was subject to frequent and debilitating flooding. J Mapu was deserving of great credit for the manner in which he had developed his difficult holding in such a short time, though his pastures lacked clover and possessed too much barley grass. Mr Kira Te Whare was an elderly man who had made a farm from scrub and ragwort country; however, he needed permanent labour to assist him. Ereura Hoera was seriously handicapped by having access to his farm impeded by a river which flooded often. Te Whitu Rareata Warbrick deserved special mention for the development that he had already completed and for the cleanliness of his cowshed and homestead. He was, however, severely handicapped by poor access and a river with no bridge. Kohika Rakaupai was making a good farm out of native bush but was handicapped by the hilly nature of the countryside. Mihi Stevens had developed a high-producing farm from swampland overgrown with rushes but badly needed permanent and reliable labour to assist her in her maintenance work. Robert Hauraki had brought scrub land into production, but had not maintained his farm as well as the others. An urgent need was the provision of better drainage from the homestead.

Mr Blake recommended that all competitors should be taken together and shown the farm of the winner so that they could see for themselves what the winning farm looked like, and perhaps what was lacking in their own farms. This occasion could be a field day, he suggested, and could do much to stimulate all entrants to better farm methods.

Presentation of the Trophy

Appropriately, for a man of Mr Callaghan's distinguished war record, he was presented with the trophy at Wairuru Pā, Raukokore, on 20 October 1948 by Lieutenant General Sir Bernard Freyberg VC, now Governor-General, who had commanded the 2nd New Zealand Expeditionary Force and the 2nd New Zealand Division with distinction in the Mediterranean during the Second World War. An address of welcome to Sir Bernard was delivered by John Waititi (formerly Major of the Māori Battalion); also present were Peter Hauraki, Sid Waititi, Moana Waititi, James Callaghan, Haukino Paora, Te Ara Ngamoki, Tohi Koopu, Paul Ngamoki and Mrs Mihi Stirling. The directors of the local Te Kaha Dairy Company were also invited. Mrs Ben Callaghan, secretary of the Wairuru Women's Marae Committee, was in charge of all arrangements.[88]

1949

Taranaki Earns a Place

Mr A V Allo, instructor in agriculture, Tauranga, was appointed as judge in 1949.

One notable feature of the 1949 placings was the awarding of second place to Waerata Harris of Inglewood, the first farmer from Taranaki to be awarded a placing. Taranaki was possibly the last of the dominant farming areas of the North Island to earn such a major placing. Generally speaking, as Richard Boast has argued, Māori dairy farming did not develop until the late 1920s, significantly pushed along by Ngata's land development schemes. As a result, a profitable and innovative industry had passed Māori by. Had the land confiscations of the 1860s not occurred, Māori farmers in Taranaki, the Waikato and Tauranga may have been better placed to take earlier advantage of the dairying industry's expansion.[89]

Winners in a Boom Year

After due consideration, Judge Allo nominated Eurera Hoera of Takahue, who had been placed third in 1945, as the winner of the 1949 trophy. Irihapeti Tuhakaraina of Tauranga was judged third and Paul Toroa of Ruatōria fourth.

In the late 1940s, major changes were occurring in New Zealand's farming sector. The value and volume of farming exports and the total trade per head of mean population reached the highest figures on record in 1948. These figures alone attested to the continuing value of farming as the major provider of the country's export income. Māori farmers were well placed to contribute to these boom export years.

A True 'Son of the Soil'

Judge Allo commended the standard of the 1949 farms.[90] The winner had 'done an exceptional job' in breaking in his farm from standing bush, Allo said. Though Mr Hoera had received minimal funding assistance over the years, he was now 'much in the credit'. Most of the enormous labour he had expended over the years had been single handed. As a consequence, his farm could be regarded with pride by any farmer, either Māori or Pākehā. The land farmed by Mr Hoera was easily the most difficult piece of land of any of this year's competitors. Despite this, the standard of his pasture management and the quality of his stock was of the highest quality. 'Mr Hoera can truly be termed a "son of the soil"', Allo said.

Other Competitors

The second and third place winners, Mr Harris and Mr Tuhakaraina, were very close behind the first-place winner. Both had brought into production areas long abandoned to weeds. The fourth-placed farmer, Mr Toroa, had done 'great work' on his farm, which he had occupied for only a short period of time. In time, said Allo, this farm would be of a high standard.

The fifth-placed farmer, Mr Hohepa Takuira of Panguru, also had a small unit that when fully developed would make a 'splendid dairy farm'. The sixth-placed farmer, Mr John Tiwha of Manunui, had a much larger unit that was more of a challenge and was seriously handicapped by having to do all of the work himself without the benefit of outside assistance. Mr Tiaki Tamaki of Pirongia, in seventh place, had a property that was badly infected with ragwort that was not under control at the time of inspection. Judge Allo noted that 'there was the prospect that this farmer would be forced out of dairying and into sheep in the near future'. However, Allo noted that Mr Tāmaki was keen to make a success of his unit and outclassed all other competitors in the amount of supplementary feed he had produced.

Placed last, the farm of Erina Hunia of Broadwood was by no means poorly farmed, but would need a considerable amount of pasture improvement and development before it attained the level of the first four competitors.

Natural Difficulties

Judge Allo noted that all competitors had significant natural difficulties to contend with, from rough topography to poor drainage or swampy and weedy ground. However, all competitors were showing considerable ability in coping with these toilsome problems. Allo commended the keenness of the Maori Affairs supervisors, 'who were doing their utmost to help the Maori farmers to farm their properties along sound modern lines'.[91]

Most farmers were employing rotational grazing in order to develop their pastures, Allo noted. However, most of the paddocks he observed were too big for the numbers of stock being carried. Subdivision should be considered a priority as soon as fencing wire and water piping become available following the restrictions following the war. In most cases, too, he considered that insufficient fertiliser was being applied. Most paddocks were well watered, though in some cases this had led to problems in pasture management.

General Comments

As many of his predecessors had done, Judge Allo cautioned about the importance of shelter and winter feed, and commented on the generally negative state of pig raising and the generally positive evidence of homesteads that were well cared for. He said that production was generally good among Māori farmers, comparing favourably with surrounding Pākehā farmers. Mr Hoera, for example, had one of the highest producing herds in his Northland district.[92]

1950

Judge Montgomery Returns

In 1950, Mr K M Montgomery returned as judge of nine nominated properties. High standards of farming were once again observed; Montgomery commented that the variation in the points he awarded 'had much to do with the adverse climate conditions of the previous four months'.

A Sound Economic Unit

The winner of this year's competition was Mr G Thompson of Ōtorohanga. Mr Montgomery commended Mr Thompson because he had had practically no supervision during his farming tenure, having had to rely on his own initiative. Although his loan account was still high, he had greatly improved his stock, pastures and chattels. The neatness of his sheds and the cleanliness of his tools was worthy of special mention. Mr Thompson's foresight in purchasing a high pedigree bull five or six years earlier had led to increasing butterfat returns, assuring this property of a good economic future. Montgomery recommended that future investment be directed towards water reticulation. Herd testing was also important; on such a small unit, every cow needed to be an efficient producer.[93]

Other Place-getters

Second-placed Mr Paul Toroa of Kākāriki, near Ruatōria, had revealed real skill in 'getting things done', Montgomery decided. The heavy rain that had fallen just before inspection day had helped to revive Mr Toroa's pastures of ryegrass, white clover and crested dogstail, which were well managed. Newly planted winter breaks with double fences would provide ample shelter and wind breaks in a few years, the judge noted. A good supply of water piping had recently arrived and would provide water for most of the farm, reducing the need to use the river water to water the stock. The stock was in very good order, and the heifers were 'an attractive line'. The piggery, however, needed remodelling.

The farm of third-placed Mr P Hei of Te Kaha was well laid out and could be described as a 'model farm', Montgomery reported, reflecting this entrant's 'cleanliness and attention to detail'. Wire fences were well constructed, and wide gateways allowed for easy vehicle and stock access. A lucerne stand was producing a good supply of grass, but some cultivation was needed to check the 'ingress of Agrostis sp'. Good fencing was also on show. Overall production per cow was not as high as Montgomery had observed in other highly concentrated farmlets: once again, herd testing was strongly recommended.

Mr K Waihi of Ruatōria, who was awarded fourth place, had made good progress in developing a troublesome swamp area. However, once the older pastures improved, this unit would compare favourably to any other in the district. The herd on this farm was not tested; this was a necessary step forward, Montgomery said. The newly remodelled piggery had also added much value to Mr Waihi's operation.[94]

Sixth-placed Mr E Pou of Kaikohe was greatly helped in his efforts by his family. Gorse and blackberry had been recently cleared, but unfortunately the piggery had been built too close to the cowshed. This would need to be moved to conform to dairy regulations. A pampas stand for winter feed had been established to complement the paspalum hay made in this district. Montgomery noted that Mr Pou's pigs were well grown and healthy. His pastures were good; those on the flat were balanced with ryegrass and white clover. The newer pastures had a good covering of brown-top, and it would be a while before fertility could be raised to allow the ryegrass and white clover to become dominant. Mr Pou was undoubtedly a good worker, but his farm had not reached the same level of development of many of the other entrants. He had, however, made good provision for winter feed, such as swede and hay, ensuring that his stock would come in well for the next season.

J Chadwick, who was placed seventh, was handling his hill country farm quite well. However, Montgomery suggested that more subdivision was required, and better provision should be made for the watering of his stock. Pasturage was relied upon solely for winter feed, which was not good farming practice. Pastures were good, but would improve with further logging up, burning, re-seeding and top dressing. For the King Country, however, a crop of roots for winter feed was most desirable.

Adverse climate conditions had affected eighth-placed competitor Mr J Savage, who Montgomery considered had received very little support from supervisors. In the judge's opinion, too much attention had been placed on his paspalum-dominant pasture, which had not produced good grasses this dry season. This had led to a low condition of cows, which would not be improved upon unless the winter was a mild one. However, a good supply of hay was on hand, which would tide Mr Savage over the winter, though some thought needed to be given to the saving of silage for those times of low pasture production.

Finally, Mr Bowman Yates, placed ninth, had been recently affected by a new road through his farm, which greatly disturbed all of his subdivision fences. Also the very dry season meant that he had had to dry off his herd earlier than normal, thus affecting his season's production returns. Given the recent weather, his herd was in good order, and was expected to come into profit during the next season 'in great fettle'. In some four years or so, when the farm had been reorganised and the piggery remodelled, Mr Yates would be a worthy competitor for the trophy.

1951

Judge Taylor Returns

Mr C R Taylor, who had recommended the introduction of a new marking system based on 'four different categories of property' when he judged the competition in 1941, returned to judge the 1951 competition.

Eight Māori farms entered, generally located between Ōtorohanga, Hicks Bay and Kaikohe in the north of the North Island. Judge Taylor visited these

Māori sheep farmers in 1951 were doing well out of limited sheep and wool sales, but many still faced water supply and pasture problems.
Three Māori men shearing sheep in a woolshed, 1954 (Alexander Turnbull Library, PAColl-6304-01)

farms in early 1951, accompanied by the respective district supervisors. The Department's inspecting field supervisor also accompanied him on seven of the eight farm visits and later filed his own report on the competition and Māori farming in general (see the text box on p. 73).[95]

Taylor introduced an updated points system, increasing the number of categories from twelve to nineteen. Assessment now covered initiative since coming under development, general management and security of commercial profit, butterfat per cow and per acre, pig meat and other production returns, pastures, fencing, water supply, shelter, sheds and yards, stock quality, stock replacement strategies and provision for supplementary feed. Also still included were the categories of 'neatness and cleanliness of house and gardens' and the keeping of good farm records. Finally, a new category was introduced dealing with 'equalising the differences in the circumstances of the competitors, such as climate, soil type, topography and other unusual features'. This category clearly allowed judges to make allowances for the 'degrees of difficulty' faced by individual Māori farmers, which had been of concern to Taylor in 1941.

A Conscientious Farmer

First place in 1951 was awarded to Mr Kopua Waihi of Ruatōria. Taylor concluded that the results of Mr Waihi's conscientious industry were well reflected in the high standard of his farm. Mr Waihi's pastures were good, although they showed a lack of top dressing, which required urgent attention. The stock on this property was generally in a fine condition though pig returns were 'half of what they could be' with an improved layout.

Mr Waihi had done well out of some limited sheep and wool sales. Fences and buildings were well maintained, though more shelter needed to be provided for. His water supply was adequate, and neatness and cleanliness were outstanding. Mr Waihi's entrant's loan account was in a very satisfactory position.

Assessing the Ahuwhenua Competition

At the completion of the 1951 competition, J H Flowers, inspecting field supervisor with the Department of Maori Affairs, reported on the Ahuwhenua competition so far. He had both positive and negative things to say about the competition itself and the state of Māori farms in general, including the following points:

- All farms the judge visited were very attractive, and no efforts had been made to 'groom' the properties prior to judging. He considered it a shame that more entrants could not have been found for the competition.
- Judge Taylor's new marking system meant that sheep and dairy could now be judged more evenly.
- The presentation to the judge of background information as to the history of farms since development was generally poor; mistakes were often made.
- Stock tallies were also often incorrect, which meant that, when comparing revenue against expenditure related to stock numbers, the judge began with incorrect data.
- All farmers should be using sheep for ragwort control – many farms inspected for the competition were badly infected.
- Records of wool sales supplied to the judge were generally patchy; important data was only obtained after asking competitors directly.
- Supervisors in District Offices should be responsible for ensuring important data was available.
- Farmers should be taught to improve their record keeping – many had no records of the pigs or bobby calves they had sold.
- In one case, a farmer could not access his farm diary because of a lost drawer key. Flowers admonished: 'An average efficient Maori farmer should be expected to keep his simple records of expenditure and receipts.'
- Pastures on inspected farms were generally poor and had not been adequately fertilised, a situation that required urgent attention; soil analysis was now imperative: 'The Maori farmer is now seeking knowledge of the more scientific aspects of farming and I think would welcome his soil being analysed and would perhaps be more keen to apply manure if he knew the deficiencies of his soil.'

Other Place-getters

Second-placed Mr C H de Thierry of Pirongia had worked hard with his family since 1939 and had converted his once gorse- and blackberry-infested property into a quality farm, with about 15 acres left to develop. Taylor considered that his pastures were generally excellent but needed constant top dressing to maintain their standard. Provision of winter feed was excellent – swedes and hay especially. Stock condition was good, though evidence of 'undershot' in the jaws of the heifers was evident. Pig production was also good and comprised a significant portion of this competitor's income; however, a new pig layout was urgently needed. Fences, gates and water supply also required attention.

Mr K A McGregor of Waihi was awarded third place. Although his farm was a part of a land development scheme, Mr McGregor was also an employee of the Department of Maori Affairs who consequently chose to do most of his farm development work himself. While he had been given the credit for this, he had of late been quite sick, and had had to rely on his wife and daughters of school age to carry on the essential work on the farm.

Certain aspects about the farm therefore were in need of attention, not surprisingly, although good work over the years was in evidence. Pastures were variable, between hay paddocks and grazing fields. Taylor suggested potash fertiliser as a necessary supplement for the pastures. Hay had been saved and was in good condition. Stock was also in good condition and showed evidence of careful handling. No pigs had been raised over the previous year, which Taylor found surprising given that this was normally one of Mr McGregor's strengths.

Water supply and shelter were good, though fences and gates were showing some disrepair, as were buildings and sheds. Ragwort was showing itself in some of the paddocks, and tools and utensils were seen to have been left lying out by the haystacks.[96]

Manu Stainton of Hicks Bay was placed fourth. Mr Stainton was a young competitor who had taken over his father's farm upon his father's death in 1946; he could not therefore be credited with having achieved the unit's development work. However, Mr Stainton had recently ploughed and sown new grass on some of the pastures, with good results. The farm suffered from flooding caused by tidal backing up of a stream running adjacent to the farm. This has had the effect of reducing the acreage available for productive work. The country around Hicks Bay was light sand dune country, which was good for the winter but difficult in other times. Taylor observed that little top dressing had been carried out, though some lime had been applied about four years ago. The stock were in a good condition, though 'undershot' jaws in heifers were pronounced and required attention, especially through the purchase of bulls. A good income had been obtained from wool and sheep, but not nearly enough from pigs. Taylor reported that salt-laden winds off the sea were affecting wire and galvanised iron on the farm. The cowshed was old but in good condition; the piggery layout was poor.

1952

Women Winners

Mr Taylor returned to judge the 1952 competition. A unique feature this year was the number of women entrants – four of eleven – and the fact that two of them were placed first and third.

The winning entrant was Mrs C R Beasley from Kaiaua, near Pōkeno. Mrs M Reid of Manganui was third. The two other women entrants were Mrs Elizabeth Clements of Ōmokoroa, who had been placed fourth in 1948, and Miss Winnie Sherlock of Tōwai. Until now, only five women had won placings in the competition since its inception in 1933.

Nominating Mrs Beasley as the winning entry, Judge Taylor wrote that her victory was 'richly deserved for her consistent endeavour, often against tremendous difficulties in the early years of development'. Mrs Beasley had farmed her property at Kaiaua for fifteen years, and had produced a farm of high quality. However, Taylor suggested further subdivision of her paddocks and electric fences for the break-feeding of pasture growth.

Taylor noted that Mrs Beasley needed to save more hay and silage when surplus growth was available, thus reducing waste. More cows might also be milked, if additional labour was available. Top dressing, while satisfactory, needed to be continued. Mrs Beasley's fences were in good condition for a dairy farm, and there was good provision of water and shelter. The quality of her stock was very good, as was her equipment and utensils. The general neatness and cleanliness of her farm were excellent, as was the condition of her records.

Other Place-getters

Mr Tapuae Rogers of Tōrere, who was a lessee of his property, was placed second. Though they had only been on the farm for two and a half years, Judge Taylor considered that Mr Rogers and his two sons had 'worked wonders' with scrub-clearing, grassing, fencing and maintaining buildings.

However, he noted that the farm had a major access problem, even by motor vehicle. Before farming his present block, Mr Rogers had cleared another block of 75 acres further up the same isolated road. In spite of his taking out heavy loans for major developments, including additions to the house, Mr Roger's accounts were actually in credit. Production figures were good in trying and difficult circumstances. Stock was well cared for and in excellent condition. Pastures were in development, but would be improved with good management and time. Pigs were a useful additional source of income. Cleanliness was very good around the cowshed and buildings and especially the home.

Taylor awarded third place to Mrs M Reid of Mangonui, whose farm was located in difficult country; as Judge Taylor commented, 'few would relish breaking in such steep hillsides covered in gorse and manuka with the ultimate view of dairying on them'. However, Taylor said that Mrs Reid had achieved this over the previous seventeen years; the endeavour had been the more remarkable because of Mrs Reid's continuing ill health.

Mrs Reid's pastures were generally very good, though her steep hillsides were always threatening to revert to subterranean clover during the dry seasons. The fences and water supplies on this property were also good, and buildings and plant were well maintained. Taylor approved of the stock and production outcomes and the 'striking' overall cleanliness of the farm, buildings and homestead.

Fourth place was awarded to Mr C H de Thierry, who had come second in the previous year. Progress since then had been good, Taylor noted. An additional

Hard-working Māori farmers were often assisted by school-age children who helped with the essential work on the farm.

Two Māori boys on horseback, Waikato, 1938 (Alexander Turnbull Library, WA-12560-G)

9 acres of noxious ragwort and gorse had been restored to first-class pasture. Thus, of the 71 acres farmed by Mr de Thierry, 56 acres were now planted in fine pasture. Supplementary feed had been separately planted, including swedes and green feed cereals. Pig production was also good, Taylor considered, but a new pig layout was urgently required. Fences and gates were reasonably good, and the farm was provided with a good water supply. Little fault could be found with the stock – Taylor noted an absence of 'undershot jawed' animals. Cleanliness around the home and farm was also very good.

Other Competitors

Mr Taylor highly commended Mr Sam Shelford of Whakapara for the quality of his pastures and stock; his property reflected 'in every detail his industry, initiative and resourcefulness'. Mrs E Clements of Ōmokoroa, who had previously been awarded fourth place in 1948, had worked hard for seventeen years with her husband, and now her son, to convert a one-time ragwort- and gorse-infested property into an excellent small dairy farm. Mr Percy Topia of Dargaville had made a good showing of a very difficult piece of low-lying swamp; given a few more years, Taylor considered he would be equal to the task of carrying off the trophy.

Mr Herewini Rewa of Ruawai, a widower, had an exceptionally clean and well-kept home. Judge Taylor observed that he was the 'most excellent cook of a hot meal, as I learned from actual experience'. Sadly, however, Taylor found his farm was in some state of disrepair, though stock returns were good. Miss Winnie Shedlock of Tōwai unfortunately 'did not stand much of a chance of securing a place this year'. However, her entry 'served to show what a capable woman could achieve through perseverance, determination and a will to succeed'.

John Chadwick of Manunui was the only sheep and cattle entry in the year's competition. Mr Taylor acknowledged that assessing his farm's 'true worth' against that of the dairy farms was problematic. He noted that Mr Chadwick's farm suffered from a lack of subdivisions and a lack of a permanent water supply in most of his paddocks. Mr Karatau Hau of Whangaruru was unfortunately away on business when Judge Taylor called, 'so he could not be tested as to his farming knowledge'. Mr Hau had been working his farm for sixteen years, breaking in the most difficult of landscapes; many attendant issues still required his attention.

1953

Isolation

It is to be noted of the 1953 competition that this year, as in all years since the end of the war, a recurring theme in Judge Taylor's report was that of farming isolation and physical difficulty. Many Ahuwhenua competitors were commended for their 'industry, initiative and resourcefulness' in the face of this challenge. The trophy winner in 1953 was a particular example. Rohe Takiari of the Ngāti Te Wehi hapū, in the Waikato, farmed at Hauturu, Kāwhia: a district that had poor roads, no electricity and mail only twice a week. The nearest post office and store were 13 kilometres away.[97]

Winner among an Outstanding Group

In complimenting Mr Takiari on his win, Judge Taylor commented that in the four years he had been judging the competition, he had not seen such an 'outstanding group as this year's list of nominees'; they exhibited 'the highest standard of farming efficiency'. The endeavours of Mr Takiari, he said, were a 'monument to his own industry and resourcefulness and an example to others of his Race'.

Mr Takiari's farm was 300 acres, 'poorly situated in a very remote area of Kawhia'. Taylor considered that his farming efforts in this difficult location showed clear thinking, reasoning power and knowledge. He noted that Mr Takiari's farm books and diary were perfectly kept. Taylor suggested that Mr Takiari consider arranging additional subdivisions to facilitate better pasture control; a more adequate water supply for the back portion of the farm; provision of more shelter and trees, especially for shade in the hot months; and soil testing, which was necessary to ascertain the need for lime and potash to supplement phosphate. Some of Mr Takiari's pastures appeared a little harsh. However, fences and buildings were in excellent order and plant, equipment and buildings were well looked after. Livestock conditions were very good, as was general cleanliness around the homestead.

Second Place

Second place was awarded to Aumihi Davis, who had taken over his farm four years earlier, at a time when it was infested with ragwort and blackberry and had few good fences. Pastures had almost been run out, and no top dressing had been undertaken for a very long time – in short, said Judge Taylor, it was 'a derelict farm'.

Sons and daughters often returned home to assume the mantle of their parents in managing the family farm when parents became older or unwell.
Māori children with buckets, c. 1910; photograph by Arthur Northwood (Alexander Turnbull Library, 1/1-010902-G)

Now, after four years of solid endeavour, a greater part of the farm had been ploughed and pastures renewed, and new subdivisions had been planned. Mr Davis had recently built a new house, cowshed and implement shed. Butterfat production had steadily risen and now stood at the excellent level of 286 lbs per cow and 184 lbs per acre, the highest figures Judge Taylor had seen in the four years he had been judging the competition.[98]

Taylor considered that stock on the farm was well cared for. The position of replacement stock was not so well developed; some planning was required to save calves to compensate for herd wastage and to maintain a reserve of heifers from which the stock might be increased without the need for purchasing high-priced stock. Pastures were generally good, though Taylor recommended soil testing to check the potash status of soils in paddocks affected by poor pasture. He also suggested some more subdivision and more shelter. He found a high standard of cleanliness and record-keeping.

Third Place

Judge Taylor awarded third place to Sam Shelford of Whakapara, who had come fifth of eleven in the previous year's competition.

Taylor reported that Mr Shelford now had 'a very nice farm in the making'. He congratulated him on his up-to-date outlook, his modern methods and his pasture management in particular, which involved break feeding, frequent harrowing and the use of electric fences. The judge suggested some extensions to his water supplies for improved pasture growth, as well as additional fertilising with potash. He also recommended soil testing. He noted that supplementary feed in the form of hay was amply provided, well stored and protected from stock.

Taylor commended Mr Shelford's cleanliness in and around the house; farm implements and equipment were also well cared for. Unfortunately, Mr Shelford's low production of butterfat was an issue. While his per-cow average of 232 lbs was satisfactory, he did not milk sufficient cows per acre, preferring instead to buy and sell stock to consume surplus growth at flush periods – a practice that could not compete with the milking of additional cows up to the full capacity of the farm's grasslands, Taylor considered. The judge suggested that Mr Shelford was running the risk of introducing disease onto his farm. Mr Shelford's loan accounts were good, though the ratio of production costs to net income was higher than it was for most of the other competitors.

Fourth Place
Taylor awarded fourth place to Miss Noti Tiopira, who had assumed control of her father's farm about three years earlier because it had been allowed to deteriorate considerably.

At that stage, Taylor noted, 'ragwort was everywhere, with fences, buildings and pastures in a poor condition'. Production had receded to a figure below 5000 lbs from 69 acres, or less than 70 lbs per acre. Since then, Taylor observed that the farm had been completely transformed: pastures restored, ragwort almost eliminated, fences rebuilt and new subdivisions initiated. The cowshed had been rejuvenated, a new pig layout had been built and water supplies had largely been restored. New grasslands had also been added to the farm, and butterfat increased accordingly to 115 lbs per acre. Stock conditions were good, though little had been done to provide for replacement stock. Good supplies of winter feed were available, and pigs supplemented the overall farm's production capacity.

Remaining Competitors
Judge Taylor found that Ilet Hemi's farm reflected the thoughtful work he had put into his property during the prior ten years, converting a derelict farm into a modern unit affected by comparative isolation. A T Hetet had occupied his property for only three years and had had to contend with virulent ragwort and blackberry, making the sowing of pastures difficult. John Savage had occupied his property for more than fifty years, and had, Taylor noted, 'battled along unaided in a most isolated place for 36 years'. Having almost no money, he had taken on roading and bush contracts in order to develop his property. Unfortunately, a greater part of his land was covered in sandy soils buffeted by high winds, detrimentally affecting his farming efforts. Hohaia Puriri had just returned from active service in Korea, reoccupying his farm of 91 acres and seeking to re-establish his badly run-out pastures. Ron King's farm was the most highly developed property prior to occupation that Judge Taylor had seen this year, though his pastures were 'fairly rough', and his cowsheds and piggery not ideal.

Early Ahuwhenua competitions emphasised economy of farm management, quality of stock and quality of produce, as well as progress in development.

Māori women harvesting kūmara, c. 1910 (Alexander Turnbull Library, 1/1-006265-G)

1954

One Competition Becomes Two

In Mr Taylor's judge's report of 1953, he had strongly recommended that two new Ahuwhenua judging categories – sheep and cattle, and dairying – be established for ease of marking but also to ensure that the marking process was fair and equitable.

In his home in Redhill, Gloucestershire, Lord Bledisloe read a copy of the current judge's report (as he did every year, often corresponding with the minister and officials on 'matters arising'), and noted the suggestion. On 2 December 1953, he wrote to the Minister of Māori Affairs, Mr E B Corbett, expressing his 'full sympathy' with the idea. Further, he offered to donate a second trophy, though he was 'none too well off these days', he added. The timing was significant, he wrote; he supported the provision of a second trophy as a means by which the Queen's coronation, and her forthcoming visit to the Dominion, 'might be celebrated amongst Maori farmers'.[99]

Accordingly, the Ahuwhenua Trophy competition in 1954 was judged in two sections. Taylor, returning as judge, noted that, in the past, judging had been difficult, with 'unlike things being hard to compare'.[100] He concluded that, this year, the separation of the competition into two distinct categories had made the judging relatively simple and was also more satisfactory from the competitor's point of view.[101]

Mr Taylor reported that the weather had been a definite factor this year, seriously affecting farm production returns.

Sheep and Cattle

First Place: The New Section's Inaugural Winner

Taylor awarded first place in the sheep and cattle section to Mr P Raharuhi of Horohoro, near Rotorua, describing him as a farmer of twenty-four years' experience who was also committed to furthering his knowledge of farming and to imparting that knowledge to Māori in his community. Taylor noted that his knowledge of the Horohoro district and its farming conditions was immense; he was also a highly skilled carpenter, as could be seen from the quality of his building improvements. Taylor commended Mr Raharuhi for 'exemplary' care and upkeep of his machinery, implements and utensils and described the management of his pastures, stock, shelter and winter feed provision as 'excellent'. The judge recommended some more subdivision, as ragwort was starting to show itself again. Some potash and phosphate were also needed on some pastures and 'a brightening up of the immediate surroundings of the house'. Taylor wrote that Mr Raharuhi's victory constituted an honour of which he could be justly proud; it was evidence of distinct achievement in practical common-sense farming, something that was 'unfortunately not always too apparent in rural life nowadays'.[102]

Second and Third Place

At the time second-placed J W Thompson had taken over his farm twelve years earlier, it had been derelict, though still carrying thirty cattle. Since then, Mr Thompson had replanted his pastures and replaced almost all of his fences. Water supplies were now much improved, Judge Taylor observed, and fields of damaging weeds had been brought under control. The property had been restored to a productive and profitable status. Originally a dairy unit, the farm had been partially converted to a sheep run because of its difficult terrain. By 1951, it had been completely converted to sheep and cattle. Stock conditions were now very good, Taylor reported, though lambs were still showing signs of having experienced a severe drought. Taylor noted that there was no woolshed or shearing equipment on the farm; a neighbour's facilities were used for the purpose.

Third place was awarded to Mr M Morehu. He controlled a large property of 296 undulating acres of pumice country, of which 248 acres were in permanent pasture. When the farm had been taken over, it was infested with bracken fern and tutu; in their place, fine ('though not first class') pasture for cattle and sheep grazing has been sown.

For the first years of his occupancy, Judge Taylor reported, Mr Morehu had battled away alone without the benefit of loans or assistance, until he had cleared and sown sufficient land to manage a herd of twenty-five cows. Ragwort had always been a costly problem. His pastures were now in a good condition, though recently a root crop planted as supplementary winter feed had failed. However, some hay was available. His fences were excellent. Taylor noted that no woolshed was on site, but Mr Morehu was able to use facilities owned by the Maori Affairs Department situated not too far distant. Taylor recommended potash and phosphate and a programme of soil testing.[103]

Other Competitors

Judge Taylor observed that competitor R Vercoe was also engaged in dairying, though the greater part of his income was derived from sheep and cattle.

Mr Vercoe's unit was 'a useful farm', although much work was still needed to convert it into a truly first-class unit. R Kingi managed a farm of 423 acres, and had recently completed some impressive subdivisions. However, his pasture growth was 'unnecessarily rank and lacking vigour', and he used insufficient fertiliser. E Pohio was farming 445 acres and had recently carried out extensive fencing and subdivision work. However, Taylor considered that his pastures were 'not as good as one would like to see', which suggested insufficient attention to the top dressing of his most difficult areas.

G Taia was a new farmer, having occupied his 450 acres for only two years and had a rather low quality of stock and low stock ratios. D Royal was farming 538 acres of well-grassed farmland with high quality stock but the condition of his pastures was disappointing to Judge Taylor. His brother, T Royal, farmed an adjoining 500-acre property with good quality stock but suffered continuing problems with a reversion to secondary growth.

Dairy

First Place

The inaugural first-place winner of the dairying section was Mrs Mihi Stevens, who had previously been awarded third place in 1947. Mrs Stevens and her husband had been working their property for twenty-two years 'in the most trying of circumstances', Judge Taylor considered, in a locality with poor access far removed from the sources of material aid – although he noted that, to some extent, this situation had now been substantially relieved.

Mrs Stevens farmed 89 acres, all of which was now in a productive condition. Twenty-two years earlier, the area had been a wīwī swamp cluttered with pūriri logs and stumps. Development had involved a tremendous amount of hard work on drainage and difficult clearing before grass could even be established or the first head of stock introduced. Upon Taylor's inspection, the area was well drained, grassed, fenced and supplied with the necessary buildings. Stock carried included seventy-four cows and replacements, three bulls, twenty heifers, five sows and twenty other pigs. Taylor found all in excellent condition, though the calves were 'hard of appearance'. He was also impressed by sheds and buildings, though he noted that the long drought had greatly impeded the provision of winter feed.

Second and Third Place

Judge Taylor awarded second place to T Rogers, who had been runner up to Mrs Beasley in the 1952 competition. Taylor saw an improvement in pasture management, fencing and water supply provision. He considered that buildings were all well maintained, though the stock was not as good as Mr Rogers' previous entry. Mr Rogers had secured a considerable income from growing kūmara and cabbages to supplement income from his mix of dairy and sheep. He was also earning a reasonable return from his pigs. Taylor reported his cleanliness around the home and milking sheds as excellent.[104]

Third place went to Aumihi Davis, who had been runner up to Mr R Takiari in the previous year's competition. Judge Taylor reported that Mr Davis' knowledge and management of modern farm practices could not be faulted. His pastures were very good and much improved, as were his fences and water supplies. Ragwort was well under control, and the upkeep of his buildings and machinery was first class. Taylor said that Mr Davis' stock was also in good condition, though his calves looked poor, possibly as a consequence of the drought. Mr Davis kept no pigs, because he was now supplying a dried milk factory. Taylor reported that his provision for supplementary feed was good, though not overly generous. Farm records were excellent and the loan account satisfactory, although there was still quite a large debt.

Potential Trophy Winners

The last competitors Taylor assessed were Mr M and Mrs D Wikaira, who farmed 98 acres of grass and crops. These entrants had taken over six years earlier, when the state of the farm had been poor. Taylor noted that much work had been done to restore the farm's productive potential, but much more needed to be done, especially the resowing of pastures and more subdivisions. This, he considered, would not only impove the general appearance of the farm but would also provide the means of increasing production to the high level necessary to win the Ahuwhenua Trophy.

1955

Assessment

After four consecutive years of judging the competition, Mr C A Taylor 'retired' in 1954.

Mr E H Arnold, assistant fields superintendent from Whāngarei, was the new judge; he had judged the 1947 competition. Judge Arnold used Taylor's points system, except that he 'rounded off' the section on 'economy of management'. Six farmers entered the sheep and cattle category, and six dairy.

In his report, Judge Arnold commented that the past decades had been noticeable for their 'progressiveness'. This had been evident in the amount of practical research being done; pasture improvement and management techniques were developing rapidly. Improved breeding and enhanced feeding of stock was now possible. The quality and volume of farm production, and increased economic efficiency, had expanded farming greatly: 'On my tour of inspection of the farms of competitors this year, I found that Maori farmers had been no less active.'[105]

Sheep and Cattle

Ex-servicemen Establishing Farms

Three of the year's competitors were devoted to fat lamb rearing, and three to wool and store production, with some fattening. All competitors were also running cattle.

Four of the competitors were ex-servicemen from World War II who had been resettled by the Department of Maori Affairs, acting for the Servicemen's Rehabilitation Board. All farms settled in this way under the Rehabilitation Act 1941 had been, as the Act stipulated, developed before being allocated. These competitors had not been long on their properties; certainly not more than three years. Judge Arnold took account of this when allocating points.

First Place

Judge Arnold awarded first place to Hekeawai Whakapahi (John Chadwick), a previous competitor. Mr Chadwick was a veteran of the First World War and was incapacitated during the judge's visit; his son acted as guide for the inspection, confirming this property as a strong 'family unit'.

The Chadwick farm was 451 acres and had been occupied for eight years, before which it was heavily bushed. Logging, burning, top dressing and seeding had all improved the unit immeasurably, Arnold observed. However, rotational grazing was impaired by an irregular water supply and inadequate subdivision. Aerial top dressing had not been possible because of the lack of a nearby airstrip. No provision had been made for supplementary feed, though some flat areas of land could have been planted in swede or other crops for this purpose. Arnold reported that the woolsheds were adequate and well designed. However, the stock was of mixed lines and was of indifferent quality.[106]

Second and Third Place

Second place was awarded to Edward Clayton Pohia, another ex-serviceman, farming 445 acres of fat lamb and cattle. Mr Pohia had vastly improved the farm's fencing, Arnold observed, and had increased the farm's water supply by laying water piping and providing sound concrete water troughs. Arnold also observed that provision of winter feed was good. However, Mr Pohia's pastures required some top dressing urgently; there was also evidence of overgrazing. His stock was generally in good condition, though the calves did not appear to be well cared for.

Third place was awarded to Robert Tu Kingi, farming a 423-acre block in close proximity to Mr Pohia; he had been farming on the block for about the same length

Four of the six entrants in the sheep and cattle section in 1955 were ex-servicemen being resettled by the Servicemen's Rehabilitation Board; thus, they had not been long on their properties when the judge arrived. The judge adjusted points accordingly.

Māori boy holding lamb in paddock, c. 1940s; photograph by Leo White (Alexander Turnbull Library, WA-25161-G)

of time. Arnold reported that Mr Kingi had expended huge amounts of effort on fencing and subdivision, though his paddocks were still too large; an extension of the water supply was therefore needed to support increased subdivision. The judge noted that aerial top dressing, ample cattle, absence of rank growth and a fresh green appearance of pasture had helped this farm to stand out. He said that more shelter and shade trees would greatly assist Mr Kingi's farming efforts. He noted that the farm records and diary had been very well kept.

Other Competitors

Among other competitors, Watchman Waaka was another ex-serviceman, farming 1012 acres of good grassland, though much of his farm was still in scrub and bush. Kelly Hawira, another ex-serviceman, was farming 792 acres with fair subdivisions, but soils were in need of replenishment. Henare and Iti Puaka were farming about 665 acres in an isolated locality 'with only a small shack to live in'; Judge Arnold observed that their good pastures and subdivision were marred by an insecure water supply that came from creeks prone to drying up in the summer.

Dairy

First Place

Five competitors in this category operated as dairy farms, while one also ran fat lambs. Over a wide variety of topographies and climate conditions (some areas had experienced severe drought), butterfat productions ranged from 82.5 lbs to 214 lbs per acre. Two of the competitors did not keep pigs.

Judge Arnold awarded first place to J W Hedley, whose efforts he commended given the impact of severe drought. His butterfat per acre was 214 lbs. Mr Hedley's farm was originally 125 acres; it was brought under the development scheme in 1938 with nineteen cows and replacements. Since then, an irrigation system had been installed and a tractor purchased, enabling ploughing. Seasonal crops were growing for supplementary feed. Arnold reported that the homestead was well sited and quite sheltered.

Second and Third Place

Judge Arnold awarded second place to Ilet Hemi who, with her husband, was farming 103 acres, 82 acres of which were in permanent pasture. The farm had been occupied for thirteen years and had required a tremendous amount of clearing. It comprised rolling farmland and some peat swamps. Arnold considered that fencing was well developed and pastures were well managed. Buildings were generally old but well maintained, and were to be replaced in the foreseeable future.

Arnold awarded third place to Foley Eru, one of the original farmers to have occupied this development block; he had been there for a good twenty years. Mr Eru's farm was 194 acres, having been increased from an original 70 acres. Arnold noted good subdivisions and adequate water and fence protections.

Other Competitors

Ratahi Tatana had occupied his farm in 1940, and for some years had lived in 'somewhat primitive conditions – cream had [been] sledged a mile across a stream subject to floods and water carried back on the return trips for shed purposes', Judge Arnold reported. Mrs R Beasley was farming a property of 152 acres and was 'managing a sound operation'. Waka Rewa was another ex-serviceman now settled on 105 acres of flat land supported by rather shallow creeks. He was working hard, Arnold considered, but needed extra labour.[107]

1956

The Twenty-first Year of Ahuwhenua

Two nominations were received in the sheep and cattle section this year; three further nominations had been withdrawn just prior to the competition. There were eight nominations in the more favoured dairy section. Judging once again was Mr Arnold, who had judged the competition the previous year and had in that report commended the appointment of judges for multiple periods because of the continuity of focus and purpose that a reappointed judge could bring.

Sheep and Cattle Competitors

The winner of the sheep and cattle section was Mr R T Kingi, who farmed a 423-acre block near Rotoiti, with 410 acres in pasture. Judge Arnold considered that excellent progress had been made in providing fencing and water reticulation, and commended Mr Kingi for sowing DDT to eliminate troublesome grass grub. An added feature of his property was his provision of supplementary feed, which meant that his flocks were well provided for. Arnold considered that Mr Kingi's stock was of a high calibre, showing excellent breeding, though his calves were 'not quite so good'. Facial eczema had affected the lambing returns.

Second place was awarded to Edward Pohia, who farmed an adjacent block of 445 acres focusing on fat lambs and cattle; much of his unit was in grass as a consequence. Judge Arnold noted that he had undertaken an enormous amount of work of late to remedy poor fencing and inadequate subdivision. He had also used aerial top dressing for the first time; a local airstrip had recently been constructed. However, Mr Pohia's farm seemed overstocked, and grass grub was a troublesome and recurring problem, as it was for his neighbour Mr Kingi. Despite this, Arnold considered that stock quality was very high, especially under a new programme of stock replacement. New lawns had been sown around the homestead, but had not been well maintained.

Dairy

First Place

Seven competitors this year were 'typical dairy farms'; one was described as a 'dairy and fat lamb' unit.

The winner was Mr Foley Eru, who had been placed third in the previous year. He was one of the original (and few remaining) Māori farmers to have occupied blocks in this area twenty-one years ago, when the area was first opened up. He had started out with sixteen cows on 70 acres; his herd had now expanded to thirty-six cows on 146 acres. He also managed a flock of 200 breeding ewes, and had 10 acres set aside in cash crops – in fact, he had three tons of potatoes in storage. Judge Arnold said that Mr Eru's farm was well subdivided, and recurring blackberry was kept at bay. His milking shed was sixteen years old, perhaps in need of replacing, but calf pens were kept spotlessly clean. Supplementary feed provision was also a strong point, and stock and pasture management were excellent.

Second Place

Second place was awarded to W J Swinton, an ex-serviceman farming 108 acres, with 102 acres in permanent pasture. Though financed through the government's rehabilitation scheme, much of the more recent development work had been completed at Mr Swinton's own expense. Production had improved over the last three years – from 6989 lbs of butterfat to 12,954 lbs, or from 68.7 lbs to 127 lbs 'per effective acre'. Judge Arnold noted that gullies and broken country had impeded further development work, though Mr Swinton had undertaken burning and spraying with some success.

Arnold considered that Mr Swinton's gates and fences were excellent, as were his pastures of ryegrass, cocksfoot and white clover. His milking sheds, which had been recently painted, were a credit to him, and his tools and equipment were well maintained. Winter feed was very good: 800 bales had been cut from 15 acres of grass. Pig production was also good. The homestead was attractive, and finances were sound.[108]

Third Place

Judge Arnold awarded third place to James Nelson of Kōpua, another ex-serviceman, farming 118 acres, of which 106 acres were in pasture and 9 acres in crops. Mr Nelson's land was flat to undulating and had at first been given over to sheep farming. With ragwort and gorse now all but eliminated, Mr Nelson had raised his butterfat output from 9301 lbs three years earlier to 13,745 lbs today. He had also recently erected ninety-six chains of new fencing and installed a new bore and deep well pump for the reticulating of water through troughs in every field.

Mr Arnold considered that the unit's pastures varied in composition, though pasture management was quite sound. Shelter and shade were good, and the milking shed was old but kept clean. Mr Nelson's dairy cows were in a fair condition, and his pigs were sold as exportable grade. Farm records and the farm diary were well kept. The house had recently been painted and was well maintained.

Other Competitors

Among the remaining entrants, Mr Sam Rare had been farming his 86-acre unit for three years; deep gullies and broken country had made his property a difficult one to manage. Tau and Robert Sheppard were 'two young Maoris' who were leasing their 140-acre farm from their father. Extensive subdivisions had been completed, but this work had encountered difficulties because of inadequate water supplies. An interesting sideline of Tau and Robert was the growing of kūmara plants for sale, or for growing as a commercial crop. Tom Horopapere was another ex-serviceman; he farmed 94 acres of undulating land including river flats. Mr Horopapere had been handicapped by a lack of finance, but had managed nonetheless to install electric power to the house and shed. P T Rika, another ex-serviceman, had been settled four years earlier on 112 acres of undulating to steep hill country; a terrain perhaps less suitable for dairying than others. Tau Iwarau Matene had farmed his property for twenty years, having been brought

Sixteen Māori farmers entered the 1957 Ahuwhenua competition: the largest number of competitors seen in a single year to date.

Hapū group of Māori shearers at Eparaima homestead, near Castlepoint, Masterton, c. 1910; photograph by George Moore (Alexander Turnbull Library, 1/2-065449-F)

into a development scheme in 1948. His extensive programme of draining, stumping, subdivision and grassing was still under way.[109]

1957

A Record Number of Nominees

According to Judge Arnold, appointed once again as judge, the number of Ahuwhenua entries had increased significantly over recent years. Arnold visited sixteen farms in early 1957: four in the sheep and cattle section and twelve dairy. Excessively wet weather in some regions, especially in the north, had made farming conditions difficult that summer; Arnold made appropriate allowances in his final determination of points and placings.

Sheep and Cattle

First Place

The sheep and cattle section for 1957 was won by Henry Mathieson Davis of Rotorua, who scored thirty-three points more than the second-place winner and thirty-five more than the third – 'in other words, a clear winner'.

Mr Davis was a new entrant in the competition. He farmed 392 acres on a 'fat lamb and cattle property'. His farm was moderately steep, 'being of an old plateau formation broken down into gullies and ridges'. He had just erected fifty chains of new fencing, dividing his unit into twelve paddocks. Judge Arnold observed that his pastures were as good as could be expected in this type of country – ryegrass, cocksfoot and white clover. The quality of Mr Davis' stock was impressive: excellent bulls and 'good type large framed ewes' were backed by a sound replacement policy. His supplementary feed was also very good, and his farm stock records were well maintained.[110]

Second Place

Second place went to John Tahuri of Ruatāhuna. Six years previously, he had taken over an isolated, run-down and overgrown farm 70 miles from the nearest railway or centre; the costs of his bringing in fertiliser and other supplies was significant. Mr Tahuri was farming 880 acres of 'flat land and terraces and easy hills', of which 350 acres were in grass and 8 in crops. Judge Arnold observed that his fences were in good condition and his yards were well planned, constructed and maintained.

He had recently added new cattle yards with a drafting race, crush pen and loading bank. His pastures and stock were of a good quality, though Arnold commented that now was the time for 'some good well bred foundation stock, both sheep and cattle, to be obtained to improve the type of stock grazed'. Perhaps the neighbour's culls could be purchased, he suggested. Supplementary feed was good, and the farm seemed well sheltered. Farm and stock records were 'of a high order'. Despite his isolation, Mr Tahuri had 'proven his capabilities on a very difficult farm', Arnold concluded.

Third Place

Third place went to Mr Desmond Royal of Lake Ōkareka, Rotorua, who farmed 568 acres, with 550 acres in good grass. The balance of his property was still given over to scrub, winged thistle, California thistle and gorse, which Judge Arnold noted risked reinfestation.

Arnold considered that fencing on Mr Royal's steep farm was very good, though a closer pattern of subdivision was needed. His top dressing with phosphate also needed to be continued to ensure sound pasture growth. Erosion in one of his gullies had become a major problem; Judge Arnold suggested that a soil conservator be asked to assist Mr Royal in dealing with this. All flocks were in a good condition, though Mr Royal's sheep had recently been shorn, 'which can be unwise so close to winter'. No supplementary feed had been allowed for, which was an issue that needed resolving. The homestead was well sited and sheltered, and the farm's records were well kept.

Dairy

Place-getters

The winner of the dairy section was Mr Rehua Cairns of Welcome Bay, Tauranga, who farmed 99 acres

As late as the 1960s, many Māori farmers still suffered from a severe lack of equipment. In 1958, Judge Allo noted that one competitor had no tractor and was doing all of the farm work with one horse.
Māori ploughing at Waikanae, c. 1908 (Alexander Turnbull Library, PA1-o-229-33-2)

of well-subdivided and well-grassed land. Further paddocks were planned in the near future. Mr Cairns' farm was undulating; Judge Arnold noted that a good water supply was 'pumped from a bore to a reservoir and then gravitated to the milking shed and every paddock'. Shelter was organised well, and pastures were well balanced, though some had been cut for hay every year, a practice difficult to avoid on such a hilly farm. Top dressing was adequate, Arnold considered, and the 'three-bail doubled-up milking shed' was in excellent condition. All stock, including pigs, were of a good quality.[111]

Judge Arnold awarded second place to Tom Haeata of Mangakino, who was farming 133 acres of rolling pumice land. Good pasture management had ensured good returns: 18,922 lbs in butterfat, or 140 lbs per effective acre. The farm now had seventeen paddocks based around a race system, with water pumped to all parts of the farm. Pastures were of a high standard, and supplementary feed was well provided for, Arnold reported. The milking shed was clean and well cared for. Dairy cattle were of a good type, and young animals were well grown. Farm records were very well maintained.

Third place was awarded to James Nelson of Te Awamutu, who was farming 118 acres with sixty-five cows, producing 14,425 lbs of butterfat. The farm was well maintained, Arnold considered, and the acquisition of a tractor had been of great assistance, especially with hay and silage making. Mr Nelson's subdivisions were also good; hedges had been planted to supplement fencing. The present milking shed, though clean, was now old and needed urgent repairs; it was also badly sited, Arnold reported.

Other Competitors

Among other competitors, Mr Swinton of Whangamatā was farming 70 acres of pasture, but was bedevilled by swamps that needed draining and clearing. Mr Alf Parker of Mangakino had occupied his 151 acres of pastures and crops for four years and had been performing particularly well, though would need additional labour to manage the physical work ahead. Mr I D Hall of Tauranga had been farming his 94-acre property for five years with some success, though the farm was of a 'poor shape', with a difficult topography to manage.

Mr C K Lingman of Te Awamutu had been farming his 176-acre unit for five years and had managed his pastures and subdivisions well, despite many setbacks. Mr Ratahi Tatana, who was farming a block of 85 acres near Kaitāia, was also commended for the quality of his 'excellent herd' and sound financial planning, though the judge noted some work was still needed to bring his pastures up to standard. Mr D Rehu of Waimiha had commenced farming his 156 acres seven years earlier, and in that time had seen his production returns grow impressively. However, the soils on his property needed constant attention.

Mr Mita Mauriohooho of Te Awamutu had taken over his 100-acre farm six years earlier in a derelict state; he had made impressive gains since, though more intensive development work lay ahead. Walter Kawiti of Waiomio had taken over his 417-acre property in 1935 and had been working ever since to convert barren scrubland into a well-producing unit. He had largely succeeded, though the country still offered many challenges. Charlie Dargaville of Panguru had occupied his 183 acres of rolling countryside in the early 1950s, and had performed creditably in bringing it into profitable production.

1958

The Sheep and Cattle Section in Recess

In 1958, fourteen entries were judged in the dairy section; there were insufficient entries received for the sheep and cattle section for the competition to proceed. Judge A V Allo, an assistant fields inspector from Whāngārei who had previously judged in 1947, inspected farms during the first half of June – somewhat later than usual.

First Place

The winner of the Ahuwhenua Trophy for 1958 was Mr Tom Haeata of Mangakino, who had been farming a 133-acre unit originally part of the Mangakino Development Block for the past five years. His farm comprised 122 acres of good grass divided into seventeen paddocks, though Judge Allo considered some of these too large. Pasture grasses were healthy, Allo observed, though there was much fog, revealing 'too lax a system of grazing'. Excellent shelter belts were in evidence, and Allo described the buildings as 'a credit to any farm'. Mr Haeata was managing a first-class Jersey herd and had 'really good' supplies of supplementary feed. In all, 1365 bales of hay had been cut, and a further 430 obtained as payment for contract work performed during the previous year. Pigs were being well wintered, Allo noted.[112]

Second Place

Second place went to W J Swinton of Whangamatā, who had occupied his 180-acre block five years

Winner of the Ahuwhenua Trophy 1958 was Mr Tom Haeata of Mangakino.
(Photo courtesy of Te Ruinga Haeata)

The replica cup and medals presented to Tom Haeata, winner of the Ahuwhenua Trophy in 1958
(Photo courtesy of Te Ruinga Haeata)

Māori 'Economic Advancement' and Farming - 1945-1961

earlier. Mr Swinton's land was undulating to hilly; sheep complemented his dairy farming. Eighteen paddocks had been established, and a further four were planned. 60 acres had been closed over autumn to save the pasture. The farm had a first-class water system, Allo observed; there was a trough in every paddock. Fences were in an excellent condition, and hedges were also being used to provide shelter.

Mr Swinton's stock was in first-class order for entering the winter: the herd average was 271 lbs of butterfat, 'which means there must be quite a number of very good cows', Allo noted. Pig returns were fair; six national hybrid sows had just been acquired. A total of 1000 bales of hay had been prepared, though no silage or cropping had been possible.

Third Place

Third place was awarded to John Peterson of Mangonui, who farmed the smallest unit in the competition: 35 acres. Mr Peterson had farmed his property for four years and had transformed a 'broken-down uneconomic unit' to one that was now yielding a good living. The land was flat and easy to farm, Allo noted, though much pasture improvement work was still needed. Fences and buildings were all Mr Peterson's work, and they were in excellent order.

Mr Peterson's farm had eight paddocks – some were now quite large. Water supply was good, Allo reported, and additional shelter was planned to overcome a shortage of wind breaks or shade. Pigs accounted for 20 percent of the farm's income. The farm possessed a 'nice herd of Jersey cows', and a good type of replacement stock was on hand. No herd testing had been done, which continued to cause some difficulties. Ample winter feed was available, and there was a good barn on site. About half of the farm had been closed for winter pasture; the negligible amount of growth suggested to Allo that the pastures had been closed too late.

Other Competitors

Among other competitors this year, James Subritzky of Ngātaki had farmed his 146-acre unit for six years, having cleared most of his difficult property of scrub and planted pastures, which still needed some improvement. John Taua of Te Mata had been working his unit for nine years, but a recent debilitating back injury had greatly impeded his development work. Mr N Warmington of Waimā had farmed his small unit of 36 acres for nine years, taking it over when it was derelict and run down. In recent years, Mr Warmington had established a good unit with mixed grasses and good returns; he was undertaking ongoing subdivision work. Mr G Kaui of Tikitere had settled on his 75-acre property eight years earlier, and had worked well to clear the land of gorse and ragwort. Though it was not an easy property to work, Mr Kaui had tidied the place up and improved his returns year by year.

James Oliver of Mangonui had been on his farm of 165 acres for the last four years, working very hard to improve a difficult property cut into three by a main road and a river. If anything, Mr Oliver had been 'trying to do too much too quickly', Allo considered, with disappointing results. E Morunga of Ōmanaia had moved onto his isolated 257-acre property nine years earlier, and with limited finance had worked hard to convert a run-down farm into an efficient dairy unit. His job was far from finished, but there was 'ample evidence of Mr Morunga's ability as a farmer, and of his energy and keenness'.

James Norman of Ngātaki had, for the previous four years, been managing a 219-acre property of mostly 'poor quality light sandy soil'. Pasture development had been slow, but Mr Norman was a 'good type of farmer who was very willing to learn', Allo noted. Mr M McMillan had a farm comprising 122 acres of good pasture, which he had developed over the previous seven years. The farm was well subdivided and stocked, but there did not seem to be supplies of winter feed.

Mathew Padlie of Awarua had been farming a 200-acre block since 1953. However, his farm had not yet reached the stage where 'a high pointing would be warranted', Mr Allo considered. Mr H Kaio of Whirinaki had been leasing his 70-acre farm for the previous twenty years; though he was a good farmer he suffered from a severe lack of equipment – 'Mr Kaio had no tractor and did all of the farm work with one horse.' The entry of J Clark of Waikeretū was 'a very difficult entry to judge' because of the almost insurmountable problems presented by an isolated property that was intersected by streams and was, at various times of the year, almost impossible to stock.[113]

1959

The Sheep and Cattle Competition Resumes

Three entries were received for the sheep and cattle section of the Ahuwhenua Trophy in 1959, enabling the competition to proceed. Eight entries were received for the dairy section. The judge appointed this year was Mr J R Murray, a farm advisory officer of Hamilton.

Sheep and Cattle

Winner

The winner of this section was Mr Jack Steedman of Welcome Bay, Tauranga, who had occupied his 529-acre block for two years. Mr Steedman impressed Mr Murray as a stockman; he had thirteen paddocks with new fencing, easing the tracks for stock movement. Difficult gorse and blackberry had been brought under control. Mr Steedman's sheep were of an excellent quality and were divided well into groups, resulting in good wool production. Murray considered that his cattle were 'rather rough' however, as a consequence of the farm being under-stocked. Water supply was good, but the farm was prone to severe flooding. Farm records were excellent and a credit to Mr Steedman; the judge was also impressed with the small office at the back of the house.

Second-equal Placing

Judge Murray awarded Desmond Royal of Ōkareka and Desmond and Pikihuia Manning of Manunui with second place equally.

Mr Royal had a large farm of 560 acres, described as 'too large for one man to handle'. His farm was limited by subdivision problems, Murray considered; he needed better control of stock and scrub regrowth. Top dressing needed to be continued, and more fences needed to be erected. Mr Royal's progress had been fair overall, though his stock numbers had not increased by much. His pastures were fair, though poorer 'the closer one got to those areas of the farm given over to bush'. Mr Royal's wool clip averaged 40 lbs per acre. The property's water supply was good, though more winter feed was needed. Some erosion was also evident.

Judge Murray considered that Mr and Mrs Manning of Manunui soundly managed their 325-acre farm, which they had occupied for only one year. The pastures on this farm were variable; top dressing was needed to make an improvement. More subdivisions were also needed to improve the wool clip of 36 lbs per acre. The judge noted that farmers of pumice soil like Mr and Mrs Manning needed to guard against overgrazing.[114]

Dairy

First Place

Judge Murray awarded first place to Mr Swinton of Whangamatā, who farmed 115 acres with sixty-four cows. Mr Swinton's subdivision and water supplies were very good, Murray thought; a double pumping system proved to be very effective. The cowshed was a great credit to Mr Swinton; the judge was especially impressed with the calf-feeding shelter, a counterbalanced drop gate with new pipe rails and new bails. The judge was impressed with the feeding and shelter provided to pigs. Hay was stored near the house, he noted; it needed its own barn. Stock was of a good type and in good condition. Winter feed was also well covered.

Second Place

Second place was awarded to J H Hedley from Hoe o Tainui, who farmed an 80-acre unit with sixty cows. Mr Hedley had previously won the trophy in 1955. After that, his butterfat production had dropped, but it increased in the 1958/1959 season, despite severe drought; at 230 lbs per acre, it was the highest of all the competitors. Pastures were well managed, Murray considered, especially in terms of regrassing. However, the cowshed needed a major overhaul; the judge noted that a new one was planned. Stock was in good condition; Mr Hedley had reverted to a Jersey herd after using Ayrshire cross. Mr Murray noted that over half of Mr Hedley's loans had been paid off.

Third Place

Third place went to W H Mauriohooho of Te Awamutu, who farmed 164 acres with 104 cows. Judge Murray considered that the appearance of the farm buildings was very good; there was a well-kept house and garden. Pastures were well managed, though a pasture renewal programme would soon be necessary. Pig returns were very high, currently at 4.28 pence per pound of butterfat for produce. The farrowing house was poor, the judge noted, and needed shifting. Much hay had been lost to the wet weather last year, and silage losses were also high. California thistle was threatening to be a problem. An irrigation plant had been purchased and was due to be installed, and a new subdivision was under way. Management and stock control were producing good results.

Remaining Entrants

T Matenga of Mangakino managed a 150-acre property with seventy-eight cows. His was a 'well laid out farm with a model subdivision' that could easily be extended, Judge Murray said. It was well managed and productive, though 'pastures were only fair and not up to standard for this district'.

Mr W R Mangu of Ōtorohanga was a hard-working competitor who was showing good development progress on his 160 acres, stocked by fifty-eight cows and 120 sheep. Mr Mangu had in recent years pushed ahead with the development of his higher potential country, the judge noted; while sensible, this had required more capital than expected.

Mr E Paki of Mangakino currently occupied a property that had once been a training farm; most improvements had already been made. His management was sound, and he was accomplishing high butterfat returns. Mr J Clark of Waikeretū managed a farm of 142 acres, with fifty-one cows. Mr Clark's butterfat returns were very high, Murray noted, and he was making improvements to the unit's subdivisions. However, little more could be done until proper access could be arranged across a creek running through the property. Mr J Te Wake of Kohukohu's farm was difficult to judge because, though he ran fifty-three cows on 198 acres, the property was clearly more suited to sheep- and cattle-rearing. Eight fields had new sown grass, but more subdivisions were required.[115]

1960

A New Decade

In 1960, Mr Murray returned as judge, for one of the largest fields of competitors ever to be entered into the competition: there were five nominees for the sheep and cattle section and an unprecedented sixteen for dairy.

Sheep and Cattle

First Place

Judge Murray awarded first place to Mr W Waaka of Punakitere, near Kaikohe. He farmed a very large 1029-acre property 'substantially on his own' and deserved much encouragement, especially as he had been on budget control for some time. Mr Waaka currently faced many problems, such as the inordinate size of his farm and wet weather that affected production for good parts of the year.

The judge noted that Mr Waaka's development work had been outstanding. Farm buildings were neat and tidy, and the unit possessed a very good water supply. However, erosion and rush growth through water seepage were recurring problems on the farm. Murray suggested semi-permanent electric fences as a solution to subdivision problems; he also advised that perhaps a blitz on fertilising would greatly progress his pasture growth. His ratio of cattle to sheep needed reducing, Murray thought, and more hay for the cattle was advisable.

Second and Third Place

Second place went to Parekura Raroa of Tokaanu, who farmed a 416-acre unit that 'needed more subdivision', Judge Murray considered. Some pastures were poor, though a good water system had been installed. Greater fertiliser application was needed, the judge thought. All stock were in good condition, though the farm was quite bare and in need of shelter, especially from the cold winds. Murray recommended that seed and fertilisers be applied to the steep hills at the back of the farm, and efforts made to ensure pasture growth was under way during autumn. A better winter crop was also necessary, he felt.

Third place was awarded to Mr Aperehama Whata of Rotoiti, who farmed a 576-acre block on difficult country. Given the amount of felled timber still lying about, Murray noted, parts of the farm were still difficult to work with. Top dressing was greatly assisting with the farm's regrowth. The condition of the stock, especially the cattle, was very good.

Remaining Competitors

Of the remaining two competitors, Mr K Grace of Tokaanu managed a 410-acre unit that had been substantially developed prior to his assuming occupancy one year earlier. Because his property could be described as 'tailor made', Murray noted, there did not exist the same opportunity for development here as 'looked for under the terms of this competition'. However, the judge commended Mr Grace for his determination to make this farm a 'really top grade farm'. The other competitor, Mr R Mitchell from Hastwells, near Eketāhuna, was a former Maori Affairs employee who had taken over 483 acres of good land originally farmed as a dairy unit, before it was conceded that the terrain lent itself better to sheep and cattle farming. Mr Mitchell managed a sound operation but had perhaps unwisely sold all of his cattle, the judge said; instead, he was grazing dairy cows for an adjoining farm, which earned him negligible grazing fees.[116]

Dairy

Winner

From the very large field of competitors, Judge Murray awarded first place to Mrs Mihi Stephens of Rangiāhua, who farmed 89 acres with sixty-four cows. Mrs Stephens' farm was exceptionally well managed, the judge said; high-quality stock were 'well catered for'. Her farm in fact had a capacity to carry more stock, he noted, dependent on seasonal issues. Mrs Stephens' reserves of hay were very high, and her crops were well stored, although the judge noted that more fertiliser was needed on her pastures, and could be 'more wisely spread'. New sheds had been built for tools and implements, and a recently installed tile drain outlet had improved drainage. The farm's heavy soil type had made pasture management difficult, and the property was prone to flooding. Overall, production had increased over the last twenty years, though it had dropped again in recent years.

Second Equal

Second place was awarded equally to J W Hedley and Mr Wallace Mangu. Mr Hedley's fifty-eight cows comprised a 'high producing farm and herd', the judge considered. Production figures had been very high – 18,648 lbs of butterfat that year, with a per-cow average of 321.5 lbs. Wet conditions had impeded development progress on this farm, as had a tall fescue weed infestation. Rush growth had also been a problem. Mr Hedley's stock were in good condition, although some of the younger animals were rather thin and showed symptoms of scouring.

Mr Mangu of Ōtorohanga farmed 160 acres of good land with 115 ewes. Good placement of fertiliser had helped Mr Mangu to develop good pastures, and his supplementary feed position was quite good, the judge noted. Young grass sown on the lower flats was coming on into good pasture. Pig returns were very high, and fencing was overall very good.

Remaining Competitors

Among the thirteen remaining competitors, Mr T Matenga managed a 150-acre property with great skill, and the judge commended him for his willingness to participate in local discussion groups focused on turning farm properties to the best advantage. Mr W K Mauriohooho of Pararewa had been ill in hospital for some time, and as a consequence, had been unable to 'carry on with farm operations as he had so ably started'. Mr R Parker of Mangakino had been doing an excellent job of rearranging fencing on his entire 156-acre unit by 'sheep-proofing' his subdivisions. Mr J Courtenay of Tauranga had worked assiduously, clearing gorse off his property and improving its access to a good water supply.

Mr S Hona of Matauri Bay, who farmed 151 acres with forty-nine cows, had seen his butterfat returns increase in recent years, though much work was still needed to improve his low-lying pastures. Mr J Te Wake of Panguru had been affected by a very dry spring that had drastically impeded production returns; however, the judge noted that most of the Northland hill country farms had dropped in production to a much greater extent. Mr H Paku of Hastwells was farming a deteriorated block that he had slowly been improving over three years.

The judge commended Mr W Morunga of Whirinaki for his efforts despite his considerable physical disability: he only had one arm. His stock were of a good quality, but his operation was marred by the type of broken country he occupied. Mr J T Wordley of Ōhaupō managed 100 acres of very good farmland that had potential, but his pastures had been turned over in the past and suffered accordingly. Mr P Toi of Ōpononi farmed 131 acres of land that was so isolated that he had once had to collect his supplies by packhorse. A recently metalled road had now changed all this. Mr Toi's farm comprised mainly broken pastures, which offered little scope for normal dairying improvements.

Mr W Watene had settled well on his 111-acre property and was fortunate in that he had had to do little development work. Mr C Dargaville of Panguru possessed a difficult and exposed farm; he needed to provide more shelter for his heavy cows. Heavy top dressing was also needed to lift the standard of his pastures, the judge considered.[117]

1961

Another Large Field

Interest in the Ahuwhenua Trophy remained high in 1961. Seventeen farmers were nominated for the competition, comprising three in the sheep and cattle section and fourteen in dairy. Mr J R Murray returned as judge.[118]

Sheep and Cattle

Winner

The winner of the sheep and cattle section for 1961 was Mr Parekura Raroa of Tokaanu, who farmed 416 acres with 916 ewes. Mr Raroa's 'heavy top dressing programme' had greatly assisted his farming efforts, Judge Murray noted; new pasture development plans were proceeding well. Top dressing needed to be continued, he suggested, at a rate of 2.5 cwt per acre as far as possible. He had an excellent cropping and pastures renewal policy, the judge reported.

Mr Raroa had recently burned out a significant section of his farm infested with gorse; hay now needed to be used in areas where burning had taken place. Shelter was also badly needed along the southern boundary, the judge noted, as well as around the home.

Second Place

Second place in this category went to Mr Aperehama Whata of Rotoiti, who was farming 576 acres with 1200 ewes. New subdivisions had been carried out since the previous year, the judge noted, with positive results. Stock was in a reasonable condition, but the judge advised that 'care must be taken not to force the breeding stock to work too hard when grazing the log areas – use the dry stock for the tougher blocks'. He also recommended new fence lines, and suggested that Mr Whata not overstock his recently developed areas if fireweed was to be controlled. Continuing with heavy top dressing was essential, he said. Hay would probably need to be purchased: 'do not attempt to go through a season without a supply on hand'.

Third Place

Third place went to George Kuru of Tokaanu, who farmed 420 acres of broken country that had required an enormous effort to clear. Once again, the judge recommended a heavy application of fertiliser: possibly a 'blitz' attack. Murray strongly recommended top dressing and further subdivisions. Mr Kuru's stock was in a fair condition, he noted, and was probably capable of a higher rate of production.

Dairy

Winner

Judge Murray picked the winner of this section as Mr Wallace Mangu of Hangatiki, near Waitomo. Mr Mangu had entered the competition twice before and had improved with every competition, the judge said. His low-lying and gorse-covered land had been cleared after considerable effort, and his dairy stock were 'easily the best in the competition'. Considering the class of country he was farming, Mr Mangu had done well in gaining the highest average for butterfat per cow. He had also produced the finest grade of cream of all competitors. Mr Mangu's shed cleanliness was outstanding, the judge reported; his pig management was sound; and his finances were commendable. He did need to pay some attention to drainage and to consider crop planting as soon as possible. Hay needed to be cut earlier, and the judge recommended arranging some shelter to protect the homestead, which was quite exposed.

Second Equal

Judge Murray awarded second place equal to J W Hedley of Hoe o Tainui, who farmed a 120-acre block of reasonably good grasslands, and Mr Tapua Pita Heperi of Rangiāhua, who farmed 101 acres carrying fifty-five cows.

Mr Hedley had been awarded second place in the competition twice before; Judge Murray said that his loyalty to the competition was admirable. He noted that Mr Hedley's early grass needed grazing off before it got too long and that he needed a pasture fertilising programme. There was evidence of weeds that needed early attention, especially tall fescue, which had taken over one whole paddock that now needed ploughing under and reseeding. Grass needed to be provided for cows just after calving, as well as shelter; 'cows should not be left on rough feeding closer than five weeks before calving', the judge admonished. More shelter was generally needed on this farm.[119]

Mr Heperi had recently suffered ill health; the judge especially commended him for his 'courage and determination' in making good his farm against tremendous odds. His annual trips to Ruakure for information and updates demonstrated his wish to keep abreast of developments in farming. Mr Heperi had designed a new elevated bail cowshed while in hospital and built it when he came out.

Mr Heperi's stock was generally good, though his pig returns were low. Flooding across his unit continued to be a problem, particularly in the recurring damage it caused to fences. Judge Murray recommended roto-cutter machinery to deal with the rushes and tall fescue problem on the farm. He said that more fertiliser was needed on outlying fields, which tended to be overgrazed. A new drainage area could be delayed for a year while the ground continued to dry out, he recommended. Tree planting and protection work could also be undertaken, in concert with Mr Heperi's neighbours.

Remaining Competitors

Among remaining competitors, Mr R Parker of Mangakino had demonstrated good herd management techniques and 'could be justly proud of his achievements', the judge noted. Murray congratulated Mr H Rika of Ōruawharo for farming a property that, especially when wet, was 'not altogether a pleasure'. Mr H P Te Hira possessed some of the best pastures in the competition and was commended for farming such a difficult soil type. Mr J Te Wake of Panguru had improved his farm immeasurably; had there been a 'prize for dairy farming in the Northland Hills, he would get it', Murray said.

Mr K Thomas of Lake Ohia had tackled a totally unimproved area 'courageously', the judge said. He was a tireless worker who had recently seen a significant area of his pasture destroyed by high tides and a gale that had blown salt water across his property. Mr C Dargaville had just had his best season to date, despite his farm being quite exposed and lacking in shelter. Mr W Te Whata of Ōpononi had done well farming a small area while under 'insecurity of tenure', the judge said (possibly in a reference to leasing). Mr K Allan of Ōtaua possessed a very attractive farm.

Mr A Savage of Whangamatā had faced significant problems in the past with drainage but had managed to install a main drain down one side of his farm, thereby 'opening up his top flat for development'. Mr N Olliver of Mangonui farmed 165 acres of 'good potential', though he had ongoing problems with gorse and bracken. Mr J Peterson of Mangonui had a small property that unfortunately could not be enlarged. His pastures were variable in quality, which seemed to restrict his stock rotation options.[120]

Occasionally, judges made the comment that Māori farms were too small or otherwise too difficult to develop; thereby, they were outside the terms of the Ahuwhenua competition.

Three young girls, a man and a toddler in a farmyard in front of a small wooden house with a corrugated iron roof and water tank, and a raupo-roofed shed, c. 1900s; photograph by William A Price (Alexander Turnbull Library, 1/2-001939-G)

Field day at Hereheretau Station, Whakakī Road, Wairoa, 2009
(Ahuwhenua Archive, Field Day No 27)

Chapter Five

The 'Spirit of Friendly Rivalry' – 1962–1972

1962

Resourcefulness

Competitors from as far afield as Tokaanu, Kaikohe, Takahiwi and Te Kōpuru took out the honours in the twenty-seventh annual Ahuwhenua Trophy competition in 1962. Fifteen entries were received this year: twelve in the dairy section and three for sheep and cattle. The judge was Mr H Macmillan Bull, a farm advisory officer from Auckland.

Judge McMillan Bull expressed his disappointment that so few entries had been received in the sheep and cattle section. In years to come, he hoped that more entries would be received while farms were 'still being developed out of the rough', rather than sheep farmers waiting until such time as they were 'more or less managing a running concern'. The development of land for farming was undoubtedly of vital importance, the judge wrote; in developing, the farmer was 'beset by many problems which called for much resourcefulness, labour and initiative'. He considered that such qualities were well supported within the 200 points on offer in the Ahuwhenua competition.[121]

Sheep and Cattle

First Place: Huge Sheep Potential

The winner of this year's sheep and cattle competition was Mr Kingi Grace of Tokaanu, who farmed 416 acres that carried 1235 ewes. Mr Grace's farm was well

Judge McMillan Bull in 1962 urged Māori farmers to compete for the Ahuwhenua Trophy while their farms were 'still being developed out of the rough', rather than waiting until they were 'a going concern'.

Māori group alongside a dwelling at Ohura, c. 1900 (Alexander Turnbull Library, 1/2-021214-G)

developed when he took it over, so his development costs were modest when compared with like properties. Maintenance requirements had been low, and the unit was located in an area of huge sheep potential. Crops and lucerne were being grown, and hay was being made. Income was derived from wool, lambs and beef, with some returns from cull sheep, netting a high annual return of 10 percent. Carrying capacity was high – 4.1 ewe equivalents per acre – as was wool production, at 37.7 lbs per acre. Mr Grace was 'thoroughly experienced with stock', the judge said. His innovative break feeding of ewes would, in time, become standard practice in pumice areas, the judge foretold. An additional area for hay-making was suggested, and more attention towards pasture fertility. Grass grub was a problem and needed watching.

Second and Third Place

Second place was awarded to Mr John Reid of Kaikohe, who farmed 553 acres and ran 2359 wethers. Judge McMillan Bull said that Mr Reid had done well to control the blackberry, tea tree, carex, bracken, ring fern, foxglove and heavy rushes that had once infested his farm – noting that the rush problem had not been entirely solved. The judge considered that Mr Reid's replacement stock situation was 'not good', but this was related to 'the particular stage of development that is required to be overcome before a ewe flock can be run with any degree of success'. Top dressing had been used successfully. Cattle needed to be wintered on the block rather than off the property, as was the custom. McMillan Bull suggested that hay could be fed out in ferny areas, which would then get trodden under with dung and grass seed. The judge noted that some of this farm's hillsides were highly dangerous to plough.

The judge awarded third place to Mr Aperehama Whata of Rotorua, who farmed 552 acres with 1200 ewes. Mr Whata was an 'excellent stockman', who managed sheep and ran cattle well. Fertiliser was an expensive item, but was essential in his farming operation. A poor burn had left much firewood, the judge noted, attracting inkweed plants that would require consistent applications of 2,4,5-T to eradicate.

Dairy

Place-getters

The winner of the dairy division was Mr W Maki of Takahiwai, who farmed 67 acres with fifty cows. Mr Maki's farm had a good balance of even and flat land, the judge noted, and his butterfat returns were very high, at 289 lbs per cow. His returns on pigs were also very high – almost the highest the judge had seen this year. However, Judge McMillan Bull considered Mr Maki's practice of buying in all of his replacement stock was unusual. He noted that an invasion of rushes on flat lands below the road was a continuing problem, though Mr Maki's plans for top dressing and feeding out would help to rectify this. Ratstail was also becoming a problem that would require consistent and expert attention.

Second place was awarded to G and R Rutledge of Te Kōpuru, who operated a 105-acre farm with sixty-one cows. The Rutledges' farm was severely exposed, and therefore vulnerable to drought. The previous year's summer had been particularly severe, and drops in production had therefore been expected, although the judge noted that the Friesian herd was in a good condition. The Rutledges had obtained a high return from pig meat, he observed; the highest equal in this year's competition. Butterfat per cow was well below the winning entry; McMillan Bull noted that a Friesian herd would always be outperformed by a Jersey herd in this regard. The pastures were good, but application of fertiliser was low.

Judge McMillan Bull awarded third place to Mr J W Hedley of Hoe o Tainui, who farmed 90 acres with sixty-four cows. The judge said that his total butterfat production, over 18,000 lbs from an effective acreage of 70 acres, had been a 'fine achievement'. Programmes to eradicate weeds had also been sound, but attention to subsequent years of regrowth was necessary. The hay paddocks needed careful fertilising, the judge noted, and long-term drainage work was necessary.[122]

Remaining Competitors

Among the competitors evaluated by Judge McMillan Bull this year, James Nokohau of Wairoa farmed 74 acres of the best soil 'of any farm inspected'. However, the judge noted that his butterfat returns were somewhat below the average. Mr A T Heta of Whāngārei was developing a very attractive well-laid-out farm, which 'originally had been a formidable undertaking'. Mr Heta was commended for his 'high application rate of fertiliser'. Kawiti Thomas of Kaitāia had done superbly well in developing his property from a totally run-down condition; Judge McMillan Bull said 'I have no doubt that he has had to overcome more difficulties than any other farm in the whole group inspected by me this year.'

Charles Berry Wells of Kaitāia was farming 131 acres with great skill, but his fertiliser application figures were 'somewhat below the average'. Rangiharuru Simeon, the youngest in this year's group, was an impressive competitor. The judge had little doubt that he would eventually develop his property into an excellent dairy farm. Charles Dargaville of Panguru had suffered quite low returns this year, providing little revenue for further improvements. Joseph Te Wake of Panguru was doing an excellent development job on one of the most difficult properties the judge inspected. He was dealing with a troublesome fern reversion problem particularly well.

Sonny Riini of Rūātoki had shown an ample grasp of good husbandry methods, and as a consequence, had achieved a rapid rise in stock and butterfat returns. Judge McMillan Bull congratulated R Stewart of Ōtangaroa for making such a success of a farm that had been severely handicapped by a lack of water. Ben Wharerau of Waimā had experienced a very low butterfat return this year, partly as a consequence of his continuing problems with grazing.

1963

Low Sheep and Cattle Entries Again

Despite Judge McMillan Bull's hopes from the previous year, in 1963, one sheep and cattle competitor withdrew just before judging began, leaving only one other entrant: Mr Alec McAllister of Kāwhia. Effectively, then, the 1963 competition was confined to the dairy section. However, Judge McMillan Bull gave some special recognition to Mr McAllister.[123]

Dairy

Winner: An Outstanding Record in Development

The winner of the dairy section was Mr Hedley of Hoe o Tainui, a frequent competitor, who farmed 90 acres with sixty-five cows. The judge described his record in having developed his difficult farm as outstanding. This year, he had reached a record butterfat production of almost 20,000 lbs. The judge noted that drainage and the eradication of tall fescue remained his main problems.

Other Place-getters

Judge McMillan Bull awarded second place to Alfred Parker of Mangakino, who farmed a 153-acre unit with ninety-three cows. Mr Parker's farm was an 'excellent unit', the judge said, comprising well-sited plantations, good subdivisions and well-kept buildings. The pasture was also very good, given the type of soil found in the area. Mr Parker had learned to work well with pumice lands. His calves, however, had looked 'unthrifty' to the judge, probably due to under-feeding or the presence of parasites. For such a big farm, the standards of butterfat per cow were very high – 263 lbs or 161 lbs per effective acre.

Third place was awarded to Bernie Parker of Ōpōtiki, who farmed 143 acres with seventy-three cows. The judge described Mr Parker's property as an excellent example of how light pumice land could be brought into production. Heavy stocking and feeding out over short periods had proved successful. Care of stock and good soil management were a feature of this farm, McMillan Bull said. However, for such an exposed property, shelter was noticeably missing and was now an urgent requirement. Butterfat production had been good but not high. Fertilising of pastures had progressed, but more was still required.

Remaining Competitors

Among remaining competitors, Judge McMillan Bull congratulated Heke Rika of Wellsford on the resourcefulness he had displayed in the management of what was 'a very difficult farm indeed'. Every winter, Mr Rika's farm had 'pugged up very badly by reason of its soil and an impervious pan below'. The cold winters had also been a problem. Clarence Simeon of Te Awamutu farmed 100 acres with sixty-four cows. The judge commended Mr Simeon for being 'the neatest young competitor' this year. His returns were good, but he was still dealing with dilapidated buildings and his hay storage practices needed improving.

The judge noted that the farm of Hare Puke Jr of Gordonton had benefitted recently from a dry autumn, additional fertiliser and improved management. The butterfat returns per acre on this farm were outstanding, but tall fescue infestation had continued to threaten many of the farmer's best pastures. Albert Savage of Whangamatā had put a great deal of hard work into his 138-acre property, but a large swamp bisecting his farm continued to cause development problems.

Hugh Nokohau of Tāneatua was farming 145 acres of land that was 'still showing visible signs of having had no fertiliser or maintenance for the previous ten years'. His stock had a 'somewhat lean and hungry look', McMillan Bull noted, which could be blamed on elements missing from the soil, rather than a lack of feed. Mr Parata Ainsley of Waihī had the best all-round pasture and the most attractive

Developing Māori land for agricultural purposes in the 1930s was arduous. Many farms were characterised by uneven contours of land, broken gullies, boulders, ragwort and blackberry.

Māori men digging a drainage ditch in the Kaitāia swamp, between 1910 and 1939; photograph by Arthur James Northwood (Alexander Turnbull Library, 1/1-010660-G)

looking stock of all the farms entered. This had been quite an achievement given the uneven contour of the land: broken gullies and boulders continued to create difficulties. Mr Ainsley also had a 'severe ragwort problem with some blackberry thrown in for good measure'.

The judge noted that Joe Peterson of Mangonui had had limited opportunities to improve his farm because of its small size and broken nature. He recommended that 100 extra acres be made available from the adjacent Lands Department's Stoney Creek Block in order to remedy the 'static situation' that Mr Peterson found himself in. John Callaghan of Ōpōtiki farmed 130 acres with sixty-six cows. His farm had almost been 'too difficult to develop' because two-thirds of his area comprised hillsides that were too steep for machinery to traverse. As a consequence, Mr Callaghan had a low stocking rate and quite low returns.

George Petricevitch of Kaitāia had done a great job of 'resurrecting his run down property in the short time of two years', the judge said. He described Mr Petricevitch's stock management was as 'remarkable'; good butterfat returns were recorded Simon Ripaki of Whakatāne was farming 145 acres in an area described as difficult to farm because of the sharply contrasting climatic conditions. Pastures on Mr Ripaki's light, sandy pumice soil burned up badly in most summers, and little growth was possible during the long winters.

Mr Hauraki Herewini of Kaikohe was fortunate in possessing such attractive and naturally fertile land. However, an extensive gorse eradication programme had been necessary on his unit, impeding development. The judge congratulated Charles Berry Wells of Kaitāia on the overall improvement of his farm, especially given a very dry summer and autumn. Mr Wells' stock had presented well, as had his pastures, though a continuing problem with gorse infestation needed urgent attention.

Mr Joseph Te Wake of Panguru occupied 160 acres of steep and broken land, with strong fern growth that needed constant attention. The dry autumn had also affected his returns. Kawati Thomas of Kaitāia had also been adversely affected by the long dry spell, which had impeded pasture growth and caused a 'serious drop in butterfat per cow'.

Mr Wananga Matenga of Rangiotu continued to be plagued by gorse, lupin and tall fescue, which were seriously limiting his development efforts. Because Mr Matenga's operation was a 'one man concern', the area to be cleared posed particular difficulties. Charles Dargaville of Panguru had successfully eliminated most of the gorse that had bedevilled his property; his farm now looked well controlled. However, a long accessible road that crossed his property continued to impede his options for subdivisions.[124]

1964

Continuing Problems with Low Sheep and Cattle Entries

The issue of low entries in the sheep and cattle section of the Ahuwhenua competition returned in 1964. Organisers worried that sheep farmers were not entering the competition because of their reluctance to be minutely assessed on their efforts to make productive some of the most difficult farming blocks in the country. In his report for this year, Judge McMillan Bull expressed concern that it had been 'particularly difficult to allocate points' to the year's two sheep and cattle competitors, as they farmed on completely different lines. One was a relatively large well-established farm with stud and flock sheep and a considerable number of cattle; the other was much smaller and running entirely from sheep. 'Such differences made it difficult when it came to marking some of the economic factors in the final summary of returns', the judge said: 'Considerable differences occurred for some of the marking categories, particularly the net income per acre and per stock unit, and also wool weight.'

'Maori Orientation', as Judge McMillan Bull Saw It

In making these observations, Judge McMillan Bull also sought to put the process of assessing Māori farmers, and the Ahuwhenua competition itself, into context. Indubitably a product of the political thinking of his time, he commented that the magnitude of the task in reorienting 'an easy-going gregarious race, fundamentally not interested in agriculture, into independent single-minded individuals with some appreciation of economics' was too depressing to contemplate. According to McMillan Bull, only when one accepted that the process of 'Maori orientation' would take hundreds of years could depression give way to an acceptance that, one day, the 'chaos of titles and uneconomic units with their bits and pieces of poor land' would be sorted out. One had only to pore over the old Ahuwhenua files to see the progress that had been made in little over thirty years, he said. The Ahuwhenua Trophy had achieved much in the past; the future challenge was for Māori farmers to win even greater achievements.[125]

Applauding the 'Tryers'

Nineteen entries were received in the dairy section of the Ahuwhenua competition in 1964, and two entries in the sheep and cattle section: this constituted the highest number of entries ever received. Judge McMillan Bull noted that the dairying entries for the year had been quite remarkable; there had been strong competition for top places, especially from Taranaki, where the only entrants from that area had come second and third. The winner, Rawson Wright, had been an outstanding competitor, leading the field by a significant margin of 80 points.

In his summary report, the judge singled out for acknowledgement those competitors who entered the competition 'year in and year out' but who, because of factors mainly out of their control, never rose above halfway. They were 'tryers', said the judge, which he felt was the main thing, and 'formed part of that hard core on which our future prosperity depended'.

Sheep and Cattle

Winner: A Frequent Entrant

The winner of the sheep and cattle section in 1964 was John Steedman, who farmed 500 acres in Welcome Bay, Tauranga. Mr Steedman had entered the competition many times, the first in 1959. His farm was an excellent unit that reflected sound management practices dealing with both land development and stock, the judge said. However, some of his ewes appeared over-conditioned, which suggested under-stocking. In reality, his ratio of cattle to sheep for the type of country he was farming was a creditable one. Mr Steedman's waterways needed more fencing to prevent stock fouling the water supplies, McMillan Bull noted. Poplars and willows needed to be moved away from the water because their leaves rotted in the water, which required constant cleaning out.

Second Place

Second place was awarded to Joseph William Thompson of Tuakau, who farmed a unit of 170 acres. The judge said that Mr Thompson's skill in managing land developments and stock was obvious. His land stood in sharp contrast to the adjacent land development block, where new developments were only just beginning. Mr Thompson's work at constructing new yards had been 'excellent', and the judge described his stock control abilities as 'unique'. His one weakness seemed to be his inability to keep written records. Also, production levels showed that his lambing percentage was low. However, supplementary feed was well managed. The judge noted that further land acquisitions were planned.

Dairy

Winner: A Former Soldier

First place in the dairy section was awarded to Rawson Wright of Tāpora, who, with his wife Wikitoria, farmed 109 acres with seventy-five cows. The judge reported that Mr Wright's farm was 'outstanding in every way and said to be one of the best farms in the north'. While they showed evidence of drought damage, his pastures were otherwise well managed. His butterfat return – 275 lbs per acre – also reflected a high standard. An up-to-date piggery had been recently constructed, contributing to a substantial increase on last year's pig meat sales figures.

Wikitoria Wright Comments on Her Husband's Ahuwhenua Win in 1964

'In the 1960s, Rawson was invited to enter the competition even though we were under the State Advances and not Maori Affairs [the State Advances Corporation was a government entity set up in 1935 through which advances on mortgages were made available on the security of farm lands and urban and suburban properties], and there was a bit of opposition to him entering because he wasn't classed as a "Māori farmer".

'We'd taken the farm over in 1954, and in the second year we started milking seventy-five heifers – can you imagine that, trying to break in 75 heifers? It was hard work. Mind you, I was never a dairy farmer; I was brought up in sheep and cattle, so it was a bit of a culture shock for me to have to milk cows, and to have babies at the same time.

'In 1964, we were the overall winners – well, Rawson was, because in those days the wife was not recognised in partnership. They brought the cup into Wellsford and put it in the BNZ. A couple of days later, at the AMP Show, the cup was presented to Rawson. We weren't allowed to take it home or anything; we were only allowed to touch it and have a photo taken. Arapeta Awatere came down from Auckland with his haka group, and Matt Rata was also there.

'I would encourage any young Māori farmer to enter the Ahuwhenua competition. There's prestige in it, and the knowledge you gain from competing against other farmers is the most amazing thing.'

Source: interview conducted by Lyn Harrison, April 2013

Wikitoria Wright with the Ahuwhenua medal and replica trophy that she and her husband, Rawson Wright, won (Dairy Category 1964)
(AHU IMG 1049)

Second and Third Place

Second place was awarded to Edward Rongomaire Tamati of Bell Block, near New Plymouth, who farmed 90 acres with eighty-nine cows. The judge noted that Mr Tamati's grazing and general management practices were complicated by his having 'odd areas of land leased around Taranaki'. However, he had done well in converting blocks infested with gorse and blackberry into fine pasture for grazing.

The judge said that grass grubs seemed prevalent in the Bell Block area, requiring a dedicated programme of DDT. However, he considered the health of Mr Tamati's animals was outstanding: the best in the competition from the point of view of size, health and general appearance. His butterfat figures too were very good – 320 lbs per cow, or 200 lbs per effective acre.

The judge awarded third place to Tongawhiti Manu of Ōeo in Central Taranaki, near Ōpunake. Mr Manu farmed 160 acres with 135 cows. His farm had the greatest potential seen this year', being flat with excellent soil. Shelter by way of boxthorn hedges was also good. Pasture production and herd improvement were high and clearly improving. Drainage was the only real problem the judge saw on this property, and perhaps some evidence of silage wastage.[126]

Remaining Competitors

Among the 'tryers' Judge McMillan Bull commended in the 1964 competition, Wiremu Mauriohooho of Te Awamutu had a particularly good farm: the judge approved of his pastures and his butterfat returns. Mihi and Karena Stephens of Ōkaihau had won the dairy section in 1960 and were farming along excellent river flats, 'carrying almost the best pasture seen in Northland'. Tiwi Black of Whakatāne was a very good pig farmer: he had the largest lay-out and pig herds of all competitors. Clarence Simeon's all-round ability left little room for criticism, although he needed more windbreak trees. He farmed 100 acres near Te Awamutu with seventy cows.

Albert Savage of Whangamatā had planted an excellent strike of new grass where a troublesome bush block had once been standing. Hare Puke's farm of 52 acres needed urgent metalling to improve its appearance and eradication of a tall fescue infestation.

Waitangi Brown of Dargaville was unique in the competition in that he was running a pedigree herd. As a consequence, his recording was excellent, covering every phase of the more complicated aspects of livestock breeding. The judge suggested that pasture renovation was necessary given the low state of his soil, but cautioned him for his 'enthusiasm for the Bevin Harrow'. Alfred Parker's Mangakino property was showing all the signs of having suffered from a prolonged 'adverse season'. Neither pasture nor stock looked well; Mr Parker had lost points 'through circumstances over which he had no control'. Tau Shepherd of Kaeo had erected new boundary fences, giving his farm a well-kept appearance. However, his herd was generally in a low condition, in part attributable to the recent drought.

Wharehuia Eri of Te Puke was 'extremely active' for a man of fifty-eight years, having done a great job in developing a farm once heavily infested with gorse and blackberry. Jim Clark's farm at Waikeretū lay well within an easy contour, and carried surprisingly good pastures. However, rocky sidings had proven difficult to clear of blackberry and gorse. Norman Warrington of Waimā farmed 82 acres with fifty-nine cows on a property with multiple land contour problems, from river flats prone to flooding to steep hillsides covered with bracken and gorse. The judge commended

Judges commended Māori sheep farmers for their first-class farming skills, while advising them to avoid overstocking, to protect their waterways from stock and to not sell hay.
K Kotua of Nelson shearing a sheep at Adams' Langley Dale farm, c. 1950s (Alexander Turnbull Library, PAColl-6303-03)

Until the 1950s, horse transport was still favoured by rural Māori.

Scene at Kāwhia showing horses by a whare, between 1910 and 1930;
photograph by Charles Whitmore Babbage (Alexander Turnbull Library, 1/1-001085-G)

Joseph Te Wake of Kohukohu for annually entering the competition in spite of the many problems he had had to overcome in developing his steep and broken farm.

Robert Smith of Waitakariri farmed 100 acres of very favourable land with an excellent balance of marine flats and dry rolling hill country suitable for winter grazing. However, Judge McMillan Bull said that his butterfat returns needed improving, and he needed to clear some noxious weeds. Areka Tapara of Waiuku had recently added an additional 30 acres to his small farm to enhance his development efforts. The judge commended Charles Dargaville of Panguru for competing every year against less isolated farmers: he had had the satisfaction of developing a farm under circumstances that would have 'daunted the courage of many contestants'.

Mr Teku Ata of Tikitere was farming 110 acres with sixty-four cows. In the past, the farm had been allowed to deteriorate to such a degree that its 'resurrection' had posed a number of formidable problems, including replacing the farm buildings, which had all reached the end of their usefulness. The farm of Arthur Subritzky of Kaitāia had suffered such a serious drought that pastures had deteriorated and the condition of stock had fallen drastically. However, his pigs were of a fair quality; if finance could be approved, more appropriate housing stock was planned.

1965

A New Judge and More Sheep and Cattle Entries

Fifteen entries were received for the 1965 Ahuwhenua competition, comprising eleven entries in the dairy section and four for sheep and cattle. Mr J R Murray, returning to judge the competition, was pleased with the numbers. He urged sheep farmers to make use of the competition, with its appropriate and fair pointing system, to assess their development progress.

Sheep and Cattle

Winner: A First-class Woolshed

The winner of the sheep and cattle section was J W Thomson of Glen Murray, who was farming 212 acres. In Judge Murray's eyes, Mr Thomson's farm bore witness to farming skills that covered most aspects of the sheep and cattle business. The general design of Mr Thomson's woolshed and yards was described as 'first class'. The condition of his stock was also high, with a stocking rate of 5.8 ewes per equivalent acre. Any weaknesses were minor and had only arisen from such issues as overstocking, the judge noted. The interest surplus on capital invested 'could have been better', but it did represent an improvement on the previous season. The judge recommended further subdivision of paddocks, and cautioned that overstocking often spoiled good development work. Hay needed to be cut earlier, and heavy long crops of poor quality needed to be avoided as winter feed. The judge advised Mr Thomson not to sell hay.[127]

Other Competitors

Second place was awarded to John Tahuri of Rotorua, who was farming a 587-acre block. Mr Tahuri's unit had not been an easy one to work, the judge noted. Pine trees were scattered all over the farm; although they offered some shelter, they were more often than not an obstruction to development work. Returns per acre per ewes wintered were the best in the competition, the judge noted, describing interest surplus on capital invested as 'quite good for this stage of development'. Murray's recommendations included a planned attack on barley grass, further subdivision, the assistance of a saw mill in clearing trees, phosphate application for pastures and selection of rams to improve wool quality.

Third place was awarded to Aperehama Whata of Rotorua, whose 576-acre block had been 'slowly turning into a good sheep farm'. Subdivisions carried out in recent years were now showing their value, the judge said, though fencing was still substandard. Logs were slowly giving way to good pasture, and good management of fertilisers had been reaping additional benefits. Murray recommended building up sheep numbers and reducing cattle, and focusing on control of ratstail. He noted that the house and farm mortgage had been substantially reduced.

There was one other dairy competitor: Mr A A Alexander. He farmed 'quite a good property' of 406 acres near Kaikohe and had clearly worked very hard to bring his unit to its current level of production. The judge noted that he had recently commenced weed control for rush and gorse, putting the rotary cutter to 'full use'. Mr Alexander's stock replacement costs had been high but would be reduced once ewes were kept as replacements, which would perhaps also lead to a better interest return on capital invested.[128]

Dairy

Winner: 'Bits and Pieces of Land'

The winner of the dairy section was Edward Rongomaire Tamati of Bell Block, near New Plymouth. Despite the awkward nature of his fragmented farm, Mr Tamati was a 'deserved winner', who had shown what could be done with 'bits and pieces of land'. All his stock were of an excellent quality; coupled with good management practices, they had brought good returns. The interest surplus in capital invested was by far the best in the competition.

Second and Third Place

Second place was awarded to Wiremu Mauriohooho of Te Awamutu, who farmed 124 acres with eighty-five cows. Despite ill health, Mr Mauriohooho had worked consistently to achieve high production returns on his farm, Judge Murray reported. More subdivision had been necessary, among other tasks, so Mr Mauriohooho's daughter had taken part in helping to run the farm. The judge noted that more cows still needed to be milked on this unit. The pastures were of a good quality, but the judge recommended an increase in fertiliser. Mr Mauriohooho's ill health had been a governing factor in what could be done on the farm; 'like many others who have been troubled and gamely come forward into the competition, this was a case where all praise was due'.[129]

Third place was awarded to Mr Tongawhiti Manu of Ōeo, near Ōpunake in Taranaki. Judge Murray described Mr Manu's 200-acre farm as a 'very good farm with high potential'. However, it was encumbered with high box thorn hedges that were expensive to trim, though they did have their uses as shelter and windbreaks. Mr Manu's production average was a reasonable 309 lbs of butterfat per cow. Top dressing rates were good, but the judge considered that, with extra fertiliser, more stock and more labour, this unit 'could be producing much more'. Hay needed to be cut a lot earlier in the year, and subdivision was also needed.

Winner of the Ahuwhenua Trophy for Dairy in 1965, Edward (Ted) Tamati of Waitara receives his award from Mrs Iriaka Ratana at Rotorua.
(Photo courtesy of Liana Poutu)

Other Competitors

Among other competitors, Mr T F Matchitt of Rotorua farmed 173 acres, and Judge Murray considered him 'capable of producing a high quality dairy unit'. Mr Matchitt's herd had given the highest butterfat quantum per cow in the competition this year, at 322 lbs. A significant herd increase was planned for the new year. Mr J Te Wake's 198-acre farm at Kohukohu was described as a typical unit located on Northland hill country; the judge congratulated him for a consistently high standard of improvements. Mr G Petricevitch of Waiharara possessed the only farm Murray had visited this year that contained 'practically all kikuyu grass'. Despite this, production returns were described as 'quite fair'.

Mr J Clark of Waikeretu had set ambitious butterfat targets on his 142-acre property and was making good progress. His herd quality was reasonably good, the judge considered, but pasture improvement needed greater attention if stock numbers were to be lifted. Mr G Dargaville of Panguru had worked hard to clear his farm of scrub and gorse and had built new fencing along his access road. His production returns had been the highest yet this year, though he still had 'a long way to go'. Mr J C Harris of Kaeo farmed 193 acres of land with a serious exposure problem to southerly climatic conditions; drainage of his 'pot holed flat' was also a major issue.

Mr J Ormsby of Rotorua managed a 'mainly peat farm' still in the early stages of development. His herd was a good one, and with further good management would produce more. Mr R Stewart of Kaeo farmed 173 acres of land described as 'quite a hard farm requiring much in the way of fencing and fertiliser'. Water supplies had long been a problem on this property, especially as the area was subject to frequent droughts.

1966

A Move Away from Established Farming Areas

Returning to judge the 1966 Ahuwhenua competition, Judge Murray expressed his disappointment that only a single candidate was contesting the sheep and cattle section.[130]

Nine nominations were received for the more favoured dairy section, all from either Northland (five), Waikato (two) or Taranaki (two). No entries were received from the older dairying areas like Waiariki or Te Tairāwhiti, where most of the earlier competitors had originated. A gradual move away from those more established farming areas had been occurring since the late 1950s.

Sheep and Cattle

The only sheep and cattle competitor was a Waiariki farmer. Judge Murray said that development and improvement on the farm of Mr J H Tahuri of Rotorua had progressed steadily; in breeding ewes, ewe equivalents per acre, wool produced per acre and capital invested per acre. Mr Tahuri's carrying capacity had also increased. The judge noted that the farmer had made a start in reducing the pine trees scattered over his property, as he had recommended last year. Yorkshire fog seemed to be recurring, Murray reported, and needed close attention. Subdivision plans needed to be continued. Ram selection each year needed to be guided by wool quality, and barley grass required constant vigilance.

Dairy

Winner: 'An Indomitable Spirit'

First place in the dairy section was awarded to Wiremu Mauriohooho of Ōwairaka, in the Waikato. Mr Mauriohooho had been a 'devoted applicant' in this competition for some years, Judge Murray noted; this year, with the production levels of his farm reaching ever higher, 'it was fairly clear that he would take some heading off'.

Despite ill health, Mr Mauriohooho had 'shown an indomitable spirit in striving to progress, helped in no small measure by his wife and daughter', Murray said. Butterfat production had increased over the previous year, and now stood at 300 lbs. An increase in fertiliser was also evident, and the judge said that an increase in cow numbers could also be considered – partly in order to control the Yorkshire fog. Hay could be cut earlier as well, to ensure better quality.

Second and Third Place

Second place in 1966 was awarded to Joseph Niwa of Pūniho, in central coastal Taranaki. High production per acre had won Mr Niwa second place, the judge said. His butterfat production of 268 lbs per acre, though not especially high, turned out to be the best in the competition because of his carrying capacity of nearly one milking cow per acre. However, the judge noted that he badly needed more shelter because of the cold mountain winds. Lack of a farm race was a drawback for this farm, and some fencing maintenance was required.[131]

Third place went to George Taurua of Poutō, Northland. His farm was prone to long dry periods. It was very long, cut into by Lake Waingata. Mr Taurua's decision to develop his front section seemed to be paying off. New fencing and subdivisions had also been of tremendous benefit, the judge said. High rates of production and low costs had led to a good net income per cow. However, pastures needed attention to prevent Yorkshire fog and cocksfoot from recurring; the judge also suggested that greater care be taken with the keeping of farm records.

Remaining Competitors

Among remaining competitors, Mr G Petricevitch of Waiharara 'deserved much more encouragement with better drainage outlets'. A good 60 acres of his property had long been lost to production because of the lack of a better drainage system. Most of the farm was in kikuyu grass, which resulted in long growth going into winter and a consequent hampering of the new season's pastures. The judge commended Mr R Simeon of Ōwairaka for turning his unit into a good dairy farm, though it seemed understocked.

Mr T Hori of Kaeo currently managed a farm with 'good potential'; he showed 'a determined spirit' in increasing production, the judge said. His per-cow production had been the highest in the competition. As had all farms in Northland, the farm of Mr H Api of Poutō had recently suffered from severe drought; ill effects were still visible. However, some good work had been done on fencing and subdivision. Mr H Roach of Ōpunake had attained good results from difficult land. He was taking a satisfactory approach to improvements, and steady increases in productivity were now possible. Mr R and Mrs I Hemi of Kaeo in Northland would increase productivity in time with additional inputs of fertiliser, fencing, labour and stock.

1967

Taking the Land – the 1967 Maori Affairs Amendment Act

The 1967 Ahuwhenua competition proceeded against a political background of continuing challenges for Māori landowners.

Even this far into the twentieth century, Māori were still dealing with Crown legislative measures aimed at compelling them to give up their lands. One such measure was the Maori Affairs Amendment Act passed this year.

The origins of the Act go back to the 1920s, when the government decided that all large-scale acquisitions of Māori land would cease. Native Minister Gordon Coates worked closely with Sir Apirana Ngata to address problems like multiple ownership and to facilitate properly funded Māori land developments. In 1929, Ngata sponsored legislation that provided for government loans to assist with such land development, and numerous development schemes commenced in the 1930s.

However, under the Maori Affairs Act 1953, Māori land could still be 'converted'; if the value of shares in it fell below £25, owners could be forced to sell it to the Maori Trustee, who would then on-sell it to other Māori owners for 'development'.

Despite significant protests, the 1967 Act intensified this process of conversion. Now, land interests worth less than $50 would be compulsorily purchased, usually by the Crown. Of course, this completely ignored the significance of Māori land to Māori themselves, as tūrangawaewae (a place to stand, or home).

At the same time, the Crown continued to take Māori land for public works in preference to other land, often without providing compensation. It was a situation that was to remain unremedied until the 1970s.[132]

Sheep and Cattle

The one sheep and cattle entry in 1967 was Mr A A Alexander, who farmed 406 acres at Ōkaihau, in Northland. Judge Murray considered that Mr Alexander's unit had 'very good potential', and decided to bestow on him 'an acknowledgement equivalent to the awarding of second place'.

Mr Alexander had a fine property and displayed careful management, Judge Murray said. Much had been done to bring his farm to its present levels of production, and the 'stage was set for more consolidated improvement'. Mr Alexander's stock

Judges often advised Māori sheep farmers to make greater provision for hay.

Horse-drawn reaper binder at work, on Māori-owned Adkin farm at Horowhenua, c. 1930s; photograph by George Leslie Adkin (Alexander Turnbull Library, 1/4-023361-G)

was in good condition, but his replacement policies, particularly in regard to cattle, needed clarifying. The judge called for better control of rushes and other weeds and said that new grasses required fertilising with heavy dressings. He also suggested a new sheep dip with a chute from the shed pens.[133]

Dairy

The winner of the Ahuwhenua Trophy for dairying for 1967 was Mr G C Hopa of Tauhei in the Waikato. Judge Murray described Mr Hopa as a hard-working farmer of lands that were the 'roughest in the region'. His stock was of the highest quality, and his management practices were sound – especially his fertilising of new pastures after eradication of troublesome bracken and gorse. Mr Hopa's buildings were well kept, and his implements were 'clean and well stored', the judge noted.

Second place was awarded to Tongawhiti Manu, from Ōeo, near Ōpunake. Mr Manu had entered the competition in prior years with some success. Like Mr Hopa, he was one of the first competitors to have been nominated from his particular area. Judge Murray described Mr Manu's farm as a unit 'with a high potential'. High-quality stock brought in good butterfat returns – 309 lbs per cow. The construction of a new herringbone shed was under way; it would accommodate the farmer's 140 cows with ease.

Murray awarded third place to Mr George Taurua of Poutō, in Northland. The judge said that high production per cow and low costs had given Mr Taurua a good net income per cow. He noted that much work had gone into necessary capital improvements. One reservation he had was the 'clumpy' nature of Mr Taurua's pastures, due mainly to cocksfoot and Yorkshire fog.[134]

1968

The 1968 Competition

Judge Murray returned for the 1968 competition. Following the pattern of previous years, only one entry was received in the sheep and cattle section.

Sheep and Cattle

The sole entrant for sheep and cattle was Mr Jack Steedman of Welcome Bay, Tauranga. Mr Steedman managed 529 acres of fine pasture with quality stock, though Judge Murray noted that the farm seemed understocked. With an increase in stock numbers, pastures 'would only improve', the judge said. Water supplies were 'quite fair', though, typically for the area, Murray noted that severe flooding had caused considerable damage to water supply ponds. The judge advised Mr Steedman to increase his subdivisions, particularly if he was contemplating additional stock. He also suggested that he 'spray gorse in late spring using 2-4-5-T', make greater provision for more hay, and discuss a top dressing programme with the local farm advisory officer.[135]

Dairy

The dairy winner was Mr Jack Karatau, who farmed 54 acres in Whangaehu, near Whanganui. Although Mr Karatau had been on his small farm for only eight years, he had achieved the highest butterfat production ever: 448 lbs per acre, or 378 lbs per cow, from sixty-four cows. Since first taking over his farm in 1960 as an ex-serviceman of the First World War, Mr Karatau had adopted sound management practices and created permanent pastures, quality shelter and a plantation, plus exceptional homestead surroundings.[136]

Second place was awarded to Tongawhiti Manu of Ōeo, near Ōpunake, who farmed 160 acres with 135 cows; this was the third time Mr Manu had received a placing. Judge Murray highly commended Mr Manu's farming management; in particular, 'the quality and vigour of his saved pasture'. This reflected excellent soil and almost flat and well-sheltered subdivisions that were conducive to strong grass growth, especially if existing fertiliser programmes were to be continued. The judge observed some waste in Mr Manu's silage stack, which must have taken many tons of valuable grass, time and labour; 'possibly a lack of appreciation of the cardinal measures employed when making good silage' was the reason.

Wharehuia Eri of Te Puke was awarded third place. Judge Murray described Mr Eri, who managed a unit of 94 acres with seventy cows, as the best stockman in the competition. He noted that Mr Eri had done well to rid most of his property of a blackberry infestation, though such issues warranted constant attention. Much of Mr Eri's farm lay in swamp, which in some seasons had impeded his grazing options; he had therefore given swamp drainage options some serious thought. Mr Eri's use of electric fences was innovative, the judge noted; it was 'unusual to inspect a farm where electric fences were so exclusively used and with such good effects'.

1969

Withdrawals

The Ahuwhenua competition for 1969 attracted two entries in the sheep and cattle section and sixteen entries in the dairy section. However, two of the dairy entries withdrew on the eve of judging: Mr Tongawhiti Manu of Ōeo because of a dairy shed reorganisation, and Mr J W Marks of the Bay of Plenty, at Judge Murray's suggestion, because of the 'sad loss of his daughter in a tractor accident earlier that year'. Mr Marks' daughter had been a great help on the farm, and this tragedy had 'disorganised his farm work', the judge noted.[137]

Sheep and Cattle

The winner of the sheep and cattle competition for 1969 was Mr Reihana F Apatu of Hastings, who managed 299 acres of farmland described as a 'particularly attractive area' that could in time become 'the envy of others'. Mr Apatu's farm was well fenced and well kept, with a neatness and tidiness 'that gave the impression of a winner', the judge said. Mr Apatu had worked hard to develop this unit, originally a 'run-off' area with no buildings or facilities to speak of, into a 'really good property'. His carrying capacity of up to eight stock units per acre was excellent and due, in the judge's eyes, to his 'forward approach'.

Second place for this year went to Mr Waka Konui of Manunui, who managed 542 acres of 'healthy and excellent stock, both sheep and cattle'. Mr Konui had 'revealed himself to be an excellent stockman', Judge Murray said. Mr Konui's lambing percentage of 100 bore this out, and was even more of a creditable achievement when one considered the old stumps and fallen timber to be found in each of his fields. The farm also had a good cattle ratio, which was paying good dividends.

In 1970, Judge Murray noted that many Māori sheep farmers were working with difficult country covered in logs and tree stumps in every field, posing issues for pasture growth and stock rotation.

Bullock team pulling logs on a corduroy road, Northland, 1920s (Alexander Turnbull Library, 1/1-006242-G)

Dairy

Place-getters

The winner for dairy was Mr T L Jones, who farmed 127 acres and had been putting into practice some very modern farming techniques and maintaining good nett returns, Judge Murray said. Provided 'he did not get too ambitious in trying to milk too many cows', Mr Jones would be able to consolidate his position and improve his pastures and the butterfat per cow and per acre. The judge also commended Mr Jones' effort to reorganise the fencing and water races in the middle section of his farm.

Second place was awarded to Mr J Edwards of Horohoro near Rotorua, whom the judge congratulated for the development of his 180 acres, particularly given the farm's contours and winter climatic conditions. Murray awarded highest marks to Mr Edwards' farm for its fencing and subdivisions; he also described supplementary feed in the form of hay as very good.

Third place was awarded to Mr G Simeon of Ōwairaka, Te Awamutu, who had achieved increased production and other improvements on his 124-acre property. He had some of the best pastures and supplementary feed 'of all the competitors', the judge said, noting, however, that ragwort was still a major problem.

Other Competitors

Among other competitors, T T Rongonui of Hāwera managed 'a well kept farm with a neat and pleasing appearance', the judge said. His production, per acre and per cow, was constantly improving. Murray commended Mr P M Carr of Hāwera for his production; he milked off 50 acres of his 80-acre farm and used the rest as a run-off for his dry stock. Grass grub infestation had impacted on his pastures, but this could be overcome with closer attention to top dressing. Mr W B Rawiri of Tinopai in Northland had earned the second highest return on capital invested of all competitors this year, though he was still down on last year. Mr Rawiri's production had not been as good as it had been, and his replacement stock were not of a high standard, possibly because of the poor quality of his hay.

Mr J T Holden of Waitomo was congratulated for his high butterfat per cow – 351 lbs – though the judge had found it difficult to get an accurate assessment of his butterfat per acre because his herd had been wintered over the sheep area. Mr T W Eri of Paengaroa, who had won third place in 1968, had suffered a slight decrease in production this year, as costs had risen, and his nett income per cow and per acre had dropped. Judge Murray commended Mr T Ruwhiu of Cape Runaway in the Bay of Plenty for increased production in a more isolated area. Although his figures were on the low side, his nett income was equal to many of the other better performing competitors.

The judge congratulated Mr R Wetere of Ngutunui, near Te Awamutu, for his improvements to an originally run-down farm, including excellent fencing. Mr D R Kirkwood of Mangatangi had been doing a good job developing another run-down property that had once been infested with ragwort and blackberry. However, some of his pastures were too long and needed grazing off, as Yorkshire fog was increasing. Mr J P Niwa of Ōkato had entered the competition three times before this year, but was still meeting with stiff competition. Had he shown more progress, he would have been placed much higher.

Mr T H Hohepa of Te Teko had recently bettered his financial position as far as nett returns were concerned, but his production figures could be much improved, Murray reported. His property was very long and narrow, which presented all manner of issues. The judge commended Mr G Ormsby of Tinopai in Northland for the way he had taken on another run-down property, which had had to be completely refenced. Mr Ormsby had stocked his farm with Friesians; given its hilly nature, a lighter animal would have been preferable.[138]

1970

Interest Remains High

The Ahuwhenua competition for 1970 attracted three entries in the sheep and cattle section and fourteen in the dairy.

Over the previous decade, entries had remained high in the dairying section, averaging about twelve nominations per year. Sheep and cattle entries had fallen away mid-decade, but were now improving. Prior to this year's competition, one of the sheep and cattle entries, Mr J T Otimi, withdrew because of road deviation work around his farm, which had disrupted his operations. Two dairy competitors also withdrew: Mr R Ormsby because he felt his farm was ill prepared, and Mr P Bell because his farm had been so badly affected by drought.[139]

Winner of the Ahuwhenua Trophy 1970, Charles Bailey of Waitara, is congratulated by Rt Hon Duncan McIntyre, Minister of Māori Affairs.
(Photo courtesy of Ra Bailey and Tiri Bailey-Nowell)

Suggestions and Follow-up Visits

An important addition to Ahuwhenua judges' reports that had been introduced by Judge Murray in 1959 was a section headed 'Suggestions': this became a standard and essential part of the reporting system. 'Suggestions' listed potential 'follow-up actions' for farmers and field staff to consider for particular farms. Within a few years, Head Office was actively monitoring these suggestions. It asked district field staff to file reports after follow-up visits with competitors, and eventually made such reports mandatory. One Head Office memorandum written in 1970 reminded staff that these reports were a requirement under 'Manual section 3.27.10', which stipulated that 'reports be submitted in duplicate'.[140]

The reports are historically interesting and tell their own story. One 1970 competitor in Whakamaara had commenced rebuilding his water race, as suggested, but finance continued to be a problem. He had attempted no further subdivisions, as had also been suggested, because of 'other circumstances which had precluded a start on this work'. However, on advice he had changed his policy of purchasing pedigree cows and was instead turning to herd testing and 'A.B.' Other competitors, following various suggestions from the judge, had attempted to clear ragwort, consulted veterinarians about stock issues and drained troublesome bogs.[141] Heavy floods had precluded follow-up work in some regions.[142] In one case, a district officer reported that 'no report had been submitted on this settler because the stock and plant had been sold and the farm was now on the open market'.[143]

Sheep and Cattle

The winner of the sheep and cattle competition for 1970 was Mr Waka Konui of Manunui, who farmed 542 acres of 'difficult country' that posed some issues for pasture growth and stock rotation. In the judge's eyes, the outstanding feature of Mr Konui's farm was the health and condition of his stock, as it had been last year. He had achieved excellent returns, as revealed by his nett income and interest return on capital. The judge said there was a 'quality and pleasing feature about the pastures, fences and stock that reflected a farmer of no mean calibre'. Second place went to Mr John Foley of Rotorua, who farmed a larger area of 684 acres 'similarly contoured to the winner's farm'. Mr Foley also had a fallen timber problem but was commended for his weed control and subdivision. Mr Foley had had to sell cattle because of the drought, but was seeking relief from the Inland Revenue Department to cover these losses.

Dairy

Place-getters

The winner of the dairy section this year was Mr Charles Bailey of Waitara, whom the judge commended for having developed a very efficient dairy farm 'with a most profitable pig diversification'. Great credit was due to Mr Bailey for his stock improvements, management practices, pasture management and conservation; he had 'the best stock seen this year' on display, the judge said. He gained top marks in all three aspects of production.

Second place was awarded to Mr Charles Berry Wells of Ōturu, Northland, who was congratulated for showing 'a real pioneer spirit' in the way he had improved his home farm and was about to begin improving a newly leased area. Mr Wells had the highest nett return per cow in this year's competition and gained the second-highest points for interest return on capital invested. His weed control and pastures had been excellent, and his farm had a neatness and appearance overall 'that caught the eye'.

Mr John Klaricich of Omapere was awarded third place for his 'hard work and perseverance' in taking over and improving a large property (400 acres) not altogether suited for dairying. His excellent work on the flats of his property had reaped great rewards, as had his very good organisational ability and appreciation of priorities.[144]

Other Competitors

Among other competitors, Mr I D Hall of Welcome Bay, Bay of Plenty, was farming 'quite an awkward shaped property – long and narrow with many sidings'. In general, his property was a neat and tidy place 'with a good house and excellent gardens', the judge said. Butterfat production had been very good, although it had been down during the most recent season due to drought conditions. Mr T W Manu of Ōeo had maintained a good standard of farming and good returns per cow, though there remained potential for further gains. Recent modifications to his herringbone shed had greatly improved his production prospects.

Mr J P Niwa of Ōkato had advanced this year in the competition due to the quality of his stock. However, drought had affected his operation, as it had other competitors. More subdivision, drainage and control of Yorkshire fog would greatly improve his unit's potential, the judge considered. Mr T T Rongonui of Hāwera, who was farming a relatively small farm (85 acres) by 'today's standards', was managing a 'neat and pleasing' property, though its production had not been as good as it could be. Drought had also affected Mr Rongonui's operations; his nett income was lower this year, and there had been no interest return on capital investment.

The judge commended Mr C Peters of Ngātaki in Northland for attempting to establish and maintain quality pastures in such a troublesome setting. Insect problems had plagued the area; Mr Peters had built starling boxes in order to recruit starlings to his defence. Mr P M Carr of Hāwera had also been affected by the recent drought; his production and returns had suffered. Urgent repairs were needed to Mr Carr's shed yards, and Yorkshire fog needed watching in his unit's 'home pastures'.

Mr D R Kirkwood of Mangatangi 'should have been placed higher in the competition', the judge said, but like other competitors he had suffered the effects of drought. However, with more land becoming available, Mr Kirkwood was looking to increase his herd size and therefore his production prospects. Mr B B Marshall of Glen Murray in the Waikato was farming 268 acres that needed closer subdivision, more fertiliser and a higher stocking rate in order to increase production. N Bishop and W King of Ōpunake were farming 114 acres, and had made a good start in reorganising a property that, being long and narrow, presented many natural difficulties, including water provision and drainage. The unit did have potential, especially in terms of good pasture growth.[145]

1971

A High Level of Interest Continues

The Ahuwhenua competition for 1971 initially attracted three entries in the sheep and cattle section, including two from the South Island. However, before judging began, the two Southland entries withdrew. 'Their withdrawal was a great pity', said Judge Murray, 'but we hope they will both come forward in future years'.

Winner of the Ahuwhenua Trophy for Dairy in 1971 was Edward (Ted) Tamati of Waitara, who had also won the award in 1965. Pictured here with Sir Paul Reeves at Owae Marae, Waitara
(Photo courtesy of Liana Poutu)

Interest in the competition meanwhile remained high among dairy farmers: sixteen entries were received. These entrants were all 'top flight competitors', Judge Murray said; they included a previous trophy winner and three previous second-place winners: 'not for many years has this section had such a line up of top grade contestants, so much so that it has been most interesting and rewarding to judge this group on this occasion'. He noted significant progress and improvement among competitors.

Sheep and Cattle

'As had occurred in 1967, so again in 1971', said Judge Murray. Mr A A Alexander of Ōkaihau East was once again the sole competitor in the sheep and cattle section. Judge Murray sought to arrive at a fair placing, in the end recommending that, again, Mr Alexander be awarded the equivalent of second place. When his production returns were contrasted over the previous three years, he noted, the results had been variable. Both stock units per acre and wool returns per acre had shown a rise to 1967 then a fall to 1971. Wool per breeding ewe wintered in lbs had shown a definite increase, as had nett income per acre and the amount of capital invested per ewe.

More broadly, Mr Alexander had improved his pasture control, but rush, moss and Yorkshire fog were proving once to be a recurring problem. Attempts had been made to obtain a more even type of wool. Mr Alexander's cattle were 'of a much better stock' than they had been, and returns were greater. Improved drainage on his bottom fields was urgently required, and more subdivisions were also necessary.

Judge Murray concluded: 'There is no doubt that improvement and progress, together with a more satisfactory financial position, indicate that this farm is a potential trophy winner.'

Dairy

First Place

The dairy winner in 1971 was Mr Edward Rongomai Tamati of Bell Block, near New Plymouth. Mr Tamati had won the trophy in 1965; this year, he won 'only marginally over the second place winner', the judge said.

Since 1965, Mr Tamati had sought to consolidate his fragmented land without success. However, by milking well over a cow to an acre, he had attained his highest ever butterfat figures, despite the impacts of drought and eczema. Good stock control and sound management were strong features of this farm, Judge Murray said; very good reserves of feed showed excellent management. Mr Tamati's returns on capital invested were among the highest in 1971.

Mr Tamati was awarded the Ahuwhenua Trophy by the local Member of Parliament, Hon David S Thompson, on 24 June 1972 at Manukorihi Marae, Waitara.

Second Place

Second place was awarded to Mr Tongawhiti Manu of Ōeo in central Taranaki, who had won second place twice before. He 'came close to first place this year', Murray said, on account of his production figures per cow and his returns on capital invested. The judge described Mr Manu's general farming standards as very high. However, his fertilising quantities had fallen short of what was necessary to ensure good pasture growth. His milking shed, though old, was of the highest quality. Mr Manu had undertaken to complete further subdivisions and rectify attendant drainage issues.

Third Place

Judge Murray awarded third place to Mr J Edwards of Horohoro near Rotorua, who was farming 180 acres in an area where 'dairying was not easy'. The judge considered that Mr Edwards had adopted good subdivision policies, and the overall management of his farm, including his record-keeping, was first class. He had made good provision for supplementary feed. The control of noxious weeds in Mr Edwards' pastures remained an important challenge; the judge suggested that he remain watchful for ragwort and blackberry. Repairs to the hay barn were necessary in order to ensure adequate cover for the hay.

Other Competitors

Among the thirteen other competitors, the judge complimented Mr W H Hughes of Tāpora in Northland on his well-laid-out farm and the quality of his stock. Mr Hughes' farm had presented a neat appearance; good pastures and successful weed control were evident. Planning for additional shelter was needed, Judge Murray noted, along with better drainage. Mr K P Ormsby of Ngutunui in the Waikato was commended on his land development progress; he had cleared substantial areas of bush and other secondary growth. Although his was in reality a mixed farm, Mr Ormsby had been dairying to a 'one man unit level' with the expectation that he could later build up the balance of his property in sheep and cattle.

Mr G B Wells of Ōturu in Northland had been awarded second place in the competition the year before. His return on capital investments remained high, and he continued to manage his property with great skill 'for this class of country', the judge said. Mr H H Rowe of Hillsborough near Inglewood had the first town milk supply farm to enter the competition; 'it had been a treat' to see such a neat and tidy farm, Judge Murray noted. Mr Rowe's returns on capital invested had dropped recently, as had all town milk suppliers because of rising costs; such farmers were recently 'losing their edge' over suppliers of factories. Mr E and Mrs R Walden of Rahotū near Ōpunake were greatly commended for establishing a dairy farm 'out of some very difficult country broken up by swamps and nobs of high ground'.

The judge commended Mr W H Roach of Ōpunake for the financial progress he had made over the years, and for his overall production. Having recently paid off his loan, he had purchased more land and was planning to move into more beef production. Mr W Toi of Ōruawharo in Northland possessed a very difficult farm to handle at the best of times, especially in the wet seasons. However, his butterfat figures were excellent. Mr R R Roach of Ōpunake had only been farming his 176-acre property for one year, though he had share-milked on site for some time. His production figures had been quite good, and he planned major subdivisions to increase his stock numbers. Mr B Tumata of Oputia in the Waikato had managed well on a farm that Judge Murray described as a 'difficult one to manage with heavy soils and constant flooding'.

The judge also commended Mr M D McConachie of Oputia in the Waikato for his planning and reorganisation of subdivisions and buildings 'to make his property more efficient for working'. Mr McConachie's pastures were good, especially those in the 'main milking section'. While N Bishop and T King of Ōaonui in Taranaki had experienced lower production figures than expected, they had made progress on improvements and pastures, and their stock was observed to be in a good condition. Judge Murray commended Mr B Yates of Takahue in Northland for his handling of a difficult property given that he was 'literally dairy farming on hill country'. Planned improvements would greatly improve his production prospects, he observed. Murray commended Mr D R Kirkwood of Mangatangi in the Waikato for his purchasing of extra lands so that he could expand his operation in pursuit of a 'more economically sound unit'.

1972

First Entries from the South Island

A pleasing aspect of the 1972 Ahuwhenua competition, wrote Judge Murray, was the rise in sheep and cattle entries, which this year had numbered four, including the first two competitors ever from the South Island. Eleven entries were received in the dairy section. 'Entrants in both sections were all keen farmers', said Judge Murray, 'and really made this year's competition one of the most interesting to judge'. Improvements and progress shown by all competitors left Murray with a 'feeling of pride in the achievements and adaptability of the Maori farmer who could hold his own with his Pakeha partner', he reported.[146]

Sheep and Cattle

The winner of the Ahuwhenua competition for sheep and cattle in 1972 was Mr A A Alexander of Ōkaihau in Northland. Judge Murray commended this 'consistent trier' for having won the trophy at last. This year, he appeared to have 'slipped back on a nett cash reserve basis, but his stocking up programme for future production had given him a 10.3% interest return on investment'. This programme represented a considerable jump and would, the judge hoped, be closely matched with drainage, subdivision and pasture improvement. Judge Murray commended Mr Alexander for his farm's neatness and tidiness.

Second place went to C H and C D Boulter of Pahia, in Southland. This father and son partnership had worked hard to develop a farm out of rough bush and scrub; they were continuing to develop clean pastures and good fencing. The farm had a general neat and tidy appearance and showed a 'spirit of development and improvement which had to be seen to be admired', Murray commented. Quite a few problems remained to be solved, but the judge congratulated the Boulter family on their work and effort.

Third-placed Mr A R Austin of Thornbury, in Southland, farmed 182 acres, which was considered to be quite a small property. The judge commended Mr Austin for his 'ability and zeal' in having made the best possible use of his property while paying off his loans in order to increase his farm equity to finance future expansion. Mr Austin's property was located on some of the best land in Southland; pastures and

Judge McMillan Bull in 1972 noted that sheep and cattle farmers were continuing to develop their farms 'out of rough bush and scrub' with an admirable 'spirit of development and improvement'.

Sheep dipping in Horowhenua, c. 1906; photograph by George Adkin (Alexander Turnbull Library, 1/4-023364-G)

stock numbers reflected high fertility. Unfortunately a sizable portion of Mr Austin's farm was prone to flooding, the judge noted, and he had no woolshed yet on site.

Mr N Hoete of Port Waikato, the final competitor, had only recently settled on his 607-acre property after 'protracted estate negotiations', Judge Murray noted. The judge commended Mr Hoete for his initiative and planning and for his early attention to fencing replacements and improved subdivisions. He had recently applied fertiliser to a good portion of the property: the first application in seven years. His buildings, house and yards were impressively neat.[147]

Dairy

Place-getters

The winner in dairy for 1972 was Mr J Edwards of Horohoro, who had been placed second in 1969 and third in 1971. Mr Edwards had demonstrated 'good all round progress', Judge Murray said; his returns had come in at very high levels. An interest return on capital of 18.3 percent was proof to Murray of sound management and wise spending. Good ragwort control was also in evidence, as was a general neatness and tidiness.

Second place was awarded to Mr Tongawhiti Manu of Ōeo, who had secured a placing four times in previous years. Achieving almost 300 lbs of butterfat per cow, Mr Manu had been the highest-performing competitor in this category, though his per-acre figures had suffered with a drop in total production (just prior to judging, he had had to reduce his herd because of labour problems).

Third place was awarded to Mr E and Mrs R Walden of Rahotū, who had achieved the highest butterfat figures per acre this year, as well as the highest interest return on capital invested, 'despite the very difficult country they were farming on'. The Waldens had surmounted many problems with great determination, the judge noted, including subdivision and drainage challenges, and there was no doubt that progress was being made – for example, a 50 percent increase in production over the five years they had occupied the farm.

Other Competitors

Among the other competitors, the judge described Mr J Te Nahu of Horohoro as a very good farmer on a property that had been difficult to work, dissected as it was with gullies that made for awkward subdivisions. Mr J Tana of Ōtūrei in Northland had risen from eighth place previously to fifth this year as a consequence of sound planning and management practices. The judge complimented Mr T Manawaroa of Ngutunui in the Waikato for his 'good, neat and tidy farm', which had made significant progress over recent years, especially in terms of butterfat figures; his were the second highest in the competition.

Mr W Watene of Welcome Bay in the Bay of Plenty had increased the size of his farm by leasing adjoining blocks of land, improving his prospects for productivity gains. Mr A Tapara of Waiuku in South Auckland had demonstrated good management of stock, wise subdivision and a concerted attack on his weed problem, with good results. Mr J W Ashby of Puketōtara in the Waikato had tackled improvements and developments on his once run-down 179-acre property with 'much energy'. Mr P Ainsley of Waihī was ably farming one of the most difficult properties possible for dairying; fencing, buildings, pastures and subdivisions on his property had all been addressed. Mr Pomare of Panguru was also dairying in a particularly difficult area; he had made many improvements over the years to buildings, pastures and fences.[148]

Field day at Hereheretau Station, Whakakī Road, Wairoa, 2009
(Ahuwhenua Archive, Field Day No 20)

Chapter Six

Challenges – 1973–2002

1973

Encouragement Needed for Sheep and Cattle Farmers

Interest in the Ahuwhenua competition among Māori dairy farmers remained high; eleven entrants contested the trophy in 1973. Originally, sixteen had entered, but five withdrew immediately prior to judging. Among them was Mr Tongawhiti Manu of Ōeo, who had been placed third once and second four times. This year may have been his best chance for winning, Judge Murray said; the judge, retiring this year, regretted his withdrawal.

Two competitors entered the sheep and cattle section. Judge Murray suggested that more effort be put into encouraging entrants in the category. This was necessary, he said, if the competition was to be continued 'on the basis of the original request of the Maori people'.[149]

Sheep and Cattle: A Competitor from Each Island

The winner of the sheep and cattle competition this year was Mr Reihana F Apatu of Ōmahu, Hawke's Bay, who had also won in 1969. Second place was awarded to C H and D C Boulter of Pahia, Southland, who had come second the previous year. These two competitors had been closely placed; 'not until the final assessment was made did Mr Apatu come out ahead of the Southland farm', Judge Murray noted.

While both farms had made significant progress in recent years, Mr Apatu stood out because of the greater expansion of his productive area and facilities, the judge said, although his operation had been severely affected by drought, leaving the property in a state that 'only time and a renovation programme could rectify'. The Apatu farm was by far the larger of the two, at 303 hectares to the Boulters' 156. Mr Apatu also had more stock units per effective hectare: 13.6 to the Boulters' 12.2.

Figures for wool per hectare and wool per ewe wintered favoured the South Island farm – 58.9 kg to Mr Apatu's 34.4 kg. The Boulters also earned a greater income per hectare ($140 as compared with Mr Apatu's $108) and per ewe ($13.23 to $11.55). However, Mr Apatu's farm had earned a greater interest return on capital invested (20.9 percent as compared with the Boulter's 14.9 percent), though he had invested less 'capital per stock unit' than the Boulters had – $38 as compared with $40.

In discussing Mr Apatu's farming operation, the judge pointed to 'a succession of droughts' the property had faced; Mr Apatu had overcome these challenges. With a much enlarged farm – it was three times the area it had been when he won the competition in 1969 – Mr Apatu had 'set the stage' for a most attractive and worthwhile property, Judge Murray said. Added facilities included a new woolshed, sheep yards and cattle yards, and further subdivisions. Mr Apatu's sound management was reflected in the farm's statistics and financial returns.

The Boulter farm had attained good production returns per acre and showed good management and general all-round efficiency, including a general neatness. Land developments were continuing, the judge noted, such as the clearing of sidings and the grassing of 'bottom blocks of land'. Increased fertilisation had reaped benefits. Wool production had increased per hectare over the last season, and the introduction of more cattle had brought improvements in overall production. The judge advised further spraying of gorse and more subdivision and said that serious pasture damage could be avoided through careful block grazing.[150]

Dairy

First Place: Setting the Bar

As he had done in the sheep and cattle section, Judge Murray equalised the points he awarded in the dairy section to account for significant differences in the operations assessed. The spread of points between first and last place this year was an astonishing 140 points. Lowest-achieving farmers had suffered in categories like butterfat per acre, nett income per hectare, nett income per cow, interest returned on capital invested and fertiliser used per hectare.

The winner in 1973 was Mr N J Ormsby from Pirongia, whose farm had immediately 'set the bar' for all others, Judge Murray said. Mr Ormsby's all-round ability and 'good productive standard' had been evident in the quality of his management. He had

Māori sheep farmers in the 1970s were enlarging their properties, building new woolsheds, subdividing new pastures and maintaining good stock quality – but droughts were continuing to impede their production.
Four men on horseback with their sheepdogs, droving sheep along the road near Awakeri, 1924; photograph by Albert Percy Godber (Alexander Turnbull Library, APG-1639-1/2-G)

succeeded in breaking in the hill portion of his farm and running some sheep and cattle in addition to his dairy herd. Drought and late calving had affected his overall production, though he had attained a good interest return on capital invested. He was now purchasing adjoining land. His supplementary feed was well catered for, and the general condition of his stock was excellent.

Other Place-getters

Judge Murray awarded second place to Mr E and Mrs R Walden of Rahotū, whose interest returned on capital invested was the highest in the year's competition. This reflected excellent all-round production despite the difficult nature of the Waldens' property, the judge said. Development work in preparation for inclement weather had proceeded well, he noted.[151]

Third place was awarded to Mr John Klaricich from Omapere, Northland, who had come third in 1970. Though Mr Klaricich was now carrying fewer cows, he had still managed to produce high levels of butterfat. The judge noted that skill was needed 'to manage a farm property like this one' – Mr Klaricich had been doing an excellent job, especially in clearing a kikuyu grass infestation on his flats and hills. His production per cow had been the best of all competitors this year.

Further Competitors

Among other competitors, Judge Murray noted that Mr C A Edwards of Ōpōtiki had 'added a refreshing note of vigour and enterprise' to this year's competition, having recently 'started from scratch' in restoring a derelict farm to profitability. The judge commended Mr W H Hughes of Taupō for good production returns on a property badly affected by dry weather. Mr R N Newdick of Tāpora in Northland had developed his property 'with a neatness and tidiness that was outstanding'. His provision of water was excellent, though nearby sandy ridges had affected his pastures.

Mr G Tauroa of Poutō in Northland had more than doubled his butterfat returns in the previous six years, and completed important development work.

Mr W H Roach of Ōpunake had recently purchased adjoining lands in order to build up a sound economic dairy unit, the judge noted. Mr K R Jackson of Whatawhata was commended for having taken responsibility for his family's property following his father's death. Formerly a forestry cadet, he had 'settled in to learn farming at short notice with the help of his mother'. Mr P Ainsley of Waihī had made good progress in eliminating ragwort from his property. W Watene of Welcome Bay had purchased additional lands and increased his stock numbers, though his practice of selling hay had not been 'in the best interests' of his own herds.[152]

1974

A New Emphasis on Community Involvement

Judge Murray was replaced by Mr A N Hall, an extensions officer in the Farm Production Division of the New Zealand Dairy Board. Judge Hall reviewed his predecessor's method of assessment and decided that some simplification of the points scale would be appropriate.

He began by noting the original directions for the competition criteria that had been set in 1933, which emphasised good husbandry, economy and profitability of management, cleanliness and neatness, and development progress, and he specified that judges take account of certain advantages entrants may have over others (like free fertiliser or access to finance). These directions, said Judge Hall, had not kept pace with changes 'both in agriculture and in the relationship of Maori agriculture to New Zealand agriculture in general'. Had the competition been established in the 1970s, he noted, emphasis would also have been placed upon such things as a particular entrant's 'influence and participation in the community'. In the 1930s, Māori rural leadership was only 'accidentally related to efficient husbandry and farming'. Nowadays, Māori farmers represented one of the few stable elements within the Māori rural population, and they made significant choices as to the extent of their community involvement.

Difficulties could arise when farmers sought to strike a fair balance between the 'husbandry of the land and the husbandry of the people', Judge Hall said – between making a property profitable and using that resource for the 'management of the family and community'. Should a competition like Ahuwhenua take note of this dilemma? Judge Hall considered that it should.

Judge Hall concluded his review of the points scale by noting that he interpreted the 1933 criterion of 'economy and profitability of management', which had hitherto been the most emphasised aspect of the competition, to be but one aspect of it.

Judge Hall's criteria therefore comprised: pasture, herd and flock management; land development; financial results; planning; equipment maintenance; household management; and community participation.[153]

Sheep and Cattle

Winners from the South

Four entries were received in the 1974 sheep and cattle competition. Judge Hall awarded first place to the Boulters of Pahia, who had come second in the previous year. Their farm had a much higher wool per hectare figure, nett income per hectare and return on capital invested than the other three competitors.

The judge commended the Boulters for their 'complete family commitment to farming' and for having been receptive to expert advice and instruction. He observed fine pasture management, which had resulted in good production returns that had been the key to profitability. The judge noted that sidling areas needed further fertilising, and that rough feed needed to be cleared away before winter so as not to hamper spring growth.

The Boulters' returns on their flock and lambing percentages had been 'satisfactory', but the flock itself needed to be improved; feeding was the key. Ewes brought in from a neighbour were clearly in a superior condition, the judge noted, through either better feeding or better genetics. The judge maintained that 'the old adage remained true – that 80% of a sheep's breeding depended on what went down its neck'. The Boulters' ewe hoggets had not been 'as big as they should have been, possibly because the feed was rough and stalky and not suitable for this class of growing animal'.

Second Place

Second place was awarded to Mr Robert T Kingi of Rotorua, a returned serviceman, who had won the trophy in 1956. In recent years, Mr Kingi had suffered from serious illness and had transferred his record-keeping to his accountant. Despite his afflictions, his entry demonstrated his keen interest in farming, the judge said. Mr Kingi's pasture management was 'the best seen this year', despite so little of his land being flat. One area had been set aside and planted in lucerne to provide hay.

Mr Kingi's herd had suffered from reduced feed in the dry summer. His prices received per kilo of wool had been quite low, and his economic farm surplus had been disastrously low, reflecting the challenges posed by difficult topography, high expenditure on labour and the need to buy in supplementary feed. Many such factors had been beyond the control of Mr Kingi and his wife, who had committed themselves to a living wage as low as $55 a week, which the judge considered 'a very meagre wage'.[154]

Third Place

Third place was awarded to Mr Norman Hoete, who had farmed the largest unit in the competition this year with 'probably the greatest potential', in Judge Hall's eyes. He described neatness and tidiness on the farm and the standard of Mr Hoete's improvements as 'exceptional'. Mr Hoete had had little formal education and had grown up at Port Waikato under the tutelage of his father, Ropiha Hoete, who had been well regarded in the district as a 'good and patient though conservative farmer'. Most of the farm's potential depended on improved management and planning, Judge Hall considered. Continued supervision was necessary to ensure Mr Hoete continued to develop his strengths, even though he was now on a fixed mortgage repayment scheme. Pasture management had largely been left to manuring, the judge noted; he considered that some time spent explaining manorial practices to Mr Hoete would be well worth while; he needed to understand 'the function of clovers in nitrogen fixation and the nitrogen cycle in general'. Hall strongly advised aerial top dressing, noting that its benefits were to be observed in the farm right next door.

The judge advised that the farmer base his culling of ewes on something more than observation, to avoid missing empty ewes. He observed that hoggets were 'on the little side' but in good order. Cattle needed to be prevented from running back into the bush during winter; winter feeding seemed to be 'somewhat haphazard'. The judge reported that, of all the stock, the late ewes were in the best order; 'but this I think may have been more good luck than good management'. He noted that farm records had been well kept by Mrs Hoete. The neatness and tidiness of the house was quite exceptional. Mrs Hoete's pantry had been stocked with row upon row of home canning and preserving; 'a great object lesson in home economics'.

Remaining Competitor

One other competitor had entered the sheep and cattle section: Mr M S Bidois of Rotorua, who at the time of his farm inspection had unfortunately been suffering from influenza. Due to Mr Bidois' sickness, his property had shown a lack of attention to buildings and stock; more than a dozen sheep were found to be cast, and most had been down for more than a day. However, Judge Hall observed that stock were generally well looked after – unlike the house and surroundings.

Dairy

First Place

Over the previous twenty years or so, observed Judge Hall, the number of dairy farmers in New Zealand had reduced from 50,000 to 20,000; many less efficient farmers had left the industry. In this period, the standards of dairy farming had dramatically increased a general trend that was also reflected in Māori dairy farming. Māori were able to take their place 'to advantage' alongside any representative group of New Zealand dairy farmers, the judge considered. He had found seven of the nine competitors in this year's completion particularly difficult to separate.[155]

The winner, Mr Monte Retemeyer of Te Awamutu, farmed 69 hectares, combining dairying with cash cropping; this had provided a 14.7 percent return on total capital invested that was the second-highest of the year's competition.

Although Mr Retemeyer's farm had a slightly untidy appearance, this largely stemmed from a major cowshed renovation that had almost been completed. An additional 96 acres of leasehold had recently been added to the farm, but was yet to contribute its full share to production. The judge observed that Mr and Mrs Retemeyer had worked very hard 'from their own resources, both financial and physical' to improve and develop their property. Mrs Retemeyer had been present during the discussions about the farm management and was obviously 'a very full partner'.

Second and Third Place

Judge Hall awarded second place to the Walders of Taranaki, who farmed 61 hectares near Rahotū and had been judged second last year. The Walders had scored the highest marks for herd management, and their production per cow had been the highest of any entrant. Their financial returns had also been good; the highest in the competition at 14.9 percent. Mr Walden had been the highest-marked entrant in both pastures and herd management and had achieved exceptional butterfat production per hectare.

The Waldens lost marks on maintenance of records, repairs and neatness and tidiness. The judge suggested that Mr Walden look to greater automation – for example, a four-wheeled tractor – in order to 'cover the ground much better' and thus reduce costs.[156]

Third place was awarded to Charles Bailey of Waitara, who farmed 78.8 hectares on a property combining dairying with pigs, resulting in good returns on capital invested. Mr Bailey's economic farm surplus per hectare was the fourth highest recorded, standing at $218 per hectare. His financial results ranked him second highest in this year's competition, though the judge noted that his income had been 'much assisted by his pig farming which had comprised almost one fifth of his total income'. The judge acknowledged the special challenges Mr Bailey faced 'by having to bring up his family on his own while farming'.

Other Competitors

Among other competitors, Judge Hall commended Mr J B Macdonald from Blenheim for successfully managing a complex operation that sat 'somewhere between a sheep and cattle farm, and a dairy farm'. However, the bulk of his stock – 180 cows – were milked for town supply. Mr Macdonald, who was seventy-two, farmed 722 acres of fertile alluvial Wairau River flatlands; his large family was involved at all levels. Mr W Watene of Welcome Bay was also older; he had served in the Second World War. Judge Hall commended Mr Watene for having entered the competition but felt his farm had been ill prepared. Mr Watene's stock had been generally in good order, but his pasture techniques were out of date.

Mr J Riini of Rūātoki had entered this year's competition to encourage others from the area to participate. His farm records were of a high standard, as was his planning and development work. The judge commended Mr Riini for the standard of his household and family management and for generally high standards of neatness. Mr and Mrs Riini were committed to the local Tūhoe community, Hall noted. Mr G Peachey was commended for his 'intelligence and interest in farming'; he had originally been a schoolteacher. He had had to learn a lot in a short time, but had been greatly been assisted by a farming consultant.

Mr Tongawhiti Manu of Ōeo, a frequent competitor, had had to contend in a strong field, 'just missing out on third by one point'. His farm was a model of neatness and tidiness, and his farm records were of a very high quality, the judge thought. Judge Hall congratulated Mr G Hohepa of Kaikohe on the quality of his pasture management: 'the best observed in this year's competition'. His herd management was rated second best.[157]

1975

A Rising Standard

Four sheep and cattle farms were nominated for the 1975 Ahuwhenua competition, and seven for the dairy. ' While it was unfortunate to see that the number of entrants had fallen', wrote Judge Hall, 'nonetheless it was pleasing to see that the standard of the entrants had risen'. So much so, he added, that the entrant awarded third place in dairying in 1974 had re-entered with much improved figures, only to be placed third once again.

Sheep and Cattle

Despite improved standards, the judge awarded no places in the 1975 sheep and cattle section.

In advising the Secretary of Maori Affairs of this decision, he described the farms that he had inspected. Two were conventional sheep and cattle properties: those of Mr Ian Perry of Tinui Valley and Mr James Morris of Ōkautete. One, Mr C Pohio's property at Lake Ellesmere, was a cropping and lamb-fattening unit. Another, Mr George Brown's 600-acre property in the same area, was an operation comprising 'grass, small seeds and, more recently, cattle'.[158] Judge Hall concluded that such different types of farm had been 'very difficult, if not impossible, to compare'.

Although judgements could be made based on financial returns, a Wairarapa sheep and cattle farm could not be compared with an Ellesmere small seeds operation because the criteria used to determine differences in profitability 'were not entirely within the control of the individual farmer'. Judges had encountered this issue before.[159] Accordingly, Mr Hall had recommended that no sheep and cattle award be made for 1975.[160]

Dairy

Place-getters

First place in the dairy competition was awarded to Mr Claude Edwards of Ōpōtiki, whose farm 'made an

Between 1950 and 1970, the number of dairy farmers in New Zealand reduced from 50,000 to 20,000. At the same time, Judge Hall noted in 1974, standards were 'dramatically increasing, a trend that was well reflected in the Maori dairy industry'.
Dairy herd in Northland, c. 1920; photograph by the Northwood brothers (Alexander Turnbull Library, 1/1-010735-G)

immediate impression' on the judge. Mr Edwards' production per cow and butterfat per hectare were far in excess of any other competitor. He had recently taken over two run-down properties and was in the process of developing them into a well-integrated 41-hectare unit.

The judge described the standard of his development work and improvements as 'a pleasure to see'. His financial results were equally satisfactory, and his farm records had been well maintained. His economic farm surplus had been higher than that of all other entrants, though their farms had been larger than his.

Second place went to Mr Tutere Hohepa of Te Teko, who had substantially improved his per-cow and per-hectare production since last year. Though this was partly due to favourable seasons, it was also attributable to hard work and consistent effort. Hall commended Mr Hohepa for having addressed important financial issues with the help of a field officer. He noted that Mr Hohepa's milking herd was in excellent order, and his dry cows were also in good condition. However, he noted that the concreting of his stock's access to the milking shed was imperative.

Mr Charles Bailey of Waitara, who was awarded third place, had substantially improved his milking cows over the previous twelve months. His cows were now in excellent order, and there were some outstanding first-calving heifers in the milking shed, the judge observed. Mr Bailey was providing adequate feed and was contemplating increasing his herd. As Judge Murray had done, Judge Hall commended him for his efforts in maintaining his farm while looking after his motherless family. Mr Bailey was now receiving help with the keeping of his grazing records, milk production records, shed sheets, calf records and heat records, the judge noted.[161]

Other Competitors

The judge commended Mr Tiwai Black of Rūātoki for farming a series of capital value lease blocks in difficult circumstances. His pasture management had been good but was complicated by the many different pasture types on his farm. He had recently planted out a large riverside area with root crops. During the previous twelve months, the Waldens of Rahotū had acquired a new house at the expense of maintaining their high management standards. As a consequence, Mr Walden's milking cows risked being underfed; sixty cows to the acre were on grass 'which could not have fed 45 to the acre for his class of cattle at that time prior to calving', the judge noted.

Mr John Taikato of Matakana Island had recently been hospitalised but had returned to take over the milking from his school-aged children. He was currently milking in an old walk-through shed; a new herringbone shed was nearing completion. Howard Moana was farming a difficult property prone to flooding and low fertility; properties adjoining his farm had long since been converted to sheep and cattle. Mr Moana's stock were assessed as being 'in a fair condition', although the judge identified several old cows and others of mediocre standard in his herds.[162]

1976

A Decline in Competitors?

In his final Ahuwhenua judge's report in 1976, Mr Hall wrote a brief report for the Ahuwhenua Trust Board in which he made a number of observations relating to the competition specifically and Māori farming in general. The number of entrants in recent years had been disappointing, he wrote, and although the standard had progressively increased, the issue as to how to generate greater interest in the competition remained.

While sheep and cattle entries had remained low, averaging 2.7 per year, dairy entries had indeed gradually fallen. Although this decrease does not seem particularly drastic in the larger context, at the time, Judge Hall looked for reasons to explain it. Perhaps the 'honour and glory accruing to the winner' were no longer sufficient to attract participants, he suggested. Some contestants had suggested that the costs of participation, which included the sponsoring of Ahuwhenua functions, had become prohibitive and a serious disincentive.

In Judge Hall's view, the contest had suffered from insufficient publicity 'at a time when every unrepresentative and disproportionate Maori option seems to have achieved wide publicity', he noted. Regretfully, he considered, New Zealand possessed a 'substantial body of racially prejudiced people'. He perceived that the Ahuwhenua Trophy presented 'a more balanced view of the Maori contribution to the economy and specifically to farming'. The honour of winning would always be the greatest attraction for entering, he maintained. 'Perhaps it is time to remind ourselves of the special contribution to Maori farming by Lord Bledisloe,' he said.

Judge Hall favoured a prize that was 'educational in nature', which might broaden 'the winner's appreciation and knowledge of farming'. However, he also noted that the winner of the year's dairy section that year had suggested that 'it could be useful to visit the farms of other contestants'.[163] He suggested that sponsorship might be advisable, envisaging a prize fund of at least $5000. Accordingly, he said, 'I enclose my cheque for $100 as a small initial contribution to such a fund.'

Sheep and Cattle

Two entries were received in the sheep and cattle section. Judge Hall awarded first place to Mr Thomas Hawira, whose 276-hectare farm was located in 'good hill country' about 12 kilometres south-west of Raetihi. The property had been partly leased from the Ātihau-Whanganui Incorporation for twenty-one years, with rights of renewal after the fifteenth and twenty-first years. Beyond the leasehold land, totalling 56 hectares, Mr Hawira also owned a 58-hectare block he had purchased from his parents in 1970 and a 97-acre block he had purchased from his brother in 1973. He managed a flock of 1700 ewes, 600 hoggets, 30 rams and 187 steers.[164]

Judge Hall found Mr Hawira's pastures and stock in a fine condition and considered the general

Ahuwhenua Competitor Numbers throughout the 1970s

Year		1970	1971	1972	1973	1974	1975	1976
Entries	Sheep and Cattle	2	1	4	2	4	4	2
	Dairy	12	16	11	11	9	7	8

upkeep and tidiness of the property noteworthy. Improvements to the water supply had cost around $1500, he reported, mainly in the 1974–1975 seasons.

A second entrant was also visited by the judge but was not awarded a place. Mr A G Cribb was managing a farm located south-east of Taumarunui. The area was freehold and comprised 208 hectares, with 1872 sheep and 175 cattle on site. Replacement sheep were bred on site, and surplus lambs sent to the works. A total of 400 of the ewes (Perendale ewes from previous farms) had been purchased from the previous owner. Mr Cribb's farm had good hill pastures and buildings, especially the new hay barn and shearers' quarters. Overall, the standard of his fencing was high.

This unit had been well managed by a previous owner, the judge noted, although subleasing in the last two years of that ownership 'had seen the farm go somewhat'. Mr Cribb had only had the farm for three years; two of those years had seen significant setbacks to sheep farming, especially in 1973 when beef prices collapsed and in 1974 when low product prices severely affected incomes. He had therefore had a limited time to make any headway. Prior to purchasing this farm, Mr Cribb had run his family's small farm, supplementing his income with shearing and fencing. Ownership of his current farm had come about as a result of sheer hard work and considerable financial assistance from the Department of Maori Affairs, Judge Hall noted.

The judge approved of Mr Cribb's good stock and financial performance (which was under budgetary control because of debt levels), but felt that in two or three years he could present his unit much more to his advantage, especially after some basic tidying up.[165]

Winner of the 1976 Ahuwhenua Trophy for Dairy, Charles Bailey of Waitara (left), is pictured here with cousin Ted Tamati of Waitara. Both Charles and Ted won the Ahuwhenua Trophy twice and were inaugural Parininihi ki Waitōtara Committee of Management members.

(Photo courtesy of Liana Poutu)

From 1970 to 1976, Ahuwhenua entries fell by about 20 percent, though in 1976, when there were ten competitors, the competition still seemed perfectly viable. Though the entries had been falling, the standards were 'progressively increasing', Judge Hall said in 1976.

Maori Agricultural College as viewed across a field, Te Hauke, Kahuranaki, Hawke's Bay, 1929; photograph by Henry Norford Whitehead (Alexander Turnbull Library, 1/1-004636-G)

Dairy

First place in dairy was awarded to Mr Charles Bailey of Waitara, who farmed 97 hectares with 150 cows. His overall production was 24,546 kg, comprising 164 kg per cow, or 253 kg per effective hectare. Mr Bailey scored well on pasture and herd management and particularly well on household and family management, given that he was raising his family alone. The judge considered that Mr Bailey's forward physical and financial planning had not been one of his strong points and did not rate his record-keeping highly. However, development of his property overall was good, he reported.

Second place went to Mr Tutere Hohepa of Te Teko, who was farming 92 hectares with forty-six cows, with an overall production of 22,722 kg, comprising 155 kg per cow, or 247 kg per effective hectare. The judge commended Mr Hohepa for his pasture and herd management, noting that his developments had progressed well. Like Mr Bailey, his forward planning needed some improving, and he needed to focus on repairing and maintaining his buildings and equipment.

The judge awarded third place to Mr Tony Edwardson of Ōpōtiki, farming 124.5 hectares: the largest property in the competition this year. Mr Edwardson managed a herd of 183 cows with an overall production of 26,562 kg, comprising 145 kg per cow, or 213 kg per effective hectare. Judge Hall commended him on the progress of development work around his property but scored his pasture management and herd husbandry lower than the first- and second-place winners.[166]

1977

Cash Prizes

On the eve of the 1977 Ahuwhenua competition, the Minister of Maori Affairs announced that a cash prize of $300 would from then on be awarded to the winners of each section, following Judge Hall's recommendation of the previous year.

Also on the recommendation of Judge Hall, the minister suggested that cash prizes be used towards travel expenses to permit the winners to study farming or to visit the farms of other entrants to look at their management.

Another major change was introduced in 1977: regional Ahuwhenua dairying competitions that would take place prior to the main national competition. The minister approved cash prizes of $50 for regional winners; each would also win the right to progress to the national final.[167]

New Judges

For the first time, two judges were appointed, with specific skills sets and appropriate farming acumen to judge each section. Mr J D McNaught was appointed to judge the sheep and cattle section; and Mr A McKenzie the dairy section.

Sheep and Cattle

First place in the sheep and cattle competition was awarded to Mr J R and Mrs S C Stewart of Wyndham, in Southland, who managed a 127-hectare farm carrying 1376 sheep and 51 cattle.

Judge McNaught noted that the Stewarts' property had 'moderate to good pastures' on free-draining soils – about 4 hectares were regrassed each year. Gorse was confined to gullies and had been well controlled. The judge described fencing around the property as generally good. The homestead stood in 'pleasant surroundings', he considered, and the unit as a whole was tidy and well maintained. The Stewarts had farmed the property for seventeen years, and in that time had installed rebuilt cattle yards, a small wintering shed, a hay barn and an impressive shed for implements, vehicles and ancillary storage.

Second place went to Mr J H Morris of Masterton, who farmed an effective area of 215 hectares with 2296 stock units on site: the most of this year's competitors. The judge commended Mr Morris for the high standard of his farm and pasture management and for creditable production returns, especially the interest returned on capital investments. Buildings, equipment and fences were all well tended, he noted.

Judge McNaught awarded third place to Mr T P and Mrs N Te Aika, who farmed a leasehold pastoral 4131-hectare property at Mason Bay on Stewart Island, comprising bush, mānuka swamp and sand dunes, as well as areas of tussock and native grasses. The Te Aikas had farmed this block for eleven years, and currently managed a flock of 908 sheep with no cattle. They had made worthwhile improvements through dedication and sheer hard work. Because of the farm's isolation, wool was the sole means of their revenue, comprising 44 percent of all receipts. The balance of their income came from possum skins and venison. 'Mr and Mrs Te Aika truly worked as a team and became farmers by their own efforts', the judge said, noting that they had been assisted by both private and 'marginal lands' finance (the Marginal Lands Act 1950 had instigated a government scheme to provide special loan facilities for farmers on marginal or deteriorated land).

The judge reported that the homestead was an 'oldish homely cottage' with no power plant. However, the buildings and shed were all in a good condition. Most of the boundaries were not fenced, though recent subdivisions had increased the number of paddocks from four to six. Overall, the farm had a restricted area of improved pasture, with a small pine plantation, an airstrip and some additional tree planting near the home, as well as a marram grass establishment. The judge noted that the Te Aikas had received assistance from the Maori Education Fund for the education of their two daughters.[168]

Dairy

Place-getters

Following a series of Ahuwhenua regional finals, which were instituted this year, four national finalists were selected to compete for the trophy. First prize was awarded to Mr Jack Karatau of Whangaehu, who farmed 43 hectares of permanent pasture, with 1 hectare set aside in homestead surroundings, plantations and shelter. Mr Karatau had won first prize in the Ahuwhenua competition of 1968 with 'some of the highest production figures on record', Judge McKenzie noted. Since then, he had greatly improved his production returns. Mr Karatau was awarded the highest possible points for financial results, the judge having compared his returns with those of the winner of that year's national ICI Farmer of the Year competition.

Second place was awarded to Mr A T Edwardson of Ōpōtiki, who the judge commended for his farm management practices and sound overall business approach. The judge observed Mr Edwardson's mature cows to be in very good order, though he needed to pay attention to younger stock. The standard of building repairs and maintenance was very high in the judge's eyes, and records of breeding, production and finance were very well kept and maintained.

Mr P and Mrs A Hemi of Kaeo were awarded third prize, in recognition of their sterling efforts over an eleven-year period in breaking in newly purchased freehold lands that required urgent clearing of gorse, bracken and tall fescue – 'I give Paul and his wife full marks for the excellence of the job they have done', Judge McKenzie said. He rated repairs, maintenance and overall tidiness on the farm very highly and observed that Mr Hemi's stock was 'better than the poor stock typical of this area of Northland'. Time, fertiliser and a more defined herd policy were now needed in order to promote production increases, he said.

Jack Karatau Remembers His Ahuwhenua Win

Jack Karatau, winner of the Dairy Competition in 1977, with Prime Minister Keith Holyoake. Jack also won the trophy in 1968.
(Karatau whānau)

'My father-in-law helped us into farming, and we brought 50 acres and milked forty-five cows in our first year. We also had two children, a boy and a girl. After the second one was born, I picked up Nola from the maternity hospital and she went straight into the cowshed that night!

'You've got to feed your cows right through the whole year, not just the time before they're milking. We bought some extra cows that two big dairy farms were culling out to bring our numbers up to 100 cows. My aim in those days was to get our production up to 400 lbs of butterfat per cow, and we did it.

'In 1968, an uncle of Nola's said, 'well, why don't you go in for that Ahuwhenua cup?' I said, 'you have to be pretty good for that, don't you?' And he said, 'you're good enough'. I went into Maori Affairs and they accepted me, so I went from there.

'I was notified by Maori Affairs that I'd won the trophy – there was no big celebration or anything like that, although it was quite a shock to the district and quite a shock to me.

'Second time around (in 1977) the Prime Minister Keith Holyoake came and presented the cup to me. I remember when I won the cup the second time we got $400 – I read recently somewhere that when you win now you get about $40,000!

'Winning the cup and the medal was something really great. Occasionally I look at it and think – that's me!'

Source: interview conducted by Lyn Harrison, May 2013

1978

Judges

In 1978, the two judges appointed for the Ahuwhenua sheep and cattle and dairy competitions respectively were Mr J D McNaught and Mr J G Simmonds.

Sheep and Cattle

First Place: The First Corporate Winner

There were five entrants in the sheep and cattle section for this year, but only two awards were made. The winner was Tiratu Station in Dannevirke, owned by three brothers: P B, R D and H J Paewai. This placing of a corporate farming entity represented a significant milestone for the Ahuwhenua competition: one that could not have been foreseen in 1933. The station was managed by one of the brothers, Ringakaha. Part of the land had been acquired as a beneficial interest; additional acres had later been purchased. The Paewai brothers, who had previously managed a successful contracting business, had also recently purchased another farm; their intention was to settle all three brothers on their own farms.[169]

Tiratu Station was 362 hectares in size, with 4345 sheep and 160 cattle. Sheep stocks were replaced from within, and wether and surplus ewe lambs were sent to the local works. Cattle numbers had decreased in recent years; replacement stock was bred from within, though weaners were brought in for fattening.

A considerable amount of tile and mole drainage had already been installed by the Paewai brothers. The judge observed that pastures were well maintained, and a sound regrassing programme was under way. Fences were generally very good, as was the water system, which originated from a spring. The farm featured a central 'pleasant unpretentious building' and adequate ancillary buildings such as woolsheds and implement sheds. 'Everything was tidy in a workman like manner', the judge noted. Scope existed for improved pastures and further drainage; it was noted that about 24 hectares 'out the back' were still in stumps.

Judges continued to commend Māori farmers for the quality of their farming management and production returns. As Judge McNaught saw it in 1978, farmers were working hard to build 'their fine operations today'.

A Māori shearing group at the Bay of Islands with a cart stacked with wool bales, 1910
(Alexander Turnbull Library, PAColl-8620)

Hepa Paewai c.1966
(Paewai family)

Mr Hepa Paewai Remembers His Ahuwhenua Win

'[When I was a child] we lived in a big marae situation, you might say, with fifteen families. We used to see all this land around us, but there were no Māori working it; it was all Pakeha people. I thought they used to own it, but we owned it and they farmed it, and I couldn't make much sense of that.

'Quite early on in our lives, our father died and our eldest brother took over as a kind of figurehead, and as much as we worked, we never seemed to have quite enough money.

'Punga went to a few lending institutions, like State Advances and others, but we weren't able to get money from them because – well, I suppose it was because we had Māori in us. So we went to the Maori Affairs and asked for money, and they gave it to us. Part of the criteria if we borrowed money from Maori Affairs was that they would automatically put us into the Ahuwhenua farming Trophy.

'Punga never told us anything until the day he said, 'Oh, we have to go down to the marae.' So we went out there, and as we drove in I said to Punga, 'What are all these people doing here?'

'Well, the Governor-General was there, Keith Holyoake, and he presented us with the trophy. I always remember what he said to us: we were good shearers, we were good farmers, we were good rugby players, so we'd have to be bloody good Kiwis.

'Winning the trophy opened doors for us in a lot of respects, like financially; we were able to go into the Farmers' Co-op and open an account because 'You guys won that competition; sure you can open an account here.'

'I would say to all the young farmers out there – go out there and do it because you get a lot of enjoyment out of it; it's very satisfying.'

Source: interview conducted by Lyn Harrison, April 2013

Second Place

Second place went to Mr M R Mohi of Waipukurau, who had managed a 213-hectare property under leasehold for the past eight years. He had taken up the farm following the death of his father-in-law, who had acquired the property as an ex-serviceman through the rehabilitation scheme. Mr Mohi supplemented his income through shearing, but intended to give this up, the judge noted.

Mr Mohi's farm comprised 2350 sheep and seventy-one cattle. Replacement stock was bred on site, and surplus wethers and ewe lambs were usually sent off to the works. The judge noted that Mr Mohi's stock was in good condition, and the farm was well run. Pastures were generally good, and good controls were in place to prevent outbreaks of thistle. Subdivision and the standard of fencing were also good, though older fences of poor quality were still being replaced. Adequate water from dams and creeks was available to most parts of the farm. Some severe erosion had caused major problems in some parts of the farm. The quality of the buildings was good, and they were well maintained.

Remaining Competitors

On the recommendation of the judge, no further awards were presented in the sheep and cattle section. Among remaining competitors, Mr J H Morris farmed 745 hectares 60 kilometres south-east of Masterton, managing 2050 sheep and 153 cattle on site. Judge McNaught commended Mr Morris for his quality management and creditable production returns. Mr C J D and Mrs A F Stewart from Wyndham farmed a 136-hectare farm carrying a stock loading of 2030 sheep and fifty cattle. The judge reported an attractive and well-laid-out farm and a positive approach to the challenges inherent in the Stewarts' locality. Mr P H Tahau of Te Hāroto farmed a property of 255 hectares carrying 2280 sheep with no cattle. He had worked very hard to build his unit into the 'fine farming operation it was today', the judge said.[170]

Dairy

The winner of the dairy competition was Mr Maurice C Anderson of Whakatāne, who owned 'a magnitude of business interests', among which this farm constituted a 'small yet important cog'. Mr Anderson scored 'near maximum points' in the stock husbandry section of this competition, though Judge Simmonds raised some queries as to aspects of his stock quality. Mr Anderson's focus on breeding was admirable, the judge reported, but pasture management was 'not one of his strong skills'. However, he noted that Mr Anderson was making every effort to lift the quality of his grasslands. With the availability of cheap feed, difficulties could arise as to appropriate stock feeding levels and pasture development, he cautioned.

The judge awarded second place to Mr W P Peachey of Ōtorohanga, whose property he described as unique in the competition in being a 'virtual self-contained traditional dairying enterprise'. Judge Simmonds noted that, in a competition like this one, this approach put the farmer at a disadvantage – it was more economic, for example, to buy in feed than to control starvation at critical periods of the season. Although Mr Peachey was an efficient self-contained dairy farmer, he could not compete against higher-producing non-self-contained units, the judge considered.

Mr C B Wells of Kaitāia was placed third. Comparing his operation had been difficult, the judge said, because of large development works being undertaken on lands adjoining his farm that were obviously affecting the efficiency of his operation. However, the judge noted that developments on Mr Wells' farm were of a very high quality; good pastures were growing that would in time foster good production returns.

1979

Interest Levels 'Disappointing'

On 7 April 1978, the Maori Trustee advised his departmental officials throughout New Zealand that the interest shown thus far among Māori farmers in the 1979 Ahuwhenua competition had been 'extremely disappointing'. Officials were even asked not to proceed with the appointment of a judge for the dairy section, as it seemed unlikely that enough entries would be received.[171]

Early in the following year, the Minister of Maori Affairs expressed similar reservations, advising his officials on 26 February 1979 that he was concerned about 'waning interest'. He urged his officials to stimulate interest through their contacts in the community.[172]

Final entry numbers were low: four in the sheep and cattle section and two in dairy.

The judge of the dairy section concluded his competition report by commenting on the continuing problem of low entries. Any award for farming achievement, he wrote, had two functions: first,

In the 1980s, judges continued to describe Māori farming as 'high quality'. Farmers were acquiring new holdings, improving pastures, modernising their water reticulation systems, embarking on comprehensive fertiliser programmes and rearing quality stock.

Two shearers at work as a woman assistant waits to lift a shorn fleece, 1949 (Alexander Turnbull Library, PAColl-6303-29)

to identify leaders or superior performers within the definition of the awards, and secondly, to encourage others to emulate these winners' achievements. Both required a level of interest among the community involved, which, in this case, was Māori farmers. If the trophy was to serve its purpose, and more entries were to be encouraged, then Māori farmers needed to know more about the trophy. Perhaps a field day and other publicity associated with the winning farms would encourage greater participation, he suggested.[173]

Sheep and Cattle

The winner of the sheep and cattle section for 1979 was Mr Reihana F Apatu of Pukehāmoamoa, 21 kilometres north-west of Hastings, who had previously won the trophy in 1969 and 1973.

Mr Apatu had solely farmed his family's 121-hectare property for fifteen years, in which time he had acquired adjoining lands on leasehold and purchase, greatly increasing his overall holdings. His stock numbers stood at 3250 sheep and 350 cattle. Replacement sheep were generally purchased, and replacement rams bred on site. The cattle on site were described as dry stock. Numbers were down on previous years, mainly because of difficult winters and the high cost of replacement stock. The judge noted that all improvements on site were of a high quality, the farm was well subdivided and pastures were generally 'good to excellent'. Barley grass was well under control, and thistle and blackberry were well attended to. Adequate water was available from a variety of sources.[174]

Second place went to the Stewarts of Wyndham, who had earlier entered the competition with some success. Their farm carried 2023 sheep and seven cattle. Replacement stock was bred on site, with some ewe purchases to upgrade stock. The judge described cattle as dry, having had no wintering. Pastures were 'good to average', and were improving with top dressing. Pasture renewal after the planting of forage crops was currently an important part of the Stewarts' pasture improvement programme. The farm was well subdivided into twenty-one paddocks, supported by good fencing. Gully gorse patches were well controlled, and an adequate water supply was available to most parts of the farm.

Two other entrants in the section were visited, but neither received an award. Mr D C Boulter managed a farm 63 kilometres west of Invercargill that was 156 hectares in size and carried 1902 sheep and sixty-seven cattle. The judge commended Mr Boulter for his good farm management practices and the fine quality of his stock. C R and P R Martin managed a farm that was located 8 kilometres east of Ōpunake. The Martins' farm was 296 hectares and carried 2599 sheep and fifty-five cattle. The property had a tidy appearance, the judge noted; buildings and plant were well maintained.

Dairy

Two entries were received for the year's dairy competition, though only one received an award.

First place went to Raumoa Amoamo, who, with his wife, was a former sharemilker who had now acquired his own property. The Amoamos' production figures had recently increased, the judge noted, though stock numbers had been reduced to provide capital for new stock purchases. The judge rated pasture management on this farm very highly. Good lucerne stands had been established, and supplements and winter feed was under good management. Previously deferred maintenance work had been factored into forward planning. The strong point of this farm was the care lavished on the stock, the judge said.

The second entrant in the section was Mr R Henry of Te Hoe, whom the judge commended for having entered the competition even though his production figures 'told a difficult story'. Mr Henry had commenced an ambitious programme aimed at increasing stock and production, but his subdivision and pasture management required urgent attention, 'on a difficult property in two pieces and subject to flooding'. Stock quality was not good, the judge noted; younger stock needed better rearing.

1980–1981

Recess

As foreshadowed by the very low entry numbers of recent years, the Ahuwhenua competition went into recess between 1980 and 1981. However, at the same time, officials continued their efforts to attract interest in the competition.

On 7 May 1980, Minister of Māori Affairs Ben Couch wrote to his officials suggesting that perhaps the competition should become biennial, 'unless there was a more enthusiastic response from Maori farmers'. He also suggested that perhaps Māori farmers themselves – or at least past winners – should be consulted about how to generate more interest.[175]

The Secretary of Māori Affairs supported the concept of a biennial competition. The efforts of field officers in canvassing for entries had been much appreciated, he wrote, 'but we must do what we can to stimulate interest in future competitions'.

1982

Ahuwhenua Returns

On 23 November 1981, the Secretary of Māori Affairs advised his field staff that it was appropriate that, given it was now fifty years since the Ahuwhenua scheme had been inaugurated in 1932, a competition should be held for the 1981/1982 year. He urged officials to encourage entries.[176]

Six months later, the secretary advised districts that applications for the Ahuwhenua competition were now to hand, and because the numbers 'were not great' (four applications had been received for sheep and cattle and ten for dairy) the minister had decided not to run district competitions, but just a national one.[177]

Sheep and Cattle

First Place

This winner of the sheep and cattle section for 1982 was Martin Kingi, who farmed 322 hectares 27 kilometres north-east of Gisborne. Mr Kingi had learned agriculture at Te Aute College and had worked in the agricultural field for seven years before being appointed to manage a farm in southern Hawke's Bay. After five years, he returned home to Whāngārā to assist on his father's farm, taking full control of it another five years later. Maori Affairs had assisted in the funding of his extensive development programme, as had the Land Development Encouragement Fund.

Mr Kingi's farm was stocked with 2420 sheep and 341 cattle. Replacement sheep stock was purchased, and replacement cattle were bred on the farm. Stock was in a good condition, the judge noted, though eczema was an ever-present threat that affected stock performance in some seasons, such as the wet autumn of 1981.

Pastures were in a good condition, the judge said, and included a significant portion of 'very good young hill pasture developed from blackberry of varying severity in the seasons 1978–1979 and 1980–1981'. The establishment of an additional 40 hectares of new pasture had not progressed to completion because of weather that had severely affected grass-sowing and cultivation. The judge reported that rush infestation had been brought under control, as had blackberry, but this needed close watching, especially in the gullies. Fences were generally good and water supply was adequate, mostly sourced from nearby dams. Pastures and subdivisions needed improving, and the judge advised top dressing.

Second Place

Second place in 1982 was awarded to the Paewai brothers for their management of 123-hectare Raupatu Station in Dannevirke. The brothers had purchased this 'run-down farm' freehold in 1977 as part of their plan to provide family members with individual properties. Hohepa Paewai now managed this station; Punga Paewai farmed his own unit as well as managing a large contracting business; and Ringa Paewai managed Tiratu Station, which had won in 1978.[178]

Substantial funding had been loaned from various sources, including the Department of Maori Affairs, for development. The property was currently stocked with 2140 sheep and seventy-three cattle. When it had first been taken over, it had stocked 700 ewes and been half-covered in scrub, with only seven paddocks, poor fencing, inadequate subdivisions and marginal water supplies.

Third Place: Keeping It in the Family

Third place went to D W W Hawkins of Te Haukē, in Hawke's Bay. Mr Hawkins had grown up on this 160-hectare freehold farm but had spent most of his early life working in the freezing works and in shearing gangs. He took over the farm from his father in 1978, working it from Hastings until a new house was built on site. While working as a freezing worker, Mr Hawkins had purchased an adjoining hill block 'to keep it in the family'.

There were 1864 sheep and fifty-five cattle on site. The farm was not an easy one to manage, comprising seven effectively independent blocks spread along the road, the house positioned midway along the site.

Dairy

First Place

The dairy competition for 1982 was won by W C and C M Edwards of Te Kūiti. The Edwardses scored strongest in pasture knowledge, finance, repairs and maintenance and progress in development; the judge did not mark their farm records and forward planning so highly.

The judge described Mr Edwards as a dairy farmer of 'well above average' ability, farming on a very difficult piece of country that had large areas of swamp, gorse and tea tree. The judge noted that he had drained a lot of swampy areas and cleared some of the difficult hill country. The farm had a ragwort problem, but the Edwards family seemed to be getting it under control.

The judge recommended that Mr Edwards keep a closer eye on his stock's condition and strive now for a better return from each animal put into the herd. He noted that the Edwardses were a close-knit family heavily involved in sport and other outside activities; fortunately, these worthy pursuits had not detracted from good care of the farm. 'The Edwards family are a good example of what the dairy industry was all about – the family unit', the judge said.[179]

Second and Third Place

Second place went to D Edwards of Ōpunake, who was farming on a 'difficult cold farm 1300 feet above sea level in Taranaki'. His farm had a problem with pastures and with 'winter pugging', the judge said, noting however that Mr Edwards was working hard to rectify the situation. Mr Edwards had a good herd and fed them well, but some of the young animals were in a poor condition, despite adequate supplementary and winter feeding. The judge concluded: 'David seemed to be a very sound dairy farmer who knows where he and his family are going, and I am sure they will make it.'

The judge awarded third place to G Brons of Reporoa, Rotorua, who scored well in pasture knowledge, repairs and maintenance and progress in development, but poorly in record-keeping and forward planning. The judge said Mr Brons had a 'very sound knowledge of farming practices in Reporoa'. He was on a very difficult block; he and his wife had achieved an increase in production through sheer effort. Stock size and condition were not of a high quality; better stock would soon see production lift, the judge foretold.

1983–1989

Another Recess

The Ahuwhenua competition once again went into recess between 1982 and 1989, during which time no annual competitions were organised.

On 11 September 1985, the Board of Maori Affairs approved in principle a recommendation that 'the Ahuwhenua Trophy should continue, following a review of the scheme headed by Sir Graham Latimer'. The secretary of the board indicated that discussions had commenced in Head Office focusing on prospective sponsorship proposals to enable this.[180]

1990

The Competition Returns Again

In 1989, Minister of Māori Affairs Koro Wetere approved the resurrection of the competition for a trial period of two years, in recognition of New Zealand's 1990 sesquicentennial commemorations. The competition would be held in conjunction with the New Zealand National Fieldays and Farmer of the Year competition.

In an agreement between the minister and the National Fieldays Society, all costs of the competition over the two-year trial were to be borne by the Society. The minister encouraged the Society to report to him upon the completion of the competition about sponsorship ideas and the feasibility of continuing with the cup in the future.[181] The 1990 Ahuwhenua competition was held for sheep and cattle only. The winners were P D and R D Paewai of Tiratu Station, Dannevirke.

1991

In 1990/1991, the Ahuwhenua competition continued for the second of its two designated trial years. One winner was announced in each section: the Parekarangi Trust from Rotorua for sheep and cattle, and Mr N and Mrs S Armitage of Atiamuri for dairy.

Droving sheep, Ōhakune, 2007
(Ahuwhenua Archive, A410 078)

The Ahuwhenua competition of 1990 was staged in conjunction with the New Zealand National Fieldays and Farmer of the Year competition. Māori sheep and cattle farmers were invited to participate.

Droving sheep at Waihau Beach on the way to the Gisborne market, 1906;
photograph by Frederick Hargreaves (Alexander Turnbull Library, 1/2-018019-F)

(John Cowpland – Alphapix)

Chapter Seven

Ahuwhenua Continuing 2003–2013

2003

The Competition Resumes

After the two 'trial' years arranged for 1990 and 1991, the Ahuwhenua competition once again sank into oblivion, suffering terminally from a lack of interest among potential competitors and a lack of the motivational drive to resurrect it among organising officials. This recess lasted for a decade, spanning the years between 1991 and 2002.

Eventually, in the early years of the new century, the critical lack of organisational motivation seemed likely, finally, to be rectified. Ms Gina Rudland and Mr Wayne Walden, board members of Meat New Zealand, urged the organisation to support a relaunching of the competition.

A new competition would take account of the changing face of Māori farming, they said; it would acknowledge the importance of incorporations and trusts in that important economic sector. The board of Meat New Zealand agreed. Thereafter, it ran the annual Ahuwhenua competitions between 2003 and 2005, before its successor organisation Meat & Wool New Zealand took over the reins.

Meat New Zealand group executive of special projects Allan Frazer was one of those who commended the re-establishing of the competition. For some time, he said, Meat New Zealand had been looking for ways to improve representation of Māori meat producers: some of the largest farmers in the country. Restoring a prestigious annual award that set a benchmark for top performances would help to do just that: 'We had two aims in the competition: to recognise excellence in Māori farming and to highlight the wider contribution of Māori farming to the national economy.'

In keeping with the aims of Sir Apirana Ngata and Lord Bledisloe in 1932, the winner of the newly established competition would receive the original Ahuwhenua Trophy and a medal based on the original trophy design, as well as a major prize package.[182]

The Critical Role of Sponsors

The relaunching of the competition in 2002 had only been possible with the substantial assistance of sponsors.

Sponsors continued to play a critical role in providing encouragement and support for the viable functioning of the Ahuwhenua competition, contributing to its facilitation of Māori farming excellence.

In 2013, the primary Ahuwhenua sponsor was the Bank of New Zealand, which had been accorded the status of platinum sponsor, with a team of 180 agribusiness partners and support staff around the country committed to continuing the BNZ's extensive role in New Zealand's agribusiness sector.

Gold sponsors also supported both competitions, contributing to the continuing viability and integrity of the competition: DairyNZ, Fonterra, Te Puni Kōkiri, the Māori Trustee, Beef and Lamb New Zealand and the Ministry of Primary Industries.

Silver sponsors were also vitally supportive: AgITO, Ballance Agri-Nutrients, PGG Wrightson and AgResearch. Bronze sponsors were also gratefully acknowledged: BDO, Allflex, AFFCO New Zealand, Polaris, LIC and Re:Gen.

Regional Finals

A series of four regional finals were arranged for 2003 for a competition that would not recognise separate categories of sheep and beef and dairy farming, as previous Ahuwhenua competitions had done. Region 1 comprised Auckland, Northland, Manukau, Coromandel and Thames; Region 2 comprised the East Coast, Poverty Bay, Hawke's Bay and the Wairarapa; Region 3 comprised the Waikato, the Bay of Plenty and King Country; and Region 4 comprised the South Island, the Chatham Islands, Taranaki, Whanganui, Manawatū and Horowhenua. Initial interest was high: twenty-nine entries were received from throughout New Zealand.

Within a few months, organisers decided to amend the regional categories because so few had been received from the northern region. The central and northern regions were therefore combined. According to Mr Walden, now the deputy chairman of Meat New Zealand, the low numbers of entries in the north could be put down to the long period that the Ahuwhenua competition had been in recess.

The first region to appoint its finalist was the southern region, which selected Whatarangi and Christina Murphy Peehi of Kahu Est. Ltd, a corporation farming 1090 hectares of 'usable rolling hill country' (688 hectares of which were freehold) with 11,800 stock units at Karioi, near Ōhakune. Recent developments had included regrassing old pastures, increasing lambing percentages, breeding and trading instead of fattening and developing new pasture blocks.[183]

The Kapenga M Trust of Rotorua was declared the winner of the revised northern/central region, and the Marotiri Farm Partnership of Tokomaru Bay won the eastern region's competition.

Following the awarding of regional winners, regional field days hosted by individual winners were held; 'highly satisfactory' attendance figures were recorded exceeding 300 people. Field days were regarded as important because they enabled winners to convey their farming and management strategies to a wider Māori farming constituency (though it was noted in a 2004 review that those who had attended were mostly Pākehā).

A Winner

An awards dinner was held at Wairākei, near Taupō, on 7 June 2003. Attending was the Governor-General Dame Sylvia Cartwright, continuing the tradition of viceregal support for the competition that had commenced seventy years earlier with Lord Bledisloe. Māori Affairs Minister Parekura Horomia was also present.[184] Dame Sylvia presented the Ahuwhenua Trophy for 2003 to the chairman of the Kapenga M Trust, Mr Sonny Sewell.[185]

The Kapenga M Trust was located 20 kilometres south of Rotorua on 1858 hectares established into three farming units: sheep and beef, dairy and deer. The trophy was awarded to the trust's sheep and beef unit.

The trust's lands had been returned to owners in 1981 and later formalised as a trust under section 438 of the Maori Affairs Act 1953. The trust itself was an ahu whenua trust under the Te Ture Whenua Maori Act 1993. It had 915 shareholders of Tūhourangi descent and functioned with a vision to 'share a commitment to caring and sharing and maintaining the land and its resources for future generations'.

Looking back on the 2003 competition, Allan Frazer said that the dinner, attended by 400 people, had been a wonderful end to the process. The competition had made a strong re-entry into the Māori and New Zealand agricultural economy, which had been essential because a significant percentage of sheep and beef production in New Zealand came from Māori enterprises, representing an important Māori commitment to the national economy.

In keeping with convention, after the dinner, the Ahuwhenua Trophy was returned to the Whanganui Regional Museum for safekeeping until next year.

2004

No National Winner

Regional competitions were once again arranged for 2004. However, because of severe flooding in the lower North Island and especially the South Island, which had caused field days to be cancelled, the previously established regions were conflated into just two. As a consequence, it was decided that there would be no national winner in 2004. Instead, the two regional winners would be announced and especially recognised.

Regional Winners

The Kuratau Trust, near Tūrangi, won the first of the two regional competitions. It farmed an extensive property of 2040 hectares, carrying 22,000 'high performing sheep and beef stock units in an area renowned for cold winters and dry summers'.

The Kuratau Trust had been working to a strategic plan compiled in 1998 and had attained a good number of their aspirations. Interestingly, the chairman of the trust, William Konui, was the son of a former trophy winner, Waaka Konui, who had taken away first prize in sheep and cattle in 1970.[186]

The winner in the other region was Te Awahohonu Forest Trust – Tararewa Station Trust, on the Napier–Taupō Road. Tararewa Station was a 2621-hectare block carrying 24,000 stock units. In 2003, the trust had achieved high percentages for its lambs (134 percent) and calves (93 percent) in an area 'subject to topographical and climate difficulties', the judge said. The trust's property was 'well advanced in its implementation programmes and adoption of technology, and was achieving good financial results'. The chairman of the trust, Mr Tamihana Nuku, accepted the win, saying it was a deserved result because 'every level of our governance and management had performed beyond expectations'.[187]

Regional Field Days

Following the announcement of regional winners, field days were held and deemed successful. In all, 155 people attended the Kuratau field day and 140 the Tarawera field day. Most of those attending were farmers. Given that the two properties were only two hours apart, it was thought unusual that the field days had not seen a 'crossover of attendees'; it was noted that the Tarawera day was largely attended by East Coast people and the Kuratau day attended by those from the Central Plateau and further south.[188]

Reviewing 2004

Following the final awarding of prizes, organisers undertook a review of the 2003 and 2004 competitions. The launch of 2003, it was noted, had occurred at a Federation of Māori Authorities conference where important personal 'one to one' contacts had been made, followed up by contacts at the operational level involving accountants and farm advisors. There was support among Māori for the reinstatement of Ahuwhenua, it was thought, leading to many entering because 'they wanted to give [the re-launched competition] a good start'.

For 2005 and beyond, the review saw several steps as essential: networking with past entrants and winners, networking with key personnel within specific incorporations and farming trusts and continuing to promote the competition among existing and expanded networks as an opportunity for Māori to promote excellence in farming and to 'participate in the delivery of strong performances in the economic, environmental, cultural and social spheres'.[189]

2005

A Focus on Sheep and Beef

Given the exponential diversification that had occurred in the farming sector during the Ahuwhenua Trophy's ten-year recess, the decision was taken in 2005 to alternate the focus of annual competitions between sheep and beef, and dairying. The 2005 competition would focus on sheep and beef.

Entry was open to Māori farms (individual farm owners, partnerships, trusts and incorporations) who were farming properties owned or leased by Māori and used mainly for beef and sheep production; entrants needed to be generating at least 80 percent of their gross farm income from beef and sheep production, and they were required to be farming at least 2500 stock units comprising sheep, beef and cattle. Farms needed also to be accounted for as a stand-alone business, separate from any other associated business enterprise (such as separately managed farming operations also owned by the entrant). Entries of more than one property owned by a trust or incorporation would be permitted, provided all of the criteria were met.

The 2005 Ahuwhenua judging criteria focused on demonstrations of excellence in Māori farming and provision of comparable value to the wider New Zealand farming community. Units would be assessed on services to the Māori farming community, quality of farm management, financial performance, governance and recognition of ngā tikanga Māori in their business practices.[190]

Commenting on the year's competition, Minister of Māori Affairs Parekura Horomia said that victory in this competition was but one aspect of it: 'The benefits of this competition accrued to all who had entered.' The Minister of Agriculture, Hon Jim Sutton, similarly applauded Māori farmers who were 'tapping into the full economic potential of their land', and congratulated Meat & Wool New Zealand on their continued administration of the Ahuwhenua Trophy, which 'celebrated the success of Maori farmers whilst locating role models who demonstrated excellence by showing us what was possible'.[191]

Judging the Regional Finals

The 2005 Ahuwhenua competition was initially staged in three regions, with an understanding that boundaries could be changed in consultation with competitors once entries had been received. The North Central region included Auckland, Northland, Manukau, Coromandel, Thames, Waikato and the Bay of Plenty. The Eastern region extended from the East Coast to Poverty Bay. The Southern region was the largest, reaching from Taranaki, Whanganui, Manawatū, Wairarapa and Hawke's Bay to the South Island, and including the Chatham Islands.

Judging of regional entrants took place during March of 2005; competitors were visited for three-hour inspections. Competitors were required to supply a certain amount of information to judges on the day of their visit: budgets for 2004/2005, business plans for 2004/2005, evidence that farmers were running their farms in accordance with business plans, evidence of long-term strategic planning, legal documentation substantiating the unit's legal

Hon Parekura Horomia, Minister of Māori Affairs, addressing the Ahuwhenua awards dinner, Rotorua, 10 June 2005
(Ahuwhenua Archive, DCSC1443)

structure and evidence of Māori descent where a farm was owned by individuals or in partnership, in which case an entrant might be asked to recite their whakapapa.

Judges assessed the efficiency with which a property was farmed relative to its potential. This was not solely based on financial measures; it could also include physical aspects such as soil type, topography and climate, as well an assessment of the farm's stage of development and financial structure. Limitations upon a unit's capacity to attract capital funding were taken into account, as were economic surpluses per hectare.

Other aspects of interest to the judges included evidence of forward planning, reinvestment practices, innovative farming systems, approaches to sustainable farming, upskilling of management, monitoring of performance and recognition of the cultural aspects of Māoridom in the context of farming practices.

Regional Finalists and the National Winner

Regional finalists were the Waihi Pukawa Trust in the Southern region; Te Pou a Kani Farms, owned by the Wairarapa Moana Incorporation, in the North Central region; and Te Aute Trust Board Farm Ltd in the Eastern region.

Competition between the operations was described by one of the judges as 'fierce but fair'. At an awards dinner held at Rotorua on 10 June 2005, the winner was announced as Te Pou a Kani Farms.

Finalist Profiles

Waihi Pukawa Trust

The Waihi Pukawa Trust farm was located on State Highway 41, 16 kilometres west of Tūrangi. The trust was chaired by Richard Fox and comprised seven trustees and two associate trustees.

Formed as a trust in 1981, Waihi Pukawa represented the hapū interests Ngāti Manunui, Ngāti Tūramakina and Ngāti Hinemihi of Ngāti Tūwharetoa. 4449 owners were registered in 2005. At the time of inspection, the trust employed fourteen staff.[192]

Development of this farm had commenced in 1947 under the supervision of the Department of Maori Affairs. Its prospects were much improved by the development of new pastures out of original pumice soils. At the time, stock decisions had been difficult to make because of rush infestations that caused serious stock bloat problems. Accordingly, large numbers of wethers were farmed; in time, large flocks grazing on red clover became a feature of the station.

Eventually, pastures became settled, and improvements like fencing, water supply, a woolshed and staff housing were completed by 1957. However, fencing remained problematic because contractors could not be employed in sufficient numbers at critical times when weed infestation was at its worst.

In 1982, the farm was handed back to its owners, and all past debts paid. At the time, Waihi Pukawa was wintering 25,000 sheep and cattle at 10 stock units per acre. Twenty-two years later, in 2004, the station was wintering 45,600 sheep, cattle and deer at a stocking rate of 14.5 stock units per hectare.[193]

At the time of inspection, the Waihi Pukawa farm was 3932 hectares in size; 602 hectares were set aside in pristine conservation bush and 186 hectares were in plantations, leaving an effective grazing area of 3144 hectares. There were 120 paddocks, to all of which an upgraded water system provided reticulated water from gravity, springs and electric pump sources.

Ahuwhenua judges commended the Waihi Pukawa Trust farm for its high-quality farm infrastructure, high gross income per hectare, cohesive governance and management practices and forward stock conditions, resulting in high per-head stock performance. Diversification into deer and merino farming also impressed the judges, as did the quality and yield of the farm's feed crops. It was noted that the trust had participated actively in the Lake Taupō nitrification developments, much to their credit.

Te Pou a Kani Farms/ Wairarapa Moana Incorporation

Te Pou a Kani Farms was situated on Scott Road in Mangakino. Its chairman was Mr Kingi Smiler, who presided over a board comprising six trustees.

The origins of Wairarapa Moana Incorporation harked back to the 1850s, when the Crown was aggressively purchasing Māori lands in the South Island, but also in Hawke's Bay and the Wairarapa. Lands were being sought to support large-scale Pākehā pastoralists, many of whom had entered into illegal leases with Ngāti Kahungunu. Governor Grey was determined to purchase these lands and went about doing so by suspect means.

In the 1850s, huge blocks of such lands were acquired in Hawke's Bay. Grey also exerted pressure on Ngāti Kahungunu to sell their lands in south Wairarapa. Fragments of these lands that remained in Māori hands would eventually be consolidated under Wairarapa Moana Incorporation.[194]

Originally a land development block set up by Māori Affairs in 1948, this property was handed back to its owners in 1983. While some development work had been completed by that time, the property was in a poor condition, with low productivity returns and significant levels of debt.[195]

In 2005, Te Pou a Kani Farms comprised 1325 hectares of sheep and beef, approximately 2870 hectares of dairy and 5000 hectares of forestry. Wairarapa Moana Incorporation comprised five elected committee members and one co-opted member. The incorporation had approximately 2783 owners. While focusing on the maximising of its commercial assets, it had also paid attention to the promotion of social programmes through a separate organisation, the Wairarapa Moana Trust, whose operations were substantially funded by the incorporation. For the previous fifteen years or so, the trust had administered annual education grants of $40,000 and marae development grants of $30,000.

The initial strategy of the incorporation's farm management had been to focus upon the trust's dairy unit, rather than its sheep and beef unit. The profitability of this dairy unit had significantly improved.

The incorporation then undertook an extensive review of the sheep and beef unit, making key personnel changes and revising livestock and forage plans to improve performance and financial returns. Following the review, 100 kilometres of new fencing were erected over 320 paddocks, a new water system was installed to provide a permanent water trough to each paddock, targeted inputs of fertiliser were introduced and building upgrades were undertaken. The review involved environmental targets, including fencing of 100 percent of gorges and streams from stock, establishment of native planting within retirement areas, establishment of a nutrient inputs budget and a water quality monitoring programme to measure nutrient losses.

In 2005, the effective area of Te Pou a Kani Farms was 1325 hectares. Soils were derived from pumice and were textually variable, free-draining and prone to leaching. Top soils were strongly resistant to water, which led at times to excessive run-off; good riparian management had therefore been necessary. The farm

Ahuwhenua Trophy winners 2005: Wairarapa Moana Incorporation/Te Pou a Kani Farms
(Ahuwhenua Archive, DCSC1502)

had been well subdivided to achieve high pasture use and good performance from sheep, bulls and dairy cattle.[196]

Judges commended managers for their implementation of intensification and development programmes on the sheep and beef unit. Though a diverse operation, they said, Te Pou a Kani Farms carried out clearly identified enterprise analyses; made use of very clear strategic planning, reporting and review processes; demonstrated a very high quality of new pastures; and had a creditable awareness of environmental issues. The judges were impressed with the calibre of the staff managing the farm and their willingness to consult outside professionals. Te Pou a Kani Farms functioned admirably, they said, taking a 'real team' approach from management up to governance level.

Te Aute Trust Board Farm Ltd

The Te Aute Trust Board Farm was located on College Road, Pukehou, in Central Hawke's Bay. The board was chaired by Mr Stan Pardoe and comprised eight trustees.

This farm dated back to the founding of Te Aute College in 1854, when 7000 acres of land were gifted by Ngāti Kahungunu rangatira Te Hapuku to support Māori education. At the time, the land was swampy and was considered too small to be farmed effectively. In 1862, the Te Aute College Trustees assumed ownership; the Te Aute Trust Board was established in 1884. The land was now general land, with the Te Aute Trust Board acting as proprietor.

In 1998, the opportunity to buy 377 hectares of neighbouring freehold land presented itself. The decision was taken to farm this and develop it as a sheep and beef unit with medium to high capital because it promised a better income than other options and meant that the land was looked after and its value ultimately enhanced.

In 2003, the Te Aute Trust Farm Ltd was established and the farm business, including stock, plant, assets and liabilities were transferred from the Trust Board to the new company. The company was 100 percent financed; it leased the farm from the Trust Board at an annual rental of $120,000, and purchased a further 100 hectares in April 2004. The farm itself comprised

784 hectares, including 73 hectares of native bush and 38 hectares comprising the school grounds, leaving an effective grazing area of 647 hectares. It leased a further 1400 hectares – in total, owning 2184 hectares. Climate was mild and temperate; soil types were mainly Matapiro and Crownthorpe light silt loam, with heavy silt loam in the form of Pukehou on 200–350 metres either side of College Road.

Current stock totals in 2005 were 3200 mixed-age breeding ewes, 1100 highlander ewe hoggets, 18 month bulls being wintered and 100 weaner beef bulls. In the past, the Te Aute farm had diversified by finishing heifers and steers and using them, as well as young bulls, to maintain pasture quality. Now, it used breeding cows in calf or with calves at foot to perform this function.[197]

The Ahuwhenua judges were impressed with a number of aspects relating to the farm: its real sense of community responsibility; its clear separation of duties between governance and management and good levels of communication; its efficiently set and reviewed budgets; its recent and ongoing successful intensification programmes (such as intensive beef and techno-systems); the recent introduction of high-fertility genetics into the sheep flock, which had significantly lifted lambing percentages; its clear finishing and stock disposal policies; its good pasture management; its stable staffing levels; and its well-maintained farm machinery.

2006

Celebrating Success in the Dairy Industry

The focus of the 2006 Ahuwhenua competition was dairy farming. With Māori contributing over 15 percent of New Zealand's total dairy export receipts and holding $100 million worth of shares in the national dairy co-operative, Fonterra, it was appropriate, said chairman of the Ahuwhenua Trophy Management Committee, Mr Bob Cottrell, to 'examine this growth phenomenon and reward excellence within the industry'.

By 2006, the new Ahuwhenua competition had already become a major marker on the Māori farming calendar.[198]

Regional Finalists and the National Winner

The three finalists for 2006 were Kōkako Trust, which farmed three dairy units in Ngātira, north of Tokoroa; Parininihi ki Waitōtara Incorporation, which managed fifteen dairy units and was the largest dairying entity in Taranaki; and the Aotearoa Trust, owned by Ngāti Raukawa ki Wharepūhunga, which farmed two dairy units in the Wharepūhunga district, south-east of Te Awamutu.

On Friday, 12 May 2006, at an awards dinner held at the Rotorua Convention Centre, Parininihi ki Waitōtara Incorporation was announced as the winner.

Finalist Profiles

Kōkako Trust

The Kōkako Trust was chaired by Mr Wano Walters and comprised four trustees.

In 1984, trustees had taken over control from the Department of Māori Affairs. There was dry stock and 208 cows on site at the time, and sharemilkers were employed. In 1987, the trust embarked on a $50,000 farm development programme focusing mainly on the dry stock block; fencing and provision of water supplies were set as the main objectives. Over the ensuing years, the dry stock acreage increased, and further substantial investments were made in fertilisers.

In 1995, a second dairy unit was established, at a cost of $500,000. Unfortunately, a severe drought in 1999 impeded development, but by 2001, the trust was producing record returns from both dairy units, sufficient to purchase two adjoining farms totalling 138 hectares. Over this time, the trust developed a four-fold strategy: to grow the farming business, to increase its total land mass, to develop farming expertise and, not least, to foster interest in farming among young people.

In 2006, the trust's three dairy units were carrying 1200 cows. The Kōkako Trust had 1700 owners; it arranged distributions annually to owners and affiliated marae, as well as making kaumātua grants and education grants and fostering participation in farming training programmes for Te Arawa young people.

The Ahuwhenua judges commended the Kōkako Trust for the impressive presentation of their farms and performance outcomes 'well above the district average'. They acknowledged the trust's attention to environmental sustainability and its cultural objectives, which they considered demonstrated pride and commitment.[199]

The Aotearoa Trust

The Aotearoa Trust was located in Wharepapa South near Te Awamutu and comprised two dairy units milking 520 cows each. A further South Island operation milked 440 cows. The trust was chaired by Mr John Edmonds and comprised five members. (One was Mr Robert Mauriohooho, a relative of Wiremu Mauriohooho, who had won the Ahuwhenua Trophy in dairying in 1966 and been runner-up the previous year.) The trust had 716 owners. Its tribal affiliation was Ngāti Raukawa ki Wharepūhunga.

The Aotearoa Trust developed out of the amalgamation of numerous small holdings totalling 4200 acres in 1953. Formerly managed as a land development block by Maori Affairs, the land was vested in the Aotearoa Trust in 1981, and the control of the trustees was formalised in June 1982. Five months later, the trust applied to the Rural Bank for loan monies to develop the first dairy unit. A second dairy unit was commenced in June 1985 after additional loan monies had been granted. In 1986, disenchantment among owners with debt issues caused 'very political troubled waters' that disturbed the trust's governance. Trustees resolved to 'unite as a tight and consistent governance board'; eventually the trust was able to regain substantial majority shareholder support.

In 1991, the trust leased an additional 105 hectares from an adjoining property, purchasing the land outright in 1996. In 1993, it purchased two South Island properties for dairy conversion, and in 1994 commenced dairying operations there. In 2000, one of the South Island properties was released in order to relieve the trust of debts and fund developments in Wharepapa South.

The Aotearoa Trust provided significant dividends to shareholders, as well as bestowing education grants and marae funding. According to the Ahuwhenua judges' report, 'their whare tupuna Hoturoa, which the Trust has restored, is a striking feature in the middle of their farm'. Key performance indicators on the trust farms were all well above district averages, the judges noted. Environmental issues were well managed, and the farms presented well.

Overall, said the judges, the dairy farms were a credit to the trust: 'The Trustees understood the challenges they faced and were taking steps to better position their Trust in order to pursue their acquisitions strategy.'[200]

Parininihi ki Waitōtara Incorporation

The Parininihi ki Waitōtara Incorporation was established in 1976 under an Act of Parliament. The incorporation was chaired by Mr Spencer Carr, who presided over a board of eight.

During 1977, 22,000 hectares made up of 364 perpetual leases with twenty-one-yearly rent reviews were transferred to the incorporation. In 1990, the first rent reviews became due, with five years to settle. One farm, Kairau Road, was sold and another purchased at Ōpunake. In 1993, another farm was purchased on Manaia Road. In 1997, the Maori Reserved Land Amendment Act was passed, reducing rent review periods to seven years (provoking Taranaki farmers to drive their tractors to Parliament in protest). In 1998, statutory rents were removed and market rentals were introduced; the first right of refusal on sales was granted to the Parininihi ki Waitōtara Incorporation. Land management plans were approved by shareholders living around New Zealand.

In 2000, the Parininihi ki Waitōtara Incorporation commenced a programme of purchasing lessees' interests; by 2005, it had acquired seventeen such interests.

In 2006, the incorporation operated fifteen farms, encompassing 801 effective hectares with 6100 cows on site. According to the Ahuwhenua judges, the corporation's governance structures reflected a 'modern and effective corporate model where performance and accountability were focussed and clear'. The incorporation's management plan and mission statement 'left no doubt as to the strategy and implementation plans to be followed'.

The farming operation submitted for judging comprised 36 effective hectares, running 136 cows with a production of 1500–1600 kg of milk solids per hectare. The judges described all aspects of this operation as admirable, especially noting evidence of comprehensive monitoring and reporting processes.

Overall, judges saw Parininihi ki Waitōtara Incorporation as 'a very polished enterprise that had made significant progress against its strategy', reflecting what could be achieved with a corporate discipline, while still delivering to the expectations of owners in a social and beneficial way.[201]

2007

New Funding; New Prizes

The relaunched Ahuwhenua Trophy competition had been made possible by sponsors like the Bank of New Zealand, Te Puni Kōkiri, Meat & Wool New Zealand, AgResearch, Ballance Agri-Nutrients, Suzuki, PGG Wrightson, AgITO and the Maori Education Trust. By 2007, as well as receiving a

Ahuwhenua Trophy winners 2007: Ātihau-Whanganui Incorporation/Pah Hill Station
(Ahuwhenua Archive, DCSC4722)

replica of the original cup donated by Lord Bledisloe in 1932 and a framed photograph of the presentation ceremony, the winners also received $40,000 worth of prizes, including $10,000 cash and a top-model Suzuki quad bike.[202]

The original sixty-three Bank of New Zealand shares of Lord Bledisloe's time had been vested with the Ahuwhenua Trust Board, established in 1933 at Apirana Ngata's behest to manage and fund the fledgling competition; however, the overwhelming portion of Ahuwhenua costs had in fact been met by the Department of Native Affairs over the years.[203]

Regional Finalists and the National Winner

The focus of the 2007 Ahuwhenua competition was sheep and beef. Regional winners were awarded medals and $15,000 in cash or farm-related items. As in previous years, the three regions comprised a North Central region from Northland down to the Waikato and Bay of Plenty; an Eastern region comprising the East Coast and Poverty Bay; and a Southern region extending from the King Country, Taranaki, Whanganui, Manawatū, Wairarapa and Hawke's Bay southwards to include the South Island and Chatham Islands.

The three finalists were the Pah Hill Station of Ōruakukuru Road, Ōhakune, in the Southern region; Tūaropaki Trust of Tirohanga Road, Mōkai, Taupō, in the North Central region; and the Matariki Partnership of Waiomatatini Road, Ruatōria, in the Eastern region.

On Friday, 15 June 2007 at the Ahuwhenua awards dinner held at the Rotorua Convention Centre, Pah Hill Station was announced the winner.

Finalist Profiles

Pah Hill Station

Pah Hill Station won the Southern region final in 2007; the Tiroa E Trust of Te Kūiti came second.

Pah Hill Station was a 1900-(effective)-hectare sheep and beef property comprising flats and easy rolling pastures with some steeper hill country. The total area of the farm was 2894 hectares; 240 hectares were set aside as conservation land. It supported 11,000 breeding ewes, 650 breeding cows and replacement stock.[204] The chairman of Ātihau-a-Paparangi Incorporation, which ran the station, was Mr Whatarangi Murphy-Peehi, who presided over a board of eight trustees.

The Ātihau-Whanganui Incorporation ran six sheep and beef stations and one dairy unit, with 5000 hectares set aside for forestry. A recent business overhaul had followed a $26.5 million Treaty settlement with the Crown, allowing the incorporation to retire most of the $31.5 million debt it had acquired when resuming control of its own lands, vested since 1903 in a series of Maori Land Board lease arrangements.[205] The incorporation's core business had been farming, though it also leased lands to other farmers. The incorporation had recently been involved in forestry developments and was also currently considering proposals from power generation companies. Pah Hill Station itself was part of an ancient Māori track that joined Rānana to Karioi, a good day's walk away.

Like ten other properties managed by the Ātihau-Whanganui Incorporation, Pah Hill Station was located on land that had been incorporated by the Māori owners in the late 1960s, when most of its land was leased out. In the 1980s, the owners decided to resume occupancy of those leased lands and run them as viable units themselves. In the 1990s, the incorporation itself took control and resumed direct farming of the land.

Pah Hill Station was affiliated to the Ātihaunui-a-Paparangi iwi of Whanganui. In 2007, the station had 7074 owners and was managed by a board of seven trustees, with a full-time staff of six managing day-to-day operations. The farm had 105 paddocks, with a natural water supply from dams and creeks. Most of the hill country consisted of rye and clover, with some native grasses.

Ahuwhenua judges commended the farm for having consistently outperformed its key performance indicators and for an ambitious programme of subdivisions, making the farm easy to manage. They were also impressed with the expertise of management staff, the health of quality stock, the introduction of cost-effective improvements and innovations in managing stock and the feed supply. The provision and standard of budgeted financial information was very high, and a good rapport between key senior management personnel had been established, it was noted. The judges also complimented Pah Hill for its innovative approach to critical environmental issues.[206]

Matariki Partnership

The Matariki Partnership won the Eastern region final in 2007; the Wi Pere Trust of Gisborne came second.

The Matariki Partnership was chaired by Wi Mackey; the board comprised seven members. The partnership operated two adjoining properties in the Waiomatatini Valley of Ruatōria. Established in 2003, the partnership combined the skills and experience of two farming entities on the East Coast – Ahikōuka A6B, established over eighty years earlier by Sir Apirana Ngata, and the Waiomatatini C Trust, established by Maori Affairs when it returned land to its owners in 1963. The profit-sharing ratio was 70 percent to Waiomatatini and 30 percent to Ahikōuka. The Matariki Partnership affiliated to Ngāti Porou and had 3000 owners.

In 2007, a full-time staff of six managed day-to-day operations. The partnership ran sheep and beef stock units on a 2480-hectare property with 1390 hectares in pasture and 150 hectares in exotic pines. Eighty hectares were also set aside in native bush, and the remaining 860 hectares was in unusable gorse- or mānuka-covered hillsides. Soils were principally sedimentary loam and silt, and the topography comprised a mix of steep hill country, easier contour and flat land. Matariki had suffered through droughts over the previous two years and had recently commissioned a feasibility study into the prospect of extracting water from the Waiapu River. As a result of recent dry conditions, it had reduced its numbers of stock. It had recently planted 80 hectares of sweetcorn and maize.[207]

The partnership's hill country consisted of forty-one paddocks averaging 27 hectares each, and its flat country contained fifty-eight paddocks averaging 4.4 hectares each. The hill country was supplied with water through dams and streams, while the flat lands had troughs supplied through springs. Most of the hill pasture consisted of rye and clover, with some native grasses. The flat land was predominantly planted out in grasses like Italian rye.

After visiting and inspecting the property, judges described the operation as 'a deliberate and considered approach to the development of an exciting venture'. They were impressed by the participation of management in local think tanks, reflecting a strong desire to obtain expertise and share knowledge, and commended the farm's governance team for their commitment to the Ngāti Porou Tapuaeroa farm project. They noted that a wide variety of communication channels was used to convey all aspects of the business to shareholders and that all trustees spoke 'comfortably and knowledgeably' about the partnership's operation.

Te Hape B Trust – stock in fine condition
(John Cowpland – Alphapix-0905131284)

Farm production and profitability were described as 'very good'; improved lambing and cattle performance were much in evidence. Good environmental policies were in place to address erosion and scrub problems, they noted.[208]

Tūaropaki Trust

Tūaropaki Trust won the North Central regional final. Second place went to Whakaue Farming Ltd of Rotorua.

The Tūaropaki Trust was registered as an ahu whenua trust under the Te Ture Whenua Māori Act 1993. The Trust Order, established by the Māori Land Court, required the trustees to administer the Tūaropaki E lands for the beneficial owners. The current chairman was Mr Tumanako Wereta, who presided over a board of seven members. The trust affiliated to Ngāti Tūwharetoa and Ngāti Raukawa. A full-time management staff of five managed day-to-day operations. The trust had 1839 owners, with a total shareholding of 16,752 shares.

The Tūaropaki Trust had its beginnings in 1952, when a group of families decided to amalgamate their land holdings into a collective body. Seven hapū were involved: Parekawa, Te Kōhera, Wairangi, Whaitā, Moekino, Haa and Tarakaiahi. The amalgamation enabled the combined lands to be developed for pastoral farming under the direction of trustees and the management of the Department of Maori Affairs. By 1979, sole responsibility for the lands was vested in the trustees – chaired by Sir Hepi Te Heu Heu.

By 2007, the Tūaropaki Trust had established a unique range of business interests, including sustainable farming, geothermal power generation and horticulture – mainly tomatoes and paprika for export.

Tūaropaki Trust lands totalled 2734 hectares, on which the trust ran 10,000 sheep and about 1200 cattle, as well as deer. It also operated two dairy units. An increase of total stock numbers to 25,000 was planned. Land that was not suitable for farming (53 hectares) was set aside for afforestation and conservation. Forestry and plantations accounted for 112 hectares; effective grazing land comprised 1926 hectares. In more recent years, the trust had taken on 650 hectares of adjacent forest, converting it to 450 acres for sheep and beef. The Tūaropaki lands overlaid most of the Mōkai geothermal field and surrounded geothermal hot springs at Ōhineariki and Parakiri. The springs were set aside with separate titles in 1891 and continued as special Māori reservations. Twenty-five hectares of land had been set aside for horticultural development (glass

Dean and Kristen Nikora and whānau, winners of the Dairy Competiton 2008
(John Cowpland – Alphapix-1054-518)

houses), and around 30 hectares for steam-field wells and pipelines and an electricity-generating plant.

The farm was located on 'yellow brown pumice silty sand', with slopes covered with Taupō ash and valleys filled with finely textured Taupō breccia. Paddock numbers had increased from ninety-eight in 2004 to 180 in 2007. Eighty percent of the farm was serviced by reticulated water; pasture was based on rye grass and white and red clover.[209]

Ahuwhenua judges commended the trust for its 'vitality and genuine desire to improve the profitability' of its extensive and attractive property. They were impressed with substantial improvements made since 2005, when the trust was last an Ahuwhenua competitor. They observed that the trust had progressively adopted a modern corporate model to manage its business; important subcommittees handled ancillary aspects of a diversifying portfolio. Trustees had taken admirable steps to upskill themselves by, for example, attending Institute of Directors courses. New sheep genetics had been utilised, the judges noted; a good use of technology showed a creditable willingness to 'try new ideas and to be innovative'. Environmental awareness was also evident: planting programmes and fencing off of waterways was in line with kaitiakitanga principles and regional policy.

2008

Seventy-five Years Strong

In 2008, the Ahuwhenua Trophy competition commemorated its seventy-fifth birthday.

Regional Finalists and the National Winner

In March 2008, the three Ahuwhenua regional finalists were announced: Dean and Kristen Nikora of Takapau in Hawke's Bay in the Eastern region; the Pārekarangi Trust of Rotorua in the North Central region; and the Hauhungaroa Partnership of Taupō in the Southern region.

On 6 June 2008, at the Ahuwhenua awards dinner held at the Energy Events Centre in Rotorua, Dean and Kristen Nikora were announced as winners.

Finalist Profiles

Dean and Kristen Nikora/Cesped Lands Ltd

Dean and Kristen Nikora were joint owners of Cesped Lands Ltd, which, like most farms in 2008, had recently experienced a serious downturn in economic fortunes. However, in their 2008 report, the Ahuwhenua judges praised the Nikoras for their innovative farming and commercial practices, noting that their growth strategy was particularly well managed, incorporating a disciplined risk analysis which nonetheless remained high.

Cesped Lands Ltd was a trust that consisted of three autonomous dairy units under a collective business and administrative structure that enabled economising on costs while maintaining earnings interest. Two of the farms, Mangatewai and Pahihi, both had farm managers, and the third farm, Ruatuki, had an equity partner as a 50/50 sharemilker. Judges described Dean's and Kristen's whānau approach to staff welfare as a strength. The manager of one of the independent units, Mangatewai, was now one of the trust's partners. As a result of a series of recent and successful investments, the trust had increased its stock holdings by 1000 cows; it now ran 3500 cows.[210]

In 1989, Dean and Kristen Nikora had left their careers at the Bank of New Zealand in Hamilton to begin dairy farming. Dean first worked as a general farmhand, after which the couple began sharemilking. In 1990, the Nikoras had purchased a minority partnership in a 1500-cow farm in Hawke's Bay; thereafter, they worked hard to grow the business, allowing for equity growth so that they could buy their own farm. In 2000, they sold their minority interest to buy Mangatewai, a farm five times larger than anything they had envisaged.

In 2008, Dean and Kristen Nikora farmed 1000 effective acres spanning five blocks of freehold and leasehold land. Their property was located near the Ruahine Ranges, about twenty minutes out of Waipukurau. The home farm, Mangatewai, was 342 hectares; the other farms ranged from 100 to 240 effective hectares. On a day-to-day basis, Dean ran the company and Kristen provided valuable office and administration support as well as calf-rearing expertise. Dean and Kristen Nikora had tribal affiliations to Ngāti Tama/Maniapoto and Ngāti Awa respectively. Cesped Lands Ltd employed a full-time and part-time staff of fourteen.

The focus for the field day following the regional final was on Mangatewai. Its effective dairy size was 440 hectares, with 100 hectares as support land. Peak cow numbers milked were 1340 in 2007, producing a total of 473,323 kg of milk solids. This represented 1075 kg of milk solids per hectare, or 353 kg of milk solids per cow.

While equity growth was a major driver of the operation, the judges noted that the Nikoras nonetheless retained an ongoing focus on the environment, through such strategies as the fencing of waterways, nutrient management, protection of wetlands and the eradication of pests to protect bush areas.

The judges were impressed with the operation's 'top end farm management in all respects', singling out grazing management that was 'simple yet well calibrated and executed'. They commended the Nikoras for their 'impressive and strong focus on asset accumulation' and for the high quality of their herd, which was described as 'quiet and in good condition'. They approved of the company's 'innovative salary packages' and its environmental protection policies; it was noted that most riparian areas had been fenced off, with a view to being fully compliant with the Dairying and Clean Streams Accord (an agreement signed in 2003 between Fonterra, the Ministry for the Environment, the Ministry of Agriculture and Forestry and regional councils) by May 2008.[211]

Parekarangi Trust

The Parekarangi Trust was chaired by Mr James Warbrick; nine trustees made up the governing board, and eight staff managed day-to-day operations. Parekarangi translated as 'the pleasing pā facing the sun'; it was from Parekarangi that the Tūhourangi/Ngāti Wāhiao people had once hunted for food and planted crops such as kūmara to provide for the people living at Whakarewarewa. Parekarangi had also provided shelter for its people during the Tarawera eruption of 1886.

The Parekarangi Trust had 173 shareholders. It farmed 665 hectares, including a 255-hectare dairy unit and a 410-hectare dry stock unit. It had been administered as a Māori incorporation between 1962 and 1981; the trust had been established in 1983 to 'promote and facilitate the use of the land in the best interests of the beneficiaries'. The trust prided itself on working well with its shareholders, and the relationship with whanaunga at Whakarewarewa remained strong, the farm continuing to provide meat for tangi and other special occasions.

After running a very successful dry stock farm, the trust moved into dairy in 1994; it now wintered 804 cows and milked 780 in its forty-bail rotary milking shed. The trust had been working towards a target of 260,000 kg of milk solids in 2008 and hoped to expand its unit to 1050 cows in the near future.

The trust had recently suffered from severe drought; the attention of trustees had turned to maintaining the quality of the stock. Stock numbers had increased in line with cow performances, resulting in a consistent growth over the previous two years. The attention had been focused on quality, not quantity, it was noted. Production had dropped from 320 kg to 266 kg of milk solids per cow during the drought, but this rate had now been restored to 330 kg of milk solids.

One interesting development was the construction of a 4-hectare youth justice facility on the trust's property; the trust was working alongside Child, Youth and Family, which was leasing this land. The land would house a fifty-five bed facility; roads and large-scale planting of native trees to provide a screen from surrounding areas would also be a feature. Supporting this facility had helped the trust meet its commitment to youth employment; there was potential for the trust also to support individual young people through the scheme.[212]

Despite economic difficulties, Parekarangi had a strong staff base of seven professionals or trainees. The unit's general environment had been much improved by the provision of new fences. Some native herbs had also been planted for the purposes of rongoa, or traditional Māori healing.

The Ahuwhenua judges commended the Parekarangi Trust for their 'strong engagement of shareholders, including tree planting and cemetery tidying days designed to bring all of the stakeholders together'. They were also impressed with the 'very close ties between the trust and its shareholders, occurring almost at the level of personal relationships'. They noted that specialised expertise was engaged where necessary. Judges described the farm as a dairy property of good scale, with adjoining support land under a very sound management regime. Management was comprehensive and well structured, they said, especially in terms of analysis of pasture and feed management and profit.

Hauhungaroa Partnership

The Hauhungaroa Partnership was chaired by Mr Howard Kahura, and had a governing board of seven members.

Reclaiming ownership of their land after 'decades of struggle' had been the greatest achievement for the Hauhungaroa Partnership, which managed 3400 hectares of land from the Hauhungaroa blocks 1D2 and 1D3 at the western end of Lake Taupō. The Partnership affiliated to Ngāti Tūwharetoa and the Ngāti Parekawa hapū; it comprised 870 shareholders.

The partnership's lands had had an early history of dairy farming; 'the old people were milking cows in the Te Aputu 1D2 block and were supplying milk to the dairy factory in Waihi back at the turn of the century'. The Department of Maori Affairs had previously administered the land; the partnership took it over in 1987, along with $500,000 worth of debt. Twenty years later, the Hauhungaroa Partnership was now very much in the black and possessed substantial assets.

The partnership's 437-hectare Taupō-based dairy unit was the part of the operation that had been entered into this year's Ahuwhenua competition; the partnership also farmed sheep, beef and deer and had investments in horticultural glasshouses. The unit had a sixty-bail rotary milking shed, built in 1999. The partnership had been engaged for three years in a $755,000 nitrate mitigation study with HortResearch and the Foundation for Research, Science and Technology to test blueberry species planted on their land. In 2008, the partnership was targeting four varieties for commercial development the following year.

Because of its closeness to Lake Taupō, the partnership had adopted a very strong environmental strategy. All of the unit's streams and wetlands had been fenced over twenty years ago, and the partnership had recently installed sixty lysimeters to monitor leachates on their dairy operation.

The partnership had recently paid attention to reliable water supplies, fencing, cropping and regrassing and had made use of off-site training to upskill staff. Milk production on the farm had fluctuated over recent years, but was on the rise again. The farm currently milked 930 cows, and in the previous year, had achieved 338 kg of milk solids per cow. Total production for the 2007/2008 season had been 257,000 kg of milk solids, down by 43,000 kg on the budgeted total due to drought. The 2008/2009 total had increased to 271,000 kg.[213] In 2009, the dairy unit had been increased by 200 hectares; the farm hoped to achieve 340,000 kg of milk solids in the 2009/2010 season.

The cowshed was fully automated: each cow carried an electronic ID recording the presence of mastitis, milk yield and the cow's weight. An in-shed feeding system stored 30 tonnes of dry pelletised feed and 30 tonnes of molasses, which could be targeted to individual cows.

The Ahuwhenua judges were impressed with the partnership's governance structure and overall processes, and in particular, with the well-balanced dividend and grant strategy developed for shareholders. They noted that a strong sense of the cultural importance of the land was well reflected in the environmental policy. The judges commended the partnership for 'lateral and innovative thinking' from the board on a whole range of issues connected with international networking. Good investment and management decisions continued to be made, and persistent efforts were taken towards solving animal health issues. The farm had strong cash flow and quality overall financial control.[214]

2009

'Agriculture Had Enabled Our Ancestors to Survive'

Welcoming participants to the 2009 field days and final judging programme, Hon Dr Pita Sharples commented that 'agriculture had enabled our ancestors to survive, and the same was now true for Maori in the modern world'. In a global economic recession especially, he said, Māori workers, managers, advisers, experts, trustees and landowners were striving for excellence, boosting production and profitability, and thereby helping to feed all peoples.

The chairman of the Ahuwhenua Trophy Executive Committee, Mr Kingi Smiler, noted that exponential growth had occurred recently in the Māori farming sector; he believed that such growth was entirely sustainable. Māori sheep and beef farmers comprised the largest group of levy payers to Meat & Wool New Zealand, he said, 'and I believe we have the capacity to make a difference that would benefit shareholders, whanau and the country as a whole'. To achieve this, Māori needed to be 'awake to consumer wants and needs while focusing on farm productivity that did not come at the cost of the environment'.[215]

Regional Finalists and the National Winner

The three 2009 Ahuwhenua regional finalists were Hereheretau Station of Whakaki, Wairoa; Morikau Station of Morikau Road, Rānana, Whanganui; and the Pakarae Whangara B5 Partnership of Gisborne. The winner of the 2009 trophy, announced at an awards dinner held in June 2009, was the Pakarae Whangara B5 Partnership.

Finalist Profiles

Pakarae Whangara B5 Partnership

The Pakarae Whangara B5 Partnership was established in July 2006, though its partners had been farming since the early 1900s. The partnership's current chairman was Ms Ingrid Collins, who presided over a

Field day at Pakarae Whangara B5 Partnership, Gisborne, 2009
(Ahuwhenua Archive, Field Day 02)

Pakarae Whangara B5 Partnership, winner of the Ahuwhenua Sheep and Beef Competition 2009
(John Cowpland – Alphapix-2500-1268)

board of five members. The partnership was affiliated to Ngāti Konohi of Te Aitanga-a-Hauiti and had 1500 owners in 2009. Eleven full-time staff managed day-to-day operations.

Creation of the partnership had seemed a logical move to partners in order to 'bring the people back together' and achieve economies of scale in both buying and selling. Whānau involved also saw it as a good solution because the individual farms were now too small to survive alone. The Pakarae Whangara B5 Partnership hoped that other incorporations and trusts in the Gisborne region would consider the same model in the future.

The Pakarae Whangara Partners had retained ownership of their own lands and livestock, while allowing these assets to be used by the partnership. Partners had also retained their own identities: all land titles remained the same, and the share registers of the two component groups remained separate from each other. However, plant and machinery was sold to the partnership.[216]

Previously, partners had farmed in a very traditional manner. The sheep flock was 100 percent Romney; a Poll Dorset ram had previously been used only on the early ewes. Romney Finn cross rams were now being used to increase productivity. Angus bulls were used across all the cows. Facial eczema-tolerant rams had been introduced to help reduce the influence of facial eczema. Terminal sires had also been bought to increase the hybrid vigour of the cattle herd. The partnership aimed to sell cattle before they had a second winter on the farm. These measures had required an extensive review of policies relating to water supply, subdivision and fertiliser.

The total area of the farm was 5448 hectares with 4544 hectares specifically set aside for grazing. Soil types included mudstone hills, pumice, ash and silt flats. The topography was flat and rolling to medium steep. In 2009, Whangara Farms was wintering 27,000 sheep and 4000 cattle.

Most of the hill country consisted of rye grasses and clover, though native grasses still had an influence on some of the pastures. The flat country had been in a cropping rotation programme.

A land environment plan assisted the farm in achieving best-practice sustainable farming practices. The partnership's social, community and ngā tikanga Māori aspirations were clearly stated.

Currently, there were 250 paddocks on the farm, and water reticulation was spread over 2000 hectares; water was also used from local dams. In 2007, Whangara Farms commenced a major development programme, the biggest part of which was the reticulated water system. By June 2011, it was planned that the whole farm would be operating under a $1 million system financed from the farm's annual surpluses.

During the previous financial year, Whangara Farms had employed four new shepherds and had been able to distribute $600,000 to business partners.

The Ahuwhenua judges commended Pakarae Whangara B5 Partnership for the 'excellent spread of skills' in the partnership's governance and management teams, saying the partnership possessed a board 'with a clear vision and independent directors'. They were also impressed with the presence of a clear strategic plan that drove a business plan and a five-year budget, which in turn drove key performance indicators.

The property was presented in a very tidy state, the judges noted; fences, weed controls, stock, crops and pastures were all developed to a very high standard. A clear subdivision strategy had been devised for improving soil nutrients, reticulating water and repasturing. The farm supported good staff training, and had integrated key senior managers into positions of governance. A nutrient budget had been designed with LandVision Ltd, and progressive fencing of waterways had much improved.[217]

Hereheretau Station

Hereheretau Station had been formed in 1922 by the Maori Soldiers' Trust to assist Māori farmers returning from the First World War; the trust had been associated with the property since that time. The station was now managed and administered by the Māori Trustee. In 1922, the Maori Soldiers' Trust had leased two Crown blocks of 220 hectares, Hereheretau 2A and 2D, for a thirty-three-year term, with the right of renewal for a further thirty-three years. It had also invested in Hoata Station in Tikitiki and the Hereheretau Station near Wairoa. Both stations suffered badly as a consequence of a depression

Field day at Hereheretau Station, Whakakī Road, Wairoa, 2009
(Ahuwhenua Archive, Field Day No 27)

just after the war, and Hoata had to be abandoned in 1925, with huge losses.

In the early 1920s, an Act was passed entrusting the Maori Trustee with the administration of the remaining assets, valued at about £12,000. These were all lost during the subsequent depression, but the Maori Trustee, having regard for the social significance of the fund, made loans to the trust to sustain it. This meant that until 1941 the Maori Soldiers' Trust had been in debt to the Maori Trustee by about £4000. When wool prices rose sharply after the Second World War, the trust quickly changed from a liability to a 'handsome asset'.

At the Sir Apirana Ngata memorial hui of 1952, attended by fifty veterans, decisions were made as to the future of the fund.[218] Given the advent of social security in 1938, there was now less need for additional funds to assist veterans. The Ngarimu VC and 28th (Maori) Battalion Memorial Scholarship Fund had been established in 1945; it was decided at the hui that the freehold of Hereheretau Station should be procured and vested in the Maori Trustee as a way of ensuring that this new scholarship fund could be sustained.

Hereheretau Station was acquired in 1952. Various parcels of land having been subdivided off, the size of the station then was 1693 hectares.

In 1998, the Maori Trustee sought a full review of the station's management and performance. The subsequent report recommended restructuring and contracting in labour and expertise. A new five-year plan was established involving new budgeting and reporting regimes, as well as a new development plan aimed at repositioning Hereheretau.

In 2003, the trust purchased a neighbouring farm of 489 hectares carrying 3800 stock units at a total cost of $2.067 million. In 2005, Hereheretau joined with five other Māori stations to participate in the Te Taumata programme, an initiative of the newly established Tairāwhiti Land Development Trust, aiming for increased farm productivity over the three-year term of the programme. In 2006, Hereheretau joined the Tairāwhiti Land Development Trust's Bernard Matthews Sheep for Profit programme, with the goal of lifting the performance of the Hereheretau flock over the ensuing three years.

Hereheretau Station had a staff of five to manage day-to-day operations. In 2009, its total area was

Pōwhiri, Morikau Field Day, 1 May 2009
(Ahuwhenua Archive, Field Day No 133)

2229 hectares, with 1839 effective grazing hectares. The farm comprised 124 paddocks ranging from 3 hectares to 52 hectares, bounded mostly by conventional fences. The soils varied from silt loam on the flats to mudstone and silt on the steeper hills, with sandy loam and ash overlay in between. Water was supplied through natural sources such as rivers, springs, dams and creeks. Pasture was typical hill country pasture: ryegrass, clover and native grasses. Stock numbers were 6763 ewes, which included 1500 ewe hoggets, and 126 rams; the total number of sheep being run on site was 9704. Total cattle stock units were 8204.

The Hereheretau Station had set environmental goals that included the permanent retirement of 172 hectares of native bush, double fencing of the Mākāretu Stream running below this native bush, development of a 24-hectare pine plantation, a gully replanting programme and a sustainable nutrient programme.

The Ahuwhenua judges were impressed with the overall governance of this property, especially the 'clear definition of the governance structure' that resulted in efficient interaction between the governance and farm teams. They also commended Hereheretau for its overall presentation; stock, pastures and infrastructures were all in good condition. They gave the station credit for the extent to which feed supplies and animal conditions were monitored and for using this information in decision making. The level of knowledge and commitment of the staff was also highly comm-ended. Clear processes and frequent assessment of performance had resulted in a high standard of financial management and performance, the judges said.

Morikau Station

Morikau Station was chaired by Ms Hari Benevides, with a board comprising seven members. In 2009, administrators of Morikau Station were looking ahead to 2010, when the station would be celebrating '100 years of farming our ancestral lands'. The station was affiliated to Ātihaunui-a-Pāparangi, and its trust had 5528 members. A staff of five managed the farm's day-to-day activities.

Parts of the Morikau 1 Block formed the core of the original farm. The Ngārākauwhakarara and Rānana properties totalled 6190 hectares, of which 1410 hectares were allotted as papakāinga and excluded from the farm lands. The station's lands had been alienated to the control of the Aotea Land Board under the Maori Land Settlement Amendment Act 1906; this had enabled land to be 'compulsorily vested in a Maori Land Board' where the lands had not been cleared properly of noxious weeds, or if the minister felt that the land was not 'properly' occupied by Māori.[219]

Despite this, a representative group of the owners met and proposed the area be incorporated and farmed. In 1911, a committee of owners was set up to exercise some oversight on the land. Within a few years, an area of 4000 acres had been planted with pastures and other crops, and Hereford cattle had been introduced. Accommodation quarters were built using timber milled on the station. Many miles of fences were built and tracks developed to improve access. By the 1930s, the farm had begun to make a profit; with the mortgage paid off, owners began to receive dividends.

In 1948, the owners asked the Minister of Māori Affairs, Walter Nash, for a greater say in the running of the farm. Nash advised them that incorporating into a legal entity would enable them to do this. But progress in 'wresting control of the farm' from the Land Board was glacial; at the time, the board incongruously reported that 'the Maori owners had no apparent desire to assume control'. However, a meeting of owners at Pūtiki Pā on 11 August 1953 agreed that the land should be re-vested with the owners and kept as one farm, and that the five blocks that made up the farm should be amalgamated into one. On 18 November 1953, the Māori Land Court consolidated the blocks into one single entity, and in 1955, it signed an order of incorporation establishing the Morikaunui Incorporation; this was only the second ever such incorporation to be set up.

The station's modern legal framework was provided by the Te Ture Whenua Maori Act 1993; it also possessed a constitution in line with the Maori Incorporations Constitution Regulations 1994.

In 2009, the total area of the farm was 4799 hectares, with 2394 hectares of effective grazing pasture. Soils were based on the Egmont ash series and yellow brown earths. Upokonui steep land soils and Mangatea hill soils tended to dominate, however, with a tendency to drain well in the winter but dry out in the summer. The farm comprised sixty-one paddocks, of which thirty had been established within the last three years. Natural streams and dams directed water into most paddocks, and there were troughs in the new paddocks. Pasture was mainly ryegrass and white clover.

His Excellency The Rt Hon Sir Anand Satyanand, Governor-General, presents the trophy to Anthony Haa, Chairman of Waipapa 9 Trust, winners of the 2010 Dairy Competition.
(AHU 2010-01)

While the station's focus had been on maximising returns, all strategic decisions and developments were required to align with its guiding principles, 'with constant reference to the past' and an acknowledgement of sustainability.

The Ahuwhenua judges commended Morikau for the extent to which board members and management had demonstrated a sound knowledge of all aspects of the business. Considerable effort had gone into establishing strategic and business plans for future growth, they noted. Brave decisions had been made in terms of staff and management changes, stock genetics and extensive property developments. The judges applauded efforts to reach out to and educate shareholders.

Judges also commended stock performance, and noted that pastures were in good condition. Regular pasture monitoring and stock weighing were undertaken. Protection of bush through the Nga Whenua Rahui initiative was also applauded. Sound decision making and investigation processes prior to investment were evident.[220]

2010

Māori 'at the Forefront'

'Our land, our waterways and our biodiversity, both natural and exotic are essential resources for our survival', wrote the Minister of Agriculture, Hon David Carter, in 2010 when welcoming all parties to the Ahuwhenua competition of 2010. Fortunately for New Zealand, he added, the country possessed some of the best environmental practices in the world.

By protecting our natural resources, said the minister, Māori believed that it was possible to sustain the ability of Papatūānuku to provide for the prosperity of all New Zealanders: 'Māori, with their views around kaitiakitanga, are at the forefront of thinking on the importance of protecting our environment in order to achieve the sustainable use of our resources.'

The minister acknowledged that the Ahuwhenua Trophy highlighted those who shared these views, and acknowledged role models for other Māori farmers. The Ahuwhenua Trophy applauded success in Māori farming, not just of the winners of the annual event: 'Those who have the courage and commitment to enter such a competition are to be commended.'

Regional Finalists and the National Winner

In March 2010, the three Ahuwhenua finalists in a competition that this year focused on dairy were announced: the Waipapa 9 Trust of Taupō; Hanerau Farms Ltd of Taipuha, Paparoa; and Rangatira 8A 17 Trust, Reporoa. On 28 May 2010, at an awards dinner held at the Taupō Events Centre, Taupō, the Waipapa 9 Trust was announced as the winner.

Finalist Profiles

Waipapa 9 Trust

The Waipapa 9 Trust lands comprised 6537 hectares, being a mix of ancestral lands developed out of cut-over bush between 1960 and 1980 and further land purchased between 1997 and 2009. The trust was chaired by Anthony Tohu Haa, who presided over a board of eight members. It was affiliated to Ngāti Tūwharetoa and Ngāti Raukawa, and had a staff of five who managed the farm on a day-to-day basis.

The original development comprised three sheep and cattle stations – Waipapa, Otanepae and Takapau – which were managed by the Department of Survey and Land Information for the trust. On 8 July 1989, the Waipapa 9 Trust took back the land and assumed full responsibility for running it. Later acquisitions allowed the growth of the trust to accelerate, developing land-based businesses alongside other commercial investments.[221]

The move to dairying began in 1996, with the development of the Ōkuhaerenga dairy unit on land of higher altitude with longer winters and a relatively short production season. Ōkuhaerenga was developed from heavy log country out of cut-over bush with poor brown-top pastures and low-fertility soils. The original operation met with mixed success, until sharemilkers Matt and Louise Pepper took over in 2005. Further investment in dairying was thereafter accelerated. In 2007 and 2008, the trust undertook the Marae Mānuka and Takapau dairy developments.

In 2010, 2400 dairy cows were being milked on Waipapa lands: the target was to increase this to 3800 by 2015. Dairying now occupied 35 percent of the trust's effectively farmed land (1400 hectares). The remainder (2632 hectares) was operated as two sheep and beef stations (Waipapa and Otanepae) wintering over 31,000 stock units. Part of the Otanepae Station was located within the Lake Taupō catchment area, and was managed under nitrogen restrictions, with a nitrogen management plan in place.

The trust had adopted a very strong environmental focus as part of its business goals and values, and over time, had retired lands, protected wetlands and waterways, planted forestry and invested in low-impact effluent systems. A cadet training unit had been established in 2006 under a separate charitable trust, to foster farming skills in young Māori.

In 2010, the Marae Mānuka dairy unit was sharemilked for Waipapa by brothers Matt and Nick Pepper. Production for the 2010 season was estimated to reach 180,000 kg milk solids from 630 cows. The contour of the land was flat to medium. The rolling to steep Ōkuhaerenga unit was also sharemilked by the Pepper brothers; its production estimate for 2010 was 250,000 kg of milk solids from 760 cows. The Takapau unit was in its first year of production, and was managed by a staff of five. Its production estimate for 2010 was 320,000 kg of milk solids from 1025 cows. It was predominantly flat on two terraces.[222]

The Ahuwhenua judges commended the trust for its clear objectives and visions, leading well into strategic business objectives. The Trust's governance structure was very good, they noted; clearly defined trustee roles assigned responsibility for individual portfolios. Succession planning was evident, and the governance team displayed strong entrepreneurial skills. Animals appeared to be in good condition, consistent with best practice. Pasture management systems, the cropping programme, rates of pasture renewal, effluent spreading and fertiliser application were all working well. The judges applauded environmental safeguards like the fencing of protected waterways.

Hanerau Farms Ltd

Hanerau Farms was owned by the Te Uri O Hau Settlement Trust, which was chaired by Mr Rawson Wright, with a board comprising four members. A staff of seven managed day-to-day affairs. Mr Wright was the son of Rawson and Wikitoria Wright, who had won the Ahuwhenua Trophy for dairying in 1964.

When Te Uri O Hau of Ngāti Whātua, representing fourteen marae around the Kaipara Harbour, became the first hapu to settle its historic grievances with the Crown in 2002, it decided that it would buy traditional Te Uri O Hau land that had been alienated over the last century whenever the opportunity arose. It also sought commercial opportunities that would maximise the return for its 7190 beneficiaries, with the ultimate aim of providing a sustainable future for its members.[223]

Hanerau Farm was the first piece of land attained by the hapū outside of the lands it had obtained through the Treaty settlement. The farm, purchased in 2004, was actually two blocks run in conjunction with each other. It comprised 176 hectares of rolling coastal country bordering the Kaipara Harbour, and included land originally owned by the hapū, including sites of cultural significance to the Ōtamatea Marae, which it overlooked. Cows from the dairy unit were being wintered there; replacement stock were grown on it; and it produced silage and hay as replacement feed.

Hanerau Farms were also developing small blocks of adjoining land on behalf of original owners, where those blocks were too small to be managed and developed individually.

Hanerau had substantially regressed its 235-hectare Taipuha dairy unit in recent years. Carbon emissions had been calculated, and an extensive seven-year programme of fencing and riparian planting of native trees was in the second year of operation. Soil type in the unit was largely heavy clay, requiring careful management to avoid damage. A 176-effective-hectare dry-stock block owned by the company but run as a separate business unit was used for grazing replacements and some wintering of dry cows.

Outside directors had been appointed to the trust for the particular expertise they brought. A five-year development plan was in place that emphasised procurement and development of Māori land, training and employment opportunities, and asset growth through capital plant and buildings.

Field day at Hanerau Farms Ltd, finalist in the Ahuwhenua Trophy 2010 Dairy Competition
(John Cowpland – Alphapix-FD-057)

Hāngi preparations for the field day at Rangatira 8A 17 Trust, finalist in the Ahuwhenua Trophy 2010 Dairy Competition
(John Cowpland – Alphapix-FD-176)

The Ahuwhenua judges commended Hanerau Farms for their management approaches, which showed understanding of and a passion for the maintenance of healthy and well-cared-for animals. They were also impressed with the strong corporate governance model, and with the trust's use of formal and informal training to upskill its people. They applauded the trust's genetics programme as innovative and well integrated. A clear and demonstrable understanding of pasture production was evident, they said; seasonal supply and seasonal pasture use were highly commended.[224]

Rangatira 8A 17 Trust

The Rangatira 8A 17 Trust was governed by a board of four chaired by Ms Diane Stockman. A staff of three managed the farm on a day-to-day basis.

The farm was stocked with 530 Friesian/Friesian-cross cows and 127 yearling heifer replacements. This was a crucial element of the farm's business strategy, which had seen the Rangatira 8A 17 Trust assets develop from a derelict farm into a multimillion-dollar enterprise for the benefit of its 1500 owners from the Rauhoto hapū of Ngāti Tūwharetoa.

The Rangatira 8A 17 Trust business story had begun in 1981, when the trust had been vested with 27 acres of hapū land at Acacia Bay Road, near Taupō. The land had been poorly maintained, and for many years had been an unofficial rubbish dump. In the following years, the trust developed the hapū assets through a programme of subdivisions and reinvestment. The trustees currently administered three land blocks within the Taupō township, kaumātua flats, a forestry block, a shopping centre and other commercial properties, and intended to purchase more land, especially farms, where it could grow the business. The trust was not looking for huge returns on capital invested, but wanted cash flows sufficient to cover debt repayment, interest and running costs so that it could invest more in the future for its people.

The Rangatira 8A 17 Trust Farm had been purchased at auction in 2002 and had a well-developed infrastructure, including a forty-a-side herringbone shed and eighty-one paddocks with a good water supply. Tracks and access to all paddocks were maintained to a high standard, with pumice material readily available from the farm quarry.

Although milk solids figures had dipped from 182,500 kg to 174,082 kg over the last two years due to low milk prices, severe weather conditions and high fertiliser costs, the farm's short-term milk solids target was 190,000 rising to 200,000 kg over five years. These projections were to a large degree weather-dependent; other restrictions included low fertility, the quality of Taupō's pumice soil and the availability of water for irrigation. The farm was fully developed; of its 221 hectares, 197 were effective, 14 were in buffer zone reserves and 10 were utilised as roads, races and buildings. The farm was 'typical volcanic plateau country', but rose to 350 metres above sea level, with stony silty ash and pumice soil.

The Rangatira 8A 17 Trust had its own irrigation system covering 170 hectares, and had obtained resource consent until 2017 to take up to 9000 cubic metres of water daily from the Torepatutahi stream, which flowed through the farm. It had been one of only two farms required to hold this water in an irrigation pond. Water for the farmhouse and the cowshed was filtered and ultraviolet treated.

The Ahuwhenua judges commended the farm's strong financial accounting, which addressed all their criteria, noting that the farm had made a good farming profit with a lower payout. They applauded the farm's record of 'solid financial performance with a single focus upon debt reduction'. Livestock was in good condition, they observed, with good separation of herds to assist the management of younger cows and in general a pasture system that worked very well. The cropping programme, rates of pasture renewal, effluent spreading and fertiliser application also seemed well balanced and integrated.[225]

2011

Applauding the Judges

Welcoming participants to the 2011 Ahuwhenua field day programme, chairman of the Management Committee Mr Kingi Smiler applauded the 'wonderful contribution' made by the judges of Ahuwhenua competitions over the many years of the trophy's existence. Each year, they had given their time and expertise not only to judge but also to inform and encourage Ahuwhenua entrants. The Ahuwhenua competition had always been fortunate in attracting some of the sector's most knowledgeable and experienced operators, he said, observing that judges were known for placing competitors under rigorous scrutiny. As a consequence, Ahuwhenua was considered to be the most comprehensively judged competition of any farming competition held in Aotearoa.[226]

Mr John Acland, who had been chief judge of the sheep and beef competition, was retiring this year, and was particularly singled out by Mr Smiler as having made 'an enormous contribution to the status that the Ahuwhenua Trophy has achieved today as the premier accolade for Māori farming success'. As chair of the Meat Board in the early nineties, Mr Acland had recognised the growing contribution and potential of Māori farming to the economy of New Zealand, Mr Smiler noted. Accordingly, he had met with Māori to explore ways in which this potential could more quickly be unleashed. He had attended hui and had heard Māori calling for the reinstatement of the Ahuwhenua Trophy competition to highlight excellence in Māori farming. He had taken this feedback to the Meat Board and had gained their support for the revival of Ahuwhenua in 2002–2003.[227]

Top: John Acland, Chief Judge Sheep and Beef Competition 2003–2009
Bottom: Doug Leeder, Chief Judge Dairy Competition 2006–2012
(Photos courtesy of Lyn Harrison)

Regional Finalists and the National Winner

In March 2011, the three Ahuwhenua finalists were announced, in a competition that this year focused on sheep and beef: Ōtakanini Tōpū Incorporation of Helensville; Waipapa 9 Trust of Taupō; and Pākihiroa Station of Tapuaeroa Valley. On 3 June 2011, at an awards dinner held at the Energy Events Centre, Rotorua, Waipapa 9 Trust were announced as the winners.

Finalist Profiles

Ōtakanini Tōpū Incorporation

Ōtakanini Tōpū Incorporation was a 2750-hectare farm bordering Muriwai Beach on Northland's west coast. Its chairman was Mr Hemi Rau, who presided over a board comprising seven members. Ōtakanini Tōpū Incorporation had between 400 and 500 shareholders, who affiliated with local hapū of Ngāti Whātua associated with the five marae in the area: most from the adjoining Haranui Marae. Returns had been spasmodic rececently, but the focus remained upon the annual dividend, as well as on distributing profits within the local committee. The farm had a staff of four who managed day-to-day affairs, plus an apprentice farm cadet.

Ōtakanini Tōpū ran stock on 1550 hectares, as well as 600 hectares of forestry and 600 hectares of unproductive mangroves and mudflats. The property had previously been administered by Crown Forestry for about thirty years and had only recently been returned to shareholders. The next forestry rotation was due in six to eight years.

Since its formation in 1951 from the amalgamation of a number of small Māori farms and the return of government-administered land, Ōtakanini Tōpū had mapped and maintained the many cultural sites within its area, including wāhi tapu, urupā and old pā sites. Its administrators had worked hard to ensure that the farm's miles of fencing and new water systems did not cut through such areas. Increased efforts were also being made to ensure that sediments did not flow into waterways and into the harbour. The farm had high hills that, when exposed to strong westerlies, caused erosion and sand-blows. Drought, kikuyu grass and coastal vegetation were local challenges.

Field day at Ōtakanini Tōpū Incorporation, finalist, Sheep and Beef Competition 2011
(AHU-OTAK-FD-0926)

Ōtakanini Tōpū Incorporation had recently rearranged its infrastructure after a much-valued member of staff, Heta Tamahori, retired after thirty-four years of service. It had appointed a new manager experienced in working with Māori incorporations, and initiated a development phase comprising subdivisions, a fishing operation and a new farming regime emphasising environmental care and the introduction of new genetics.

The contour consisted of flat to rolling land with some steeper slopes; the soil type was pinaki and loamy sand, with some peat on the eastern side. Ōtakanini Tōpū wintered 16,000 stock units: 8500 sheep and 7500 cattle. The core business of the incorporation was sheep and beef farming, which had largely been the case since 1951.[228]

The Ahuwhenua judges commended the Ōtakanini Tōpū Incorporation committee of management as an 'enthusiastic group' with a good mix of skills. Clear lines of communication between levels of authority were evident, they said. Stock health and conditions were very good: ewes in particular were in a good condition. The judges also commended the Ōtakanini Tōpū Incorporation for initiating, in collaboration with the Auckland Regional Council, a sustainable land and environment plan that incorporated land management units and land use capabilities.[229]

Waipapa 9 Trust

The Waipapa 9 Trust had won the competition in the previous year for its dairy farming. In 2011, it had 1208 owners and was governed by a board of six, with a staff of seven to manage day-to-day operations.

The trust's Waipapa Station was a sheep and beef property incorporating a terminal breeding and finishing operation, a bull beef finishing operation and winter dairy support. It was located on easy to medium hill country, with some steep country.

The total area of the farm was 1534 hectares, with an effective grazing area of 1100 hectares. There were 274 paddocks, and water supply was by reticulation over the whole property, being part community scheme and part internal farm supply. Pasture comprised mainly ryegrass and clover, but some significant areas of brown-top-dominant pasture remained. There were 5860 ewes, 591 hoggets, 78 rams and 1207 cattle in July 2010. Highlander composite ewes were managed as a terminal flock, mated to terminal Primera rams. Replacement ewes were purchased from the Otanepae breeding unit; no hoggets were reared on site. Independent veterinary professionals reviewed an animal health plan for all stock annually.

Kingi Smiler, Chairman, Ahuwhenua Trophy Management Committee (AHU12-257)

Waipapa's environmental goals were based upon striving to have 'minimal impact on the environment'; this was managed through increased awareness, strategic investment and sustainable management practices. Environmental policies included protection of wetlands, new lined effluent ponds, annual soil testing and nutrient budgeting and retirement of farm land (300 hectares) as part of the Lake Taupō protection programme. Social, community and ngā tikanga Māori goals and strategies were important to the trust, which funded education grants, kaumātua grants, marae maintenance and participation in the Āwhina Group (a collaborative alliance with a number of other Māori economic authorities) for the Volcanic Plateau.

The Ahuwhenua judges commended the Waipapa 9 Trust for its well-organised and innovative approach to governance, noting that trustees were allocated specialist tasks and that high productivity was evident; efforts were being made to ensure that the best genetics were in place, and stock was performing well and was in good condition after a number of difficult years. A long-term stable workforce of staff and advisors was also an impressive feature of this operation. Careful planting to prevent soil erosion in particular was noted.

Pakihiroa Station

Pakihiroa Station was described as a rugged 3140-hectare property of medium to steep hill country on the East Coast. It was held in trust for Ngāti Porou, whom it represented through whakapapa, by Te Rūnanga o Ngāti Porou. The property was administered by a board of seven directors (three representing the shareholders and four independent appointees), chaired by Mr Selwyn Parata. The farm employed a local shearing gang and when necessary brought on contractors for fencing and capital works. The station had a staff of six who managed day-to-day operations.

The purchase of Pakihiroa by Te Rūnanga o Ngāti Porou occurred in 1991, thereby guaranteeing all Ngāti Porou tūrangawaewae within their tribal homeland. The purchase also returned the sacred mountain Hikurangi to the people, as well as the resting place for Nukutaimemeha, Māui's waka.[230]

Pakihiroa was farmed by a subsidiary company of the Rūnanga, Pakihiroa Farms Ltd. The farming company had a long-term vision: to develop a commercial focus, 'leaving its shareholder to deal with social issues'.

The property ran 10,700 stock units on 1256 effective hectares. It also had 325 hectares of pine

Pōwhiri for Pakihiroa Field Day, Sheep and Beef Competition finalist 2011
(AHU-Paki-FD 02491)

and eucalyptus; the balance of the property included Hikurangi maunga and part of the Raukūmara Range. Pakihiroa was a lynchpin in a Ngāti Porou economic aim to supply high-quality food to consumers based on the best products sourced from 'our seas and lands'.

Pakihiroa had a finishing farm, Puanga, located 10 kilometres north of Gisborne, where stock was finished for processing.

Pakihiroa had a good reliable summer rainfall, without the challenge of toxins that affected lower altitude farms. As a result, the farm benefitted from a good summer pasture production well suited to sheep and cattle breeding, with some finishing of lambs as the season permitted. Other challenges did exist, such as a high erosion risk, shallow topsoils, occasional snowfalls and relative isolation.

Pakihiroa had been a stand-alone company since 2007, resulting in much better lines of responsibility and communication between all interested parties.

The Ahuwhenua judges commended the Pakihiroa Station board for its 'excellent mix of youthful energy and wise counsel to govern this operation'. Also impressing the judges were appropriate stocking rates, which provided a good platform for concentrating on per-head performance, and the adoption of a land environmental plan focusing on significant investment in land stabilisation: low-input farming was seen as 'appropriate for this tough piece of country.'[231]

2012

Lord Bledisloe's Warning

On 10 July 1939, former Governor-General Lord Bledisloe wrote from his Lydney Park Estate in Gloucestershire to the New Zealand Prime Minister, M J Savage, thanking him for having sent a copy of the judge's report for the 1939 Ahuwhenua competition. 'I am very glad to learn', he wrote, 'that this annual competition, which I inaugurated and endowed seven years ago, is helping to encourage more enlightened and profitable methods of farming among the Maori people.' Lord Bledisloe was particularly glad to note the emphasis placed by the judge on herd testing 'in regard to dairy cattle and the importance of selecting bulls with a good milk record behind them'.

In this letter, Lord Bledisloe also issued a warning: 'New Zealand Dairy farming will never achieve stability as long as the price of milk and the value of dairy cattle are based upon butter-fat only and

The Governor-General His Excellency Lieutenant General The Rt Hon Sir Jerry Mateparae and Roku Mihinui, Chairman of Kapenga M Trust, winner of the Ahuwhenua Trophy Dairy Competition 2012
(AHU-Awards-337)

not the other milk solids (casein and sugar).' Such an emphasis placed undue premium upon Jersey cattle, he said, 'and pulls down severely the value of New Zealand cheese which (under your climate and pasture conditions) ought to be the best in the world'. Farmers, he insisted, should be remunerated for all the valuable ingredients in their milk and not butterfat only.

Lord Bledisloe concluded by thanking Prime Minister Savage; 'the Maori race deserves all the help and encouragement you are giving them', he said.[232]

Regional Finalists and the National Winner

In March 2012, the three Ahuwhenua regional finalists were announced: Kapenga M Trust of Rotorua; Tauhara Moana Trust of Taupō; and Waewaetutuki 10 – Wharepi Whanau Trust of Te Puke. On 8 June 2012, at an awards dinner held at the Skycity Convention Centre, Auckland, Kapenga M Trust was announced as winner.

Kapenga M Trust cows feeding
(Ahuwhenua Archives)

Finalist Profiles

Kapenga M Trust

The Kapenga M Trust was chaired by Mr Roku Mihinui, with a board comprising seven members; a staff of six managed day-to-day affairs. Its 915 shareholders were of Tūhourangi descent. It had won the trophy in 2003.

The Trust comprised a 330-hectare property running a mixed Jersey and Friesian herd of 1020 stock units; its strategy centred on developing high-breeding-worth cows that could harvest high levels of pasture.

Between 2008 and 2009, the Kapenga M Trust farm's total milk production had increased from 241,441 kg to 371,169 kg, although the farm's herd had only increased by nine cows. Production per cow had gone from 246 kg to 372 kg.

The trust owned a total of 1858 hectares, which included a sheep and beef farm and a dairy farm. Two sharemilkers had been in place for the previous four years, operating under a 50:50 sharemilking arrangement that provided incentives for the sharemilkers to invest in the farm and its herd. This strategy had been positive for both of the parties concerned, although it was noted that the trust usually tried to employ staff of Tūhourangi descent, where it could identify or train relevant expertise.

The trust was also currently harvesting 100 hectares of forestry woods that had been planted twenty-eight years earlier. Proceeds from this had encouraged owners to purchase lands with ancestral connections to the people as they became available. In 2008, the trust sold one of it farms located 60 kilometres away and purchased a 250-hectare dairy farm adjoining its Waikite Valley Holdings.

Historically, the area had always been known as Tūhourangi's 'food bowl', due to its abundant supply of birds and other wild foods, and later its network of gardens. Kapenga M was farming land with which it had a long whakapapa connection. Tūhourangi had historically settled inland from Maketū; more especially since the eruption of Mt Tarawera in 1886.

The key to the success of Kapenga M's operation today was the enforcement of mana whenua ownership and guardianship of the land.

The Ahuwhenua judges were impressed with the Kapenga M Trust's clear understanding of the role of governance. They saw evidence of a clear operational plan and a strong ethos of introducing new ideas, noting that plans were in place to introduce 'trainee trustees'. Milk production was high at 977 kg per hectare, especially considering the difficult farm contours and altitude. The judges commended stocking rates and pasture management; cow genetics, considered important, remained a strong focus area. They approved of high soil fertility levels and a good infrastructure overall. The quality of the housing on site was excellent, they noted.[233]

Tauhara Moana Trust

The Tauhara Moana Trust was chaired by Mr Toby Rameka, with a board comprising seven members. It was tribally affiliated to Ngāti Tūwharetoa and Te Arawa and had 2037 owners. A full-time staff of thirteen managed day-to-day operations.

The Tauhara Moana Trust comprised 715 hectares of ancestral land (the Tauhara North 3B Block) and 363 hectares of land purchased in 2004 (the Parariki Block). It had also recently secured a long-term lease for a further 952 hectares of neighbouring lands.

The original Tauhara Moana Trust development had been a sheep and beef unit managed by the Department of Survey and Land Information. In 1984, the trust took back full responsibility for the management and operation of the business, and continued with the sheep and beef operation. In 2006, the trust decided to move into dairying. It developed a 500-cow dairy unit and, at the same time, arranged for the piped irrigation of approximately 55 hectares of land from a nearby factory. Originally established as a sharemilking unit, this dairying initiative did not succeed; costs always exceeded income. Consequently, the farm was leased by a farming group that unfortunately ended up in receivership. The trustees negotiated with the receiver to take back the farm in March 2010.

The trust had a fifty-five-a-side herringbone shed with a rotating breast rail that enabled the expansion of its dairy operation from 500 cows to 1700 cows. In the 2011/12 season, 1600 dairy cows were milked: this figure was expected to increase to 3000 by 2015. The trust had a strong environmental focus, and had over time retired farmlands, protected waterways and wetlands and invested in low effluent systems. The trust was originally a supporter of Fonterra, but in 2011, opted out to become a foundation supplier of Miraka Ltd, the first Māori-owned dairy processing plant.

The Ahuwhenua judges commended the Tauhara Moana Trust for their thorough system analysis and governance actions, which were in accord with the wishes of their people. They noted that the trust used advisory trustees as a means of 'risk strategy'.

Field day at the Tauhara Moana Trust, finalist in the Dairy Competition 2012
(Ahuwhenua Archives)

The trust had managed a significant amount of growth well over the past few years; specialised committees had addressed specific issues. Tauhara Moana was a regional focus farm, which meant intense scrutiny and analysis of its operations. Reproductive performance in the herd was improving significantly, and there was strong rural professional support available to a strongly performing management team.[234]

Waewaetutuki 10

Waewaetutuki 10 was a 71-hectare dairy unit owned by the Wharepi Whanau Trust, chaired by Rehua Smallman and comprising seven trustees.

The Wharepi Whanau Trust was established in 1996 to retain and farm ancestral land. The trust administered the land for its fifty-nine shareholders, all of whom had links to Ngati Pukenga. As all of the owners descended from one of the five children of the tupuna Wharepi, it was decided that the governance board should represent all of these five children. Between 1996 and 1999, this decision was formalised through the Maori Land Court and was reflected in the current makeup of the trustees.

The farm was made up of two substantial blocks, which were leased before being amalgamated in 2000 to form Waewaetutuki 10. At that stage, the trust entered into a ten-year lease on the understanding that in time the trust would take over the running of the farm. This was achieved in 2009, and the trust was then able to purchase cows, plant, machinery and Fonterra shares. The Maori Trustee assisted and still currently helped with the management of the farm.

Waewaetutuki 10 had thus recently completed its third year of operation; in that time its herd of Friesian milking cows had increased from 178 to 200. Milk production had also increased from 45,000 kg of milk solids in 2008 to an expected 72,000 kg in 2012.

In 2010, the Wharepi Whanau Trust leased a further 21 hectares to Waewaetutuki 10 to cater for increases in stock numbers. Waewaetutuki 10 had 60 acres in pasture, 6 hectares in maize production and 5 hectares in waterways. It had upgraded its effluent system and worked closely with the New Zealand Transport Agency to repair road

Finalists Waewaetutuki 10, Wharepi Whanau Trust Field Day 2012

(John Cowpland – Alphapix)

Tangaroa Walker, inaugural winner of the Ahuwhenua Young Māori Farmer of the Year Award 2012 for dairy farming

(AHU-Awards-Dinner-396)

frontages along the highway in order to prevent serious flooding.

The Ahuwhenua judges commended Waewaetutuki 10 on team work observed at all levels. The trust's business plan was clear and well summarised, they said, and its 'no frills' approach to farming seemed to have worked very well. Soil fertility was high, and there was evidence of a dramatic increase in productivity. Waterways were well fenced, and water quality had improved significantly.

2013

Eighty Years of Ahuwhenua

By 2013, the Ahuwhenua competition had been in existence for eighty years, during which time the trophy had been contested fifty-four times. It was ten years since the trophy had been successfully relaunched in 2003, 'reflecting the changes made in the Maori Agribusiness sector and the increasing importance of Maori incorporations and Trusts'.

The current Ahuwhenua trustees are the Minister of Māori Affairs, the Minister of Primary Industries and the chief executive of Te Puni Kōkiri (the equivalent three roles were signed up by Ngata in 1933: the Minister of Native Affairs, the Minister of Agriculture and the Under Secretary of Native Affairs). The present trustees have delegated the management of the competition to the Ahuwhenua Trophy Management Committee, which since 2007 has been chaired by Mr Kingi Smiler.

Recent changes to the competition have included the introduction of the Ahuwhenua Young Māori Farmer of the Year Award. This was first won by Tangaroa Walker (Ngāti Ranginui, Ngāti Pūkenga) in 2012. In June 2013, Jordan Smith from Kearins Brothers Ltd's farm in Te Kuiti was announced as the inaugural Ahuwhenua Young Māori Sheep and Beef Farmer of the Year.[235]

Regional Finalists and the National Winner

The three Ahuwhenua regional finalists in 2013 were Te Uranga B2 Incorporation – Upoko B2 of Taumarunui; Te Awahohonu Forest Trust – Tarawera Station of Te Hāroto, Hawke's Bay; and Te Hape B Trust – Te Hape Station of Te Kūiti. On 7 June 2013, at a dinner held at the Pettigrew Green Arena, Taradale, attended by over 850 people, Te Awahohonu Forest Trust – Tarawera Station was announced as the winner of the year's competition, which had focused on sheep and beef.

Finalist Profiles

Te Uranga B2 Incorporation – Upoko B2

Te Uranga B2 Incorporation – Upoko B2 was chaired by Ms Traci Houpapa, and comprised four members. It had 760 shareholders.

Established in 1910, Te Uranga B2 Incorporation had been trading in its own right since 1958. The Incorporation's sheep and beef unit – Upoko B2 – comprised 1123 hectares, and was one of four integrated business units, which included dairy and forestry operations. The property had been placed under Maori Affairs control in 1958 until 1981, when it was returned to its owners.

For the next ten years, owners focused on the retention of ancestral lands, along with the purchasing of neighbouring freehold blocks as they became available. They also aimed to protect regenerating forests, and had negotiated forestry planting of pine trees with Carter Holt Harvey. A programme for conversion to dairy by 2010 was also instigated. By 1999, Te Uranga B2 Incorporation was receiving incomes from milk, forestry, wool, sheep and beef cattle.[236]

The Upoko B2 unit comprised yellow brown earth, loam and pumice. It was currently stocking 140 Angus breeding cows, 170 trading steers, 200 heifers, 300 weaner cattle and 5500 mixed-age

Pōwhiri for field day at Te Uranga B2 Inc., finalist, Sheep and Beef Competition 2013
(John Cowpland – Alphapix- 2304131617)

Romney/Coopworth ewes. Grazing support was provided for 200 heifers and 400 winter cows. In March 2013, Te Uranga B2 won three Ballance Farm Environment Awards. It was now positioned as a significant regional economic partner, and was active in supporting local communal, social and sporting occasions. It also supported local schools, especially with forestry programmes, and made grants to marae and kaumātua for tangihanga and health and education purposes.

The Ahuwhenua judges commended Te Uranga B2 Incorporation for its strong leadership and its engagement with key professionals. The incorporation had a clearly established commitment to cultural growth and preservation, they said, as well as to local schools. The Upoko B2 unit was applauded for its consistent and monitored financial performance: targeted results were in the top 20 percent of Beef and Lamb New Zealand. Gross farm revenues had been good, reflecting improved feed conversion efficiency and improved productivity. The judges highly commended Te Uranga B2 Incorporation's environmental and sustainability goals.

Te Awahohonu Forest Trust – Tarawera Station

The Te Awahohonu Forest Trust – Tarawera Station was chaired by Mr Tamihana Nuku and comprised seven members; there were 1150 shareholders. It had been a regional winner of the Ahuwhenua competition in 2004, when no national winner was announced.

The Tarawera Station was described as 'a relatively minor portion' of the Te Awahohonu Forest Trust's 21,000-hectare land-based asset, but remained 'its oldest business'. The farm had been developed by Māori Affairs between 1965 and 1987, when it was returned to Ngāti Kahungunu and Ngāti Hineuru owners and the trust took control.[237]

In 2013, Tarawera Station ran nearly 30,000 stock units on 2865 effective hectares, including recently leased lands adjoining the main farm. The unit's highlander composite flock included 16,000 ewes and 5000 replacement ewe lambs. High lambing percentage records and carcass quality derived from good feeding and improved genetics. The Red Stabilizer beef herd comprised 1000 breeding cows plus heifer replacements and finishing steers and bulls. Calving was around 85 percent, with all surplus

Te Awahohonu Forest Trust Field Day, 2013
(John Cowpland – Alphapix-0205131313)

cattle finished prime. Fertiliser was applied annually across the whole farm. Feed budgeting ensured economic feed supply and allocations. Stock numbers were expected to be down this year because of the recent drought.

Tarawera Station maintained over 10,000 hectares in native forest, which was home to protected birds.

The Ahuwhenua judges commended the trust for their strong governance, focus on developing new trustees and excellent engagement with professionals. It also applauded its social, communal and ngā tikanga Māori policies; especially its support for wānanga to promote successful Māori role models. They described the trust's financial performance as 'consistent and excellent'; environmental and sustainability goals were equally impressive. The judges noted that trustee succession planning had been commenced and that staff had regularly entered competitions to benchmark the farm's ongoing performance.

Te Hape B Trust – Te Hape Station

The Te Hape B Trust was chaired by Mr Hardie Peni and comprised six members. It had 854 owners.

Te Hape Station comprised the largest farmed area by one entity in the Waitomo District. Located 500 metres above sea level, it contained the head waters of three major rivers. It had 1222 hectares of native bush under Nga Whenua Rahui covenant, as well as maturing pine trees.[238] The land comprised some significant sacred sites.

Originally surveyed in the 1890s into milling blocks, the land was taken over by Maori Affairs in the 1950s, eventually being handed back to owners in 1974 encumbered with debt. Since then, improvements to subdivisions, pasture management and genetics had elicited positive results.

As a breeding and finishing unit, Te Hape Station was the largest operation in the Trust's 10,000-hectare land portfolio. It ran 31,000 stock units, of which 62 percent were Perendale cross sheep and 38 percent were cattle, mainly Angus. Because winters on the station were rough, as many stock units as possible were finished in order to reduce capital stock numbers; cropping was used to improve live weight gains.

Ahuwhenua judges were impressed with the trust's plans to make the business operate efficiently and to allocate profits, describing this as a 'very professional approach and a true credit to the Trustees'. The trust's approach to social, community and ngā tikanga Māori goals was highly commended, as were its environmental and sustainability strategies. Relations between trustees and staff were observed to be very good.[239]

Pōwhiri at Te Miringa te Kakara marae, hosted by Te Hape B Trust, finalist, Sheep and Beef Competition 2013
(John Cowpland – Alphapix-0905130553)

Conclusion to Part One

'Administering the Policy Effectively'

I trust (we will) solve the real problem, how to administer the policy effectively so as to satisfy every business requirement and yet carry with it the enthusiastic goodwill of the human elements for whose benefits it has been conceived.[240]

Sir Apirana Ngata, 1934

The 'Human Element'

In early 1934, the two co-founders of the Ahuwhenua Trophy briefly exchanged letters. From his official quarters in Auckland, Governor-General Lord Bledisloe wrote to his colleague, Native Minister Sir Apirana Ngata, asking for advice on a range of issues affecting Māori. As a side note, he expressed concern over the future of Ngata's land development schemes, given recent controversies over funding and management. He wished Ngata well, expressing confidence in the policies.

In reply, Ngata expressed his gratitude to Lord Bledisloe and pointed to what he felt was the real issue – how to 'administer the policy effectively' – by which he meant how to ensure that all business imperatives were met while never losing sight of the 'human elements' for which the land development schemes were launched in the first place.

Sustainability

Almost eighty years later, Ahuwhenua Trophy Management Committee chairman Kingi Smiler touched on a similar theme when introducing the 2013 Ahuwhenua finalists: the issue of how to attain business performance measures while sustaining those human elements.

'We have three fantastic finalists for 2013,' he said, 'with all finalists performing extremely well under the criteria that have been set for the competition.' The governance and leadership of all three finalists was 'very strong'; each farm had met the dairying industry's financial benchmarks, thereby representing strong Māori exemplars for the whole of the industry in New Zealand to see.

However, beyond such measures, 'the role of each finalist as kaitiaki in terms of the environment was also extremely strong'. Each finalist represented major whānau, and their communication with shareholders was exceptional, as shareholders demonstrated with pride at every field day.[241]

Ms Tracy Houpapa, the chairperson of one of the 2013 finalists, Te Uranga B2 Incorporation, reiterated this view: 'If we win the Ahuwhenua competition, it will be an acknowledgement of the strategic visions of our tupuna,' she said. Success would also affirm the work of previous management committees and years of support from hapū and whānau. Irrespective of who won, it was perfectly clear, she said, that 'Māori primary industry had arrived'.[242]

Ngata's Vision
Apirana Ngata's vision for the land development schemes of the 1930s had its genesis in his earlier years, when his father, Paratene Ngata, was seeking to stem the loss of Māori lands while maximising their economic potential. He saw that the way forward was through land consolidations – especially of fragmented and multiply owned titles – and incorporations.

Even earlier, Wiremu Pere was also an advocate of land consolidations and expressed concerns about the rate of land dispossession. At a meeting at Waipatu in 1898, Pere had addressed Premier Richard Seddon, saying, 'I have not many words to say by way of introduction. You are perfectly aware of the matters I shall place before you.' In recent years, he said, Pākehā had acquired 60 million acres of Māori land, leaving Māori with perhaps only 4 million acres. Māori were now asking that all land sales be stopped, he said. All land remaining to Māori should be 'absolutely reserved to Maori, forever and ever, until the end of time'. Land purchase agents also needed restraining, he thought, because they were coming among unsuspecting Māori, 'throwing corn for the fowls to eat'.[243]

Modern-day Ahuwhenua: Winning the Competition in 2005
Maintaining the land and consolidating its economic potential in order to sustain coming generations have remained critical tenets of Māori communities since the days of Ngata and Pere. Speaking of the success of Te Pou a Kani Farms in winning the Ahuwhenua Trophy in 2005, Kingi Smiler said that the 'biggest buzz' had been the pride of shareholders themselves at the farm's success. The success had given the incorporation's governance teams the confidence to continuing looking for 'new things that needed to happen', he said.

After winning the competition, Te Pou a Kani Farms went out of sheep farming and concentrated on establishing a viable dairy milking platform: 'In 2005, we were doing approximately 2.2 million kg of milk solids, whereas in 2012, we are now doing 4.5 million kg of milk solids, with our cow herds increasing from 7000 cows to 10,000.' It diversified its outputs potential by investing in Miraka Ltd, in order to 'export our product direct to international markets'.

Such developments had been an important part of growing and sustaining the economic potential of the farm, thereby 'satisfying the business requirements' of the incorporation, Kingi Smiler noted. However, 'the developments I've been talking about are there for one purpose – to sustain future generations, especially our mokopuna'.[244]

2006: A 'Massive Achievement'
Winning the Ahuwhenua Trophy in 2006 was a 'massive achievement' for Parininihi ki Waitōtara, according to present chairman Dion Tuuta. It represented the culmination of ten years of hard work consolidating the business: 'Even today, seven years after our success, we often reflect on the fact that we won the trophy.' Since winning the focus of the board had been very clear: 'PKW must be more active in the management of our properties.'

Since their win, Parininihi ki Waitōtara had moved towards a more diversified management structure: 'so, for example, instead of one hundred percent 50/50 sharemilking, we now have a range of managed farms where we employ staff directly ourselves'. The land was the basis of 'who we are', Tuuta said; 'Taranaki Maori were anchored into this land by their whakapapa.' The land was not a commodity to be bought and sold. The ultimate aim of Parininihi ki Waitōtara was 'to look after the land sustainably for our future generations'.[245]

2007: Pride of the Shareholders
Mr Dana Blackburn, Director of the Ātihau-Whanganui Incorporation, which farmed Pah Hill Station, said that their winning the trophy in 2007 had met with an 'incredible response and celebrations'; among the most significant of the benefits it had brought was pride among shareholders, owners and staff. In 2007, Pah Hill Station was one segment of a much larger

'The land is the basis of who we are,' said Dion Tuuta, chairman of Parininihi ki Waitōtara, winners in 2006. (Mount Taranaki and landscape viewed from Eltham, 2013; photograph by Bryn Thomas)

organisation, which had now merged into 'an integrated, one farm system instead of each operating independently – we all now work more closely together for the common cause'.

To those Māori farmers thinking of entering the competition, Mr Blackburn said 'I would recommend it, not only for the outcomes you can achieve but for the advantages of just participating, which were immense'.[246]

2008: A Reward for Years of Effort

'Winning the Ahuwhenua Trophy for us was the culmination of twenty years of hard work running the family farm together,' said Dean Nikora, who, with his wife Kristen, had won the trophy in dairying in 2008. Judging had been intense, he said, 'but that was what we liked about the competition'. The Ahuwhenua competition criteria were stringent, but Dean Nikora said that putting the business before such scrutiny was ultimately rewarding for the sake of the feedback, which had been greatly beneficial. He appreciated the independence and expertise of the judges.

2009: Fantastic 'For All of Māoridom'

For the Pakarae Whangara B5 Partnership, winning the trophy in 2009 had been 'absolutely fantastic, not only for us as the holders of this land, but for all of Māoridom,' said Ingrid Collins, chairperson of the partnership. Since winning, Pakarae Whangara B5 Partnership had instituted two major changes: the purchase of a new farm for providing alternative grazing in summer and reticulating the entire operation with 'good clean water for our stock to drink all year round'. Winning the trophy had helped the partnership in obtaining additional finance – 'winning the competition constituted an expression of confidence in our governance and management, which the banks took note of'.

Kahurangi performers, Ahuwhenua Trophy awards dinner 2013
(John Cowpland – Alphapix-0706133205)

2010 and 2011: 'Still a Buzz'

The thrill in winning back-to-back trophies 'was still a buzz now', according to Anthony Tohu Haa, chairman of the Waipapa 9 Trust, who won the trophy twice – for dairy in 2010 and for sheep and beef in 2011. Three or four years down the track, he said, the cup still held the prestige that it did when it was won: 'Winning the trophy opened up new horizons for us to secure and grow our business, and this we have managed to do.'

2012: All About Looking Ahead

Winning was 'all about looking ahead', said Roku Mihinui, chairman of the Kapenga M Trust, which also won the trophy twice – for sheep in 2003 and for dairy in 2012. 'The biggest thing is to keep looking ahead ... you must refocus on the land and its potential.' As a result, for example, Kapenga M Trust now had an intense and continuous environmental sustainability programme, 'and much of what we do, we had learned from other participants'. Participating in Ahuwhenua had been a 'huge plus' for the trust, he said. To those who were contemplating entering the competition, Mr Mihinui's advice was simple – 'go for it!'[247]

2014–2021

Part Two

Ahuwhenua Trophy
Competitions Continue

(John Cowpland – Alphapix)

Introduction

Growing the Trophy – 2014–2021

Celebrating Eighty Years

In 2013, *Ahuwhenua: Celebrating 80 Years of Māori Farming* was published by Huia Publishers and the Ahuwhenua Trophy Management Committee. In 2014, *Ahuwhenua* was also a finalist in the Ngā Kupu Ōra Māori Book Awards presented by Massey University, attesting to the significance and relevance of the inspiring Ahuwhenua Trophy story.

Ahuwhenua: Celebrating 80 Years of Māori Farming coincided with the commemoration of eighty years of the Ahuwhenua Trophy competitions. The book was launched by Hon Dr Pita Sharples, Minister of Māori Development, and Mr Kingi Smiler, Chairman of the Ahuwhenua Trophy Management Committee on 3 October 2013 at the Annual Conference of the Federation of Māori Authorities in Hastings.

Ahuwhenua examined the development and expansion of the Ahuwhenua Trophy competition from its inception in 1933. Earlier, during the 1920s, Sir Apirana Ngata (Member of Parliament, Eastern Māori) had presided over the Native Lands Consolidation Commission where he had consistently argued for greater state financing and support for Māori land developments. At that time, Māori communal land holdings had egregiously declined, with a bleak future looming for pā and papa kāinga.

However, once appointed as Native Minister on 10 December 1928, Ngata was able to introduce the Native Land Amendment and Native Claims Adjustment Act 1929, which offered extensive state assistance for Māori to open up and occupy their land holdings scattered about the native districts.[248]

Increasing numbers of Māori men and women – many quite inexperienced but no less determined – were soon moving back into these isolated areas, working the land under the most difficult circumstances, enduring penury, sacrifice and loneliness, contributing to the slow yet inexorable process of advancing Māori economic security. They were also restoring and maintaining the ancestral lands upon which tūpuna once trod.

The long process of developing such land-based economic security for Māori was but one of the innumerable aspects of the story of the prestigious Ahuwhenua Trophy. When Sir Apirana Ngata and Governor-General Lord Bledisloe launched the first Ahuwhenua competition in Rotorua in 1933, Māori who wished to farm the land were faced with a prodigious set of obstacles.

Guided by Ngata's single-minded determination, and Lord Bledisloe's wish that Māori be supported in these endeavours, the Ahuwhenua Trophy emerged as a significant benchmark to which Māori could aspire, in a 'spirit of friendly rivalry', as they set about clearing, burning, ploughing, planting, watering, fertilising and grassing their challenging lands in pursuit of economic, social and cultural sustainability.

A New Decade

Ahuwhenua: Celebrating 80 Years of Māori Farming described the Ahuwhenua years from 1933–2013, through the many challenges faced by Māori farmers, whānau and administrators alike.

This new section, *Ahuwhenua: Trophy Competitions Continue 2014–2021* carries on this historical analysis by describing the Ahuwhenua Trophy competitions from 2014–2021. Between these years, the trophy progressed from strength to strength, emphasising what was admirable and most sustaining about the Māori agricultural industry whilst it adapted to rapid and, at times, unforseen change. In 2012, an Ahuwhenua Award for a Young Māori Farmer of the Year was also introduced, since then attracting young Māori contestants of exceptional quality.

Originally limited to the categories of dairy, and sheep and beef, the Ahuwhenua Trophy was extended in 2018 to include horticulture, with the inaugural Ahuwhenua Trophy for Horticulture presented in 2020. This represented a significant expansion of the Ahuwhenua competition 'and not before time', said the Chairman of the Ahuwhenua Trophy Management Committee, Mr Kingi Smiler.

Mr Smiler said he was delighted that horticulture had joined the Ahuwhenua whānau through the staging of an award especially dedicated to recognising Māori excellence in horticulture. Māori had always been proficient horticulturalists with many demonstrating great skill and innovation. It was therefore fitting that, whilst still embracing the vision and values of the trophy's two founders, Sir Apirana Ngata and

Horticulture Ahuwhenua Trophy Haumiatiketike introduced in 2020
(John Cowpland – Alphapix)

Ahuwhenua Horticulture Trophy winners 2020, Te Kaha 15B Hineora Orchard. Chairman Norm Carter holds the trophy, with Governor-General The Rt Hon Dame Patsy Reddy (third from right) and Hon Damien O'Connor, Minister of Agriculture (right).

(John Cowpland – Alphapix)

Lord Bledisloe, Māori growers and horticulturalists now had the opportunity to be publically recognised for their efforts, in the same way as Māori in the sheep and beef, and dairy industries, he said.[249]

Ahuwhenua: Trophy Competitions Continue opens with three brief historical chapters. The first chapter (Chapter Eight) describes what might be called the 'origins of modern Māori horticulture', between the years 1845 and 1860 as Māori cultivators and growers came to terms with rapidly changing agricultural and economic circumstances. The second chapter (Chapter Nine) accounts for the significant land dispossessions that egregiously affected Māori between 1858 and 1912.

The third chapter (Chapter Ten) moves forward to 1933, briefly revisiting the first thirty years of the Ahuwhenua Trophy competition, highlighting some of the innumerable challenges and immense difficulties faced by those earliest Ahuwhenua pioneers as, in the spirit of Sir Apirana Ngata's vision, Māori struggled to recover and re-establish their economic, social and cultural foothold on the land.

That they suceeded is evident in the strength and vitality of the Ahuwhenua Trophy today. The following chapter (Chapter Eleven) recounts the eight years of continuing Ahuwhenua Trophy competitions between 2014 and 2021. A final chapter (Chapter Twelve) describes a new award for young Māori farmers, which was introduced in 2012 to celebrate exceptional young Māori working in the agricultural sector. In 2020, Ahuwhenua recognition was extended to include a Young Māori Grower of the Year.

As Mr Smiler observed at that time, the exceptional calibre of all Ahuwheua finalists continued to demonstrate the strength of the Māori agribusiness sector and, through the Ahuwhenua Trophy, Māori were able to develop their attributes further by showcasing the quality of their farming enterprises.[250]

(John Cowpland – Alphapix)

Chapter Eight

Origins of Modern Māori Horticulture – 1840–1860

Establishing Māori Farming

When the Ahuwhenua Trophy competition was launched in 1933, its purpose was to encourage 'friendly rivalry' amongst interested Māori farmers, of which there were many, the first competition attracting over seventy entries.

From 1933 to 1953, Māori farmers were adjudged with all of their livestock included in the mix. At that time, most Māori farms comprised varying combinations of sheep, cattle, dairy and pigs, with some goats and poultry seen on occasion. In 1954, however, at the suggestion of judges, the Ahuwhenua competition was divided into two specific categories: dairy, and sheep and cattle. These two new categories recognised how Māori farmers were increasingly focusing on dairy; or sheep and cattle.

Thereafter, both categories were adjudged every year, until 2005 when, with diversification between the two ever increasing, it was decided that the two categories would alternate each year. The focus for 2005 would be sheep and beef, with dairy the following year. And, in 2020, a third competition category was introduced, that of horticulture, with all three categories each to be adjudged in three-year alternate cycles.

In Chapter One of this book, a number of notable features relating to the rise of Māori sheep and cattle farming are discussed. These features include the growth of Māori sheep and cattle farming in Hawke's Bay and the East Coast after 1850, contributing to the growth of Napier as the hub of an ever-expanding national sheep industry.

Chapter One also discusses Māori participation in dairying, possibly dating from 1908 when Kurupu Tareha purchased a dairy farm – a first for Māori – near Taradale. This is not to discount the commendable efforts of Rāwiri Taiwhanga who in 1826 was making butter for sale to Bay of Islands merchants, possibly becoming New Zealand's first commercial dairy farmer.[251]

However, despite the promising beginings, Māori agriculture declined rapidly after 1860 because of land loss, war and consequent economic uncertainty, lasting at least to the land developments era of the 1920s and 1930s.

With dairy, and sheep and cattle, canvassed in Chapter One, the purpose of this chapter is to briefly discuss early Māori participation in the third Ahuwhenua category, that of horticulture.

Modern Māori Growing and Horticulture

Across the nineteenth century, Māori constantly sought government assistance and finance to develop their extensive land holdings. As James Carroll (Member of Parliament, Eastern Māori) argued in 1891, Māori were turning to sheep farming and stock raising, wishing to become profitable producers like their Pākehā neighbours. Māori desired 'to become useful settlers and contribute to the wealth of the country'. Māori should therefore be 'afforded facilities for rendering productive the lands they already possess'.[252]

As James Carroll often emphasised, Māori had once thrived on the land, sustaining a trading economy, providing produce and supplies for fledgling Pākehā communities and, given the chance, Māori would do so again. Carroll was referring to the earliest years of Pākehā settlement between 1840 and 1860 when Māori horticulturalists and growers had developed and expanded their endeavours to attain a position of almost total market dominance.

The years from 1840–1860 have been described as 'the heyday of Maori agriculture' with 'large quantities of Māori grown produce [playing] a significant part in feeding the European population'. In parts of the country, especially in the Auckland province, Māori produce was available in such quantities as to generate significant revenues.[253] By their success, Māori growers and farmers were also able to contribute substantial sums to local provincial economies, in Auckland and elsewhere.[254]

Across such a short time-span, Māori growers embraced new seeds, implements, technologies and (not least) market conditions as they sought to emulate the economic success of Pākehā farmers. The years between 1840 and 1860 were notable, then, because this was the period when Māori horticulturalists learned their trade and made their mark, managing their cultivations, conveying their produce to market, somewhat undermining Pākehā presumptions of supply and demand, and, not least, coming to grips with the challenges posed by the evolving Pākehā economy.

Māori Learning the Trade

After the 1850s, the Auckland province contained almost one-half of the Māori population of New Zealand. This included Northland, the Waikato and the Thames-Coromandel areas, which together comprised approximately 42 percent, or 24,000 Māori, out of the total Māori population of approximately 59,000 people in New Zealand at that time. Most of these Auckland Māori – about 20 percent – lived in the Waikato.[255] The Pākehā population of Auckland in 1860 has been estimated as approximately 23,732, far exceeding the other major Pākehā population centres of Wellington (13,837), Nelson-Marlborough (12,000), Canterbury (15,370) and Otago (12,691).[256]

Partly because of these population numbers, advances in Māori horticulture in Auckland exceeded those experienced in other native districts across the country, especially since Auckland included the fertile valleys of the Waikato where horticulture flourished. Hundreds of cultivations could be seen lining the banks of the Waikato River, especially in areas where Māori had managed to retain legal title.

Throughout New Zealand, by the 1850s, some of the more traditional Māori forms of land utilisation had changed. But other forms of customary land use had survived, consistent with Māori retaining other customary tenets across all aspects of their lives, like traditional dress, cultural protocols, social customs and community regulators of behaviour and religion. Nonetheless, patterns of Māori horticulture were changing quickly in line with the new technologies underscoring the introduction of exotic crops, like wheat and potatoes.[257]

Crops such as these had been available to Māori since the days of the earliest Pākehā arrivals. However, it was the missionaries who introduced most of the new seeds in sufficient quantities to make large-scale planting possible. The missionaries also directed their evangelism of Māori towards farming. With new seeds and techniques available, it was thought, Māori could eventually attain economic self-sufficiency, which was always a missionary priority.

One way to impart this Māori economic self-sufficiency, and thus raise Māori living standards, was the agricultural training farm. An example was the model farm established at Waimate where Rawiri Taiwhanga was based, learning his trade selling butter to Pākehā in Kororareka.[258]

Partly as a consequence of missionary efforts such as these, Māori horticulture developed quickly, especially once the mission societies began to expand southward into remote areas of the country. By the time that large-scale numbers of Pākehā began to arrive, Māori were able to provide regular supplies of crops and meat to the fledgling settler enclaves.

In Auckland, new settlers often observed that Māori wished 'to farm like the white settlers'. Large numbers of Māori moved nearer to the colonial settlements, gaining employment as labourers on Pākehā farms in order to further their knowledge, as well as to earn a living. Māori were quick to learn and soon became 'excellent ploughmen, reapers, bullock-drivers and shearers'. Pākehā farmers welcomed 'this source of cheap labour', which was in short supply, thereby easing the entry of Māori into this early labour market.

The government of Governor George Grey also wished to assist Māori onto the land and into economic self-sufficiency through the expansion of horticulture. If Māori could be encouraged to farm, thought Grey, with their cultivations needing to be constantly manned and protected, then Māori might be persuaded to desist from engaging in warfare. The government also published a periodical in te reo Māori and English entitled *The Māori Messenger*, which devoted pages to imparting up-to-date farming information.[259]

Grey consequently enacted his 'flour and sugar policy', providing hospitality to rangatira, along with food, seeds, livestock, implements like harnesses and ploughs and the means to build flour mills.[260] Grants were also made to mission societies and schools, encouraging the teaching of agriculture, especially if the school had a farm attached, as was often the case.

In 1857, Crown officials reported on such a school at Kai Iwi, near Whanganui. Operated by the Wesleyan Church, the school occupied 300 acres purchased from the Crown in 1853. According to a report filed by three local Justices of the Peace, 'between sixty and seventy acres have been enclosed and subdivided and the greater part of these laid under grass and other crops'. Stock on hand consisted of six working bullocks, nineteen cows, an 'excellent wheat mill capable of supplying the wants of a large station', a hand threshing and a winnowing machine, a bullock cart, ploughs 'and all other implements ordinarily used on a farm'.

Further south in Whanganui, the Church of England had established an Industrial School 'enclosing 40 acres, 14 of which are under grass, wheat, potatoes and other crops'. When the school had officially opened, on 7 February 1854, twenty-four 'Native scholars' had been admitted.[261]

Establishing Horticulture

From the 1850s, most Māori farming operations comprised the growing of crops. Some livestock were maintained but in the main only for community consumption. Raising livestock in sufficient numbers to render the operation economically viable did not arise until much later, though such developments as raising stock were experienced unevenly across the native districts. Undoubtedly, sheep and cattle farming were an early feature of some tribal landscapes, like that of the East Coast.

By the time of the earliest Pākehā arrivals, the only traditional crop still being harvested by Māori was the kūmara. Other native crops, like the taro and the gourd, had largely declined, given the demanding growing conditions prevalent across New Zealand, especially in the south.

In 1830, J S Polack observed Māori constructing special 'kumara-houses', which were 'expressly built for the storing of that favourite vegetable'. The care, neatness and ornate workmanship involved equalled the workmanship normally reserved for much superior houses, reflecting the kūmara's continuing cultural value. In fact, writes Mere Roberts, Māori interpreted much of their world through the whakapapa of organic things, like the kūmara. Recognising such approaches, she argues, as utilised by early Māori growers, makes it possible to understand the nature of Māori knowledge at large and how it related to the organic environment, 'particularly with regard to the management of valued resources'.[262]

The floors of kūmara houses were raised, sides were boarded up and a verandah with pillars supported the roof. 'On the roof a wooden image [was] generally fixed, remarkable for latitude of tounge and protuberance in the abdomen'.[263]

The kūmara was 'the most invaluable food possessed by the New Zealander', writes Polack. Many varieties were observed, arising from the influence of an ever-changing climate, as well as variable soils and exposure. But though still a staple of the Māori diet, the quality of kūmara seemed to be rapidly deteriorating, leading to its decline. Some species were now planted, it seemed, for ceremonial purposes only, with the special huts erected for careful storage. 'On planting the kumara, the land becomes tapued, as also [did] the planters in sowing the seed.' As a consequence of its diminishing vitality, the kūmara declined across the 1850s as a popular economically viable plantation crop for Māori.[264]

European Crops

As European crops were introduced, they were quickly picked up by Māori, rapidly spreading across the country. Of these, wheat and potatoes were the most successful, accounting for the bulk of Māori gardening. The potato in particular rapidly replaced the kūmara as a staple of the Māori diet because it was easy to grow and harvest. Other crops were equally successful, like corn and maize, although they were not produced on the same scale as wheat and potatoes. Nor were otherwise-popular vegetables like cabbages, turnips, marrows and beans, alongside fruit like apples, peaches and quinces.

Generally speaking, the quality of Māori produce was sufficient to guarantee continuing markets, though this was not always the case. Māori-grown tobacco for example was always inferior to the imported varieties and therefore did not feature as an important trade item for Māori, instead being grown mainly for home consumption.

In the Waiariki and Bay of Plenty districts in 1857, the following acreages of crops grown by Māori were recorded:[265]

Māori Crops Acreage 1857 Waiariki and Bay of Plenty

- Wheat — 3450 acres
- Potatoes — 3050 acres
- Maize — 1735 acres
- Kūmara — 1230 acres
- Tobacco — 16 acres
- Sown grass — 7 acres
- Barley — 4 acres
- Other crops — 110 acres

European Farming Methods

Following the 1850s, Māori methods of planting combined traditional Māori implements with new tools introduced by Pākehā, although it was not long before Pākehā tools prevailed. Generations earlier, a ko (digging stick) would have been used by Māori to clear the ground, but this was a rare sight by the 1850s. Traditional implements developed by Māori over time 'were of the most primitive nature, hence any form of earthwork was tedious and laborious'. Some digging implements like the rapa maire (a form of spade) and the koko (scoop) were generally only used when constructing palisades.

Even so, though rare, the following implements were still observed in use on Māori plantations; hengahenga (a form of hoe), kaheru (wooden spade), ketu (a hand cultivating tool), patupatu (a heavy clod breaker) and a timotimo (a form of grubber). In some parts of the country, like Te Urewera, where Pākehā encroachment was rare, such traditional tools were still being used as late as the 1890s.[266]

However, metal implements inevitably became more common, like mattocks and axes, for good reason, given the comparative inefficiency of traditional tools. Ploughs were also in common use, not only for planting, but also for digging up root crops like potatoes in order to retrieve the root systems for replanting. Ploughs were pulled by oxen. Though they were owners of innumerable horses, Māori rarely used them as beasts of burden.

Metal tools like the sickle were also used to harvest wheat and maize. However, reaping was rarely timely or efficient, with much of the crop lost. Threshing wheat with a flail was also common, with some tribes purchasing threshing machines in the late 1850s.

Fertilising the soil was also rare. Māori farmers were often exhorted via *The Māori Messenger* to use bonedust or other manures, sound advice that Māori seldom heeded. This created problems for Māori soils, which were soon depleted, especially where the same crops were planted, year after year. Rotation of crops, or resting the soil, were concepts Māori seemed unable, or unwilling, to grasp.[267]

This possibly relates to the continuing importance of cultural edicts and contexts. One method of averting an impending diseased crop, as observed by Pākehā, was the reciting of certain karakia and the interring within cultivations of bones belonging to a revered tūpuna. On occasions, ethnologists like John White recorded such extensive preparations, especially where culturally important kūmara crops were seen as deficient.[268]

Perhaps as a consequence, some Māori farmers demonstrated a marked aversion to the use of fertilisers, especially those derived from animal products. Accordingly, after a few years of intensive gardening leaving soils depleted, Māori would move on, clearing more land by burning, seeking to establish new cultivations. This was a tried and trusted method of land clearance employed by Māori since the days before the advent of Pākehā.

The environmental costs of such land clearances, however, were substantial. In the post-Pākehā period, discarded cultivations now left fallow would soon acquire dock, thistle and other introduced noxious weeds which, growing in profusion, began spreading into the new cultivations being established. As Māori moved their cultivations, they also continued their pre-Pākehā practice of establishing several cultivations at once, with all sites well separated from each other. The origins of such a dispersing of cultivations lie in the distant past, when iwi and hapū sought to conceal their plantations from hostile attack. Since the advent of Pākehā, Māori were still observed to be creating more than one plantation, sited well away from pā and papa kāinga, to ensure the safety of tribal communities from attack.

On occasions, this created considerable problems with the division of labour. As in the pre-European days, all Māori participated in the planting and harvesting. However, ensuring that all plantations received the intensive care required undoubtedly taxed Māori communities.

Abandoned cultivations were not used thereafter by Māori, despite Pākehā advisers urging that they be replanted with a variety of English grasses to support the raising of livestock. Some Māori did attempt to plant out their discarded cultivations in grass, but Māori appeals to the government for the necessary seed to be supplied in significant quantities invariably went unanswered. As a consequence, a large portion of the 'wastelands' often referred to by Crown officials was in fact tribal land that had once been used, and now abandoned, as cultivations.[269]

Fencing in Wandering Stock

Another fairly straightforward yet demanding issue to bedevil Māori cultivations was that of wandering stock and the need for adequate fencing. Given their customary practice of 'cropping different lands in rotation, partly to emphasise continuing ownership of the lands', Māori resisted building and rebuilding costly fences, as cultivations moved every few years.

According to a Civil Commissioner reporting in 1861, the task of determining where, and how, fences were to be erected around cultivations often

Early Māori owned large numbers of livestock, especially horses, though horses were not used as farm animals.
(Danny Keenan)

fell to the local village rūnanga. Such rūnanga were set up by Resident Magistrates as a means of gaining tribal compliance in the maintenance of law and order. The rūnanga also set the terms under which roaming stock like pigs could be restrained within any given village. Because no two village committees were the same, wrote the commissioner, such rūnanga determinations varied greatly across tribal regions, at times seeming to represent 'a formidable heterogeneous mass without order'.[270]

In fact, the problem of fencing and stock trespass in the native districts seemed to be a difficult one to solve, wrote the Civil Commissioner. There were two issues, he said. One related to the fencing of cultivations in compliance with edicts issued by the local rūnanga or Resident Magistrate. The other issue related to the 'wastelands' – unused land holdings – which, though often extensively grassed, were always unfenced.

Trespass between pā and papa kāinga had been a problem since well before Pākehā had arrived, with redress, then as now, being of a very summary nature. If appeals to owners to pay costs and repair damage went unheeded, trespassing animals could be seized and slaughtered 'and eaten with great relish'. Owners of the confiscated stock would then seek redress as soon as possible on the basis of a quid pro quo, 'doing the same in return on the first opportunity and so the balance was struck and good temper maintained', though such practices, albeit culturally sound, undoubtedly impeded economic growth.

This was fine between Māori. However, if Pākehā were involved, things were different. Pigs found wandering off papa kāinga and damaging Pākehā lands, for example, would soon be 'impounded and detained'. After long discussions, Māori might be prevailed upon to pay damages by way of property transfer. However, once Pākehā introduced substantial numbers of cattle and sheep into the mix, 'the evil was proportionately increased' as such stock holdings spread across the country.[271]

Responsibility for Trespass

When erecting fences to protect their cultivations, wrote the commissioner, 'it may be observed that the Maoris erect their fences, not so much as a matter of obligation to their neighbour as of convenience to themselves'. Māori argued that their crops did not go to the cattle; the cattle came to the crops. The onus to restrain wandering cattle therefore lay with the cattle owners.

Māori therefore objected to erecting 'substantial fences'. Their own fences, they argued, did perfectly well, keeping their pigs out of the cultivations because, facing injury, pigs were disinclined to burrow beneath their sharp-pointed fences, whereas they were known to burrow beneath a stone wall with impunity. Māori objected to providing costly fences for Pākehā convenience. Pākehā farmers were responsible for restraining their stock; it was not the responsibility of Māori.

As for horses and cattle, these broke through Māori stick fences with ease, and therefore required restraining, as did sheep capable of leaping over cultivation fences. 'Hedge and ditch' fences were of only marginal value, unless the ditch was very wide and deep. 'Log and stone' fences were preferable, especially against horses and cattle, but again, they could be easily leapt by sheep.

On Waiheke Island, where substantial cultivations were located, only minimal fencing had been installed by Māori because of the absence of stock on the island. However, Māori were objecting to Pākehā farmers from Auckland bringing their stock to the island and without authority letting them loose to graze at will. As a consequence, such stock frequently wandered into the cultivations. The same often happened if Pākehā farmers adjoining Māori lands did not restrain their stock. 'Bad feeling [was] the consequence.'

In the Coromandel, Māori cultivations remained unfenced, though they were situated in the very centre of the district, because so few Pākehā cattle were found in the neighbourhood. On the eastern side of the Thames River, 'the unfenced cultivations [were] extensive, with no European stock'. However, where fences did exist, it was not uncommon for them to extend around more than one proprietor's plantations.

'And so, with respect to nearly every other part of the colony, where natives are still indisposed to erect suitable fences, the evil will increase with every increasing year until some stringent provisions be made to apply to all classes of the community'.[272]

Livestock

Livestock farming did not initially attract Māori interest, largely because Māori lacked pasture. However, some figures do exist for Māori livestock holdings in 1857, in the Waiariki and Bay of Plenty area.

Māori Livestock 1857 Waiariki and Bay of Plenty

- Pigs — 5000
- Cattle — 207
- Goats — 34
- Horses — 990
- Sheep — 144

Source: Hargreaves, 'The Maori Agriculture', p. 68

Pigs were the most lucrative of livestock options because pastures were not required. Instead, pigs could be left to scavenge in the villages and nearby bush. Some tribes also ran cattle, but, like the pigs, they were generally allowed to roam and scavenge, becoming what was termed 'bush cattle', though some tribes did provide pasture and other crops for feeding. Some Māori also had cows, but these were for domestic use only, like milking. Sheep were generally unpopular because they could not survive in the wild or be left to scavenge.

Māori also found sheep to be prone to disease that was exceedingly difficult to treat, and there was the constant problem of wild dogs roaming the bush. That dogs were a particular problem was also noted by Pākehā surveyors who were then cutting their lines through areas farmed by Māori. Sightings of such dogs were common with packs of dogs that had escaped from villages now roaming through the bush and kiekie scrub. Surveyors were often advised not to venture out alone.[273]

However, of all the tribes that did show some interest in sheep farming, the most successful were possibly the tribes around Taupō who possessed a significant number of open pastures. Though of poor volcanic quality, the pastures nonetheless could support limited numbers of sheep. In 1859, Ngāti Tūwharetoa owned some 170 sheep. By 1862, this number had increased to 2000. However, with roading so undeveloped, it was not possible to take the wool to the coast to be shipped to market.[274]

Flour Mills

Initially, Māori who went into flour production employed small steel handmills for grinding wheat.

Such small handmills were a common sight amongst flour producing tribes during the 1850s. However, in the end, their output was limited and they tended to break down often. This entailed carting the mill to Auckland for repairs. Flour Mills powered by water were soon recognised as more efficient and easier to manage.

The earliest Māori-owned flour mill was constructed at Aotea in 1846. The mill was managed by a Pākehā millwright and could grind wheat at the rate of two bushels an hour. However, the mill had not survived for long, soon running into disrepair.

The next flour mills appeared near Te Awamutu in 1849. By the end of that year, six flour mills were owned and operated by Māori in the Waikato region. Within four years, a further ten mills had been built and were operating. Pākehā were employed to do the building, though this led to workmanship that was of a variable quality, with some mills closing down soon after commencing.

In order to encourage Māori milling, the government appointed an Inspector of Native Mills whose task was to assist Māori in the management of their mills, providing training in the hope that in time Māori would manage the mills themselves.[275]

The Auckland Market

Throughout the 1850s, Auckland was the largest centre for Māori produce, as the largest population centre and also the largest exporting port. 'Crowds of Maori vendors thronged the streets selling their wares', always ready to haggle and hold out for the best sales possible. Until all their produce was sold, most Māori occupied an area in Commercial and Mechanic's Bay, while others moved to the outskirts of town.

It is not possible to evaluate the amount of Māori produce sold in Auckland because returns, such as the volume of canoe traffic bearing goods, give an incomplete picture. Māori produce was also brought into the town and on-sold by Pākehā merchants. In 1855, some forty-nine coastal vessels were registered as Māori owned, transporting Māori produce to market. By one estimate, each shipment could expect to earn between £200 and £500. Māori also utilised an equal number of smaller craft carrying supplies to market.

Figures are available that account for the volumes of produce being shipped into Auckland by coastal vessels. Whilst Pākehā and Māori produce is undifferentiated, it seems certain that most of this produce would have been produced by Māori since, at this time, Pākehā had not encroached far beyond Auckland into the Waikato, where the large-scale growing areas had been established.[276]

Quite apart from what these figures convey, vast quantities of produce were also delivered to market by road. Māori also commonly carried their produce to market themselves. It has been estimated that each Māori bore up to 60 lbs of produce. According to one estimate, the total value of all produce brought to Auckland by Māori amounted to at least £16,000 annually. However, these estimates varied, depending on market conditions. In 1853, the canoe trade alone was worth £13,000, and £16,000 the following year, before declining sharply thereafter.

Whatever the values, and they were undoubtedly significant, it was commonly recognised throughout

Māori Produce Imported to Auckland 1852–1856

		Wheat (Bushels)	Maize (Bushels)	Potatoes (Tons)
1852	Coastal Vessels	30,633	10,959	188
	Canoes	1674	2157	235
1853	Coastal Vessels	37,541	14,556	850
	Canoes	2454	4139	282
1854	Coastal Vessels	65,833	18,569	535
	Canoes	3715	1123	94
1855	Coastal Vessels	82,228	29,250	1175
	Canoes	1372 bags	1398	212
1856	Coastal Vessels	56,930	7873	279
	Canoes	3557	774	125

Source: Hargreaves, 'The Maori Agriculture', p. 74

Auckland commercial circles that the trade in food and provisions was firmly in the hands of Māori. In fact, the supply of horticultural produce had been left to Māori, who had shown themselves to be able to produce significant quantities of food relatively cheaply. Pākehā farmers had therefore opted for large scale pastoral runs, developing pastures in the process, tacitly leaving the supply of food and provisions to Māori.

Much of this trade was also destined for export. Of all the items produced, wheat remained the most valuable export, while pigs and maize were also important. However, from an export high in 1854 of 80 percent, the Auckland share of the national total of wheat exports declined to almost nil by 1860. In addition, the pork and bacon sent to Australia did not enjoy a high reputation, also leading to an eventual collapse in the market, partly caused by Māori feeding their pigs on fish. Potatoes also sent overseas were eventually seen as inferior because they were often diseased, poor in quality and badly packaged.

Towards the end of the 1850s, as the wheat trade was seen to decline sharply, Māori began to hold back their wheat in the hope of forcing better prices. For a time, once resumed, this proved successful, but in the end the trade in wheat entirely collapsed, most notably in the face of increased wheat imports from Australia.

By 1860, 45,348 bushels of wheat were being imported into Auckland to feed a rapidly expanding Pākehā population. With the Land Wars in Taranaki having commenced, this population now included a sizable military component. By 1861, this quantum of imports had grown to 55,910 bushels and, in 1862, to 188,931 bushels.[277]

The Challenges Ahead

The challenges that would now face Māori farmers and horticulturalists after the 1850s were undoubtedly formidable. As discussed in the next chapter (Chapter Nine), the Land Wars and land dispossessions that occurred after 1860 would egregiously disrupt Māori farming initiatives. As we will see, significant losses occurred through the aegis of the Native Land Court. A stagnant national economy, and an absence of government assistance, also severely impeded Māori farming developments, with impacts that would be far-reaching.

(John Cowpland – Alphapix)

Whangara Farms
(John Cowpland – **Alphapix**)

Chapter Nine

Dispossession of Māori Lands – 1858–1912

Māori Land Loss

In the early days of Pākehā settlement, Māori horticulture thrived with whānau and hapū successfully harnessing their efforts to supply rapidly expanding markets, especially in Auckland. However, after 1860, Māori efforts to maintain their cultivations and trade in foodstuffs, much less developing larger farming enterprises, were progressively undermined by the progressive and enormous loss of tribal lands.

This chapter briefly analyses the rates of Māori land loss between 1858, when the Crown first attempted to extinguish Māori land titles, and 1912, when the significant era of Liberal government land purchases came to an end.

By then, the government had acquired some 15 million acres of Māori land in the North Island, with only approximately 27 percent of the North Island left in Māori hands. By 1912, the government had also acquired virtually all of the South Island. Thereafter, from 1912–1928, Māori land purchases tapered off, constituting nonetheless a further 1.4 million acres. By 1929, the rates of land loss had almost reduced to a trickle, though, given the negligible amounts of land still retained by Māori, the continuing losses were of course significant.[278]

As well as briefly quantifying the losses of customary land to 1912, this chapter will also briefly describe aspects of the iniquitous legal apparatus assembled by the government in order to render these losses possible. A range of Māori responses to these measures is also briefly canvassed.

Ahuwhenua Success

When the Ahuwhenua Trophy competition was established in 1933, all Māori – not least the trophy's co-founder Sir Apirana Ngata – were well aware of the long history of Māori land loss and dispossession that had devastated tribal economies and communities since the earliest days of colonial settlement.

Earlier, in 1931, Governor-General Lord Bledisloe had visited a number of land development schemes throughout Rotorua and the Bay of Plenty. Inspired by the commitment and hard work of the Māori farmers on view, Bledisloe had joined with Ngata in launching the Ahuwhenua competition.

The inaugural winner – William Swinton – was announced on 27 March 1933, winning in a creditable field of seventy-one entries, all from Waiariki, with at least ten other farms also visited by Judge W Dempster 'as a courtesy'. Swinton's success was all the more remarkable, said Judge Dempster, because he had completed extensive regrassing, along with the building of an impressive cow shed, which provided space for the storage and care of his farm machinery and utensils.[279]

Swinton's striking success was also in part made possible by the Māori land development schemes established by Ngata in 1929. After decades of urging the government to act on Māori land loss and economic marginalisation, Ngata had managed to secure government funding for tribal landowners to develop their lands and supporting infrastructures.

Owners were also authorised to amalgamate their land titles into single, consolidated entities. Land development and consolidation schemes quickly followed as Māori seized this new opportunity to apply themselves to the economic, community and cultural regeneration now on offer, much facilitated and encouraged thereafter by the Ahuwhenua Trophy competition.

When presenting his first report to Parliament in 1931, covering the first two years of land developments, Ngata gave details of the lands and expenditures involved, along with an extensive analysis of historical, psychological and economic aspects that underpinned these important development schemes. By this time, for example, almost three-quarters of a million acres were under development, providing for more than 8000 Māori.[280]

The schemes represented a 'last-ditch effort' by Ngata to restore and regenerate Māori lands remaining in Māori hands. By 1933, when the Ahuwhenua Trophy competition was launched, Māori had lost almost all of their original tribal estates. Much earlier, during the 1850s, widely dispersed tribal communities had still possessed most of their lands, though significant acreages had been lost throughout the Far North, Wairarapa, Wellington and especially Hawke's Bay.

But after 1863, war and land confiscations, and Native Land Court rulings, had led to the dispossession of vast tribal holdings. Aggregating and never-ending government legislation expedited the losses, especially between 1890 and 1912. By 1922, when the first Māori Trust Board was established by Ngata (Te Arawa), only 8 percent of New Zealand remained in Māori hands. By the 1930s, Māori land holdings had continued to diminish, compelling Ngata to launch his development schemes in order to stem the losses whilst facilitating the development of lands remaining to Māori.

Origins of Land Loss

The long process of wholesale Māori land dispossession really began in 1858 when Governor Gore Browne agreed to relinquish his authority over monies allocated by Parliament for Māori purposes, instead transferring authority to a restive Pākehā General Assembly. A new portfolio was established, Minister of Native Affairs, occupied by Colonial Treasurer Christopher William Richmond on 27 August 1858.

Soon afterwards, the Native Territorial Rights Bill 1858 was introduced into the General Assembly. By this measure, the government would have the power to extinguish all customary Māori land titles. However, the Bill was overruled by the British Parliament and never implemented. But the settler politicians had signaled their intentions and would return to the strategy of extinguishing Māori land titles in 1862.

At that time, Māori still retained approximately 80 percent of the North Island, comprising approximately 23.2 million acres (or 9.4 million hectares). Most of the 6 million acres owned by Pākehā at that time had been sold to the Crown which, since 1840, had possessed a sole right of purchase arising from the pre-emption provision contained in Article Two of the Treaty of Waitangi. In other words, only the government could purchase Māori land.

In Auckland, for example, Māori still possessed about 2.4 million acres, or about 58 percent of their original estate. In Hauraki, the tribes possessed 735,000 acres, or 90 percent of their original holdings. In the Waikato, Tainui still retained 2.2 million acres, or 91 percent of their customary lands. And in Taranaki, the eight iwi of Taranaki Whānui still possessed a combined 1.8 million acres, or 96 percent of their original holdings.[281]

In the South Island, the Crown took advantage of the Treaty pre-emption provisions by negotiating nine substantial land purchases with Ngāi Tahu between 1844 and 1863. The largest of these purchases were the Otago purchase of 1844, comprising 400,000 acres, and the Canterbury purchase, transacted in 1848, which comprised some 20 million acres. By 1865, the entire South Island had been purchased, except for 175,000 acres of reserves and other lands exempt from sale. Additional reserves were created for Ngāi Tahu in 1906, but some 95,000 acres of this land was subsequently lost through alienations between 1910 and 1939.[282]

Māori Responses

Māori were quick to respond to the threat of extinguishment and increasing land losses. A Māori gathering at Manawapou, South Taranaki, in April 1854 determined that unified political action was needed to stem the losses.[283] Two young rangatira who were present, Tamihana Te Rauparaha and Matene Te Whiwhi from Ngāti Raukawa, proposed the formation of a Māori King movement. Thereafter, they travelled the country, seeking an appropriate candidate. After some years of consultations, Potatau Te Wherowhero of the Waikato was appointed Māori King at Ngāruawahia in 1858.

Māori support for the Māori King's determination to resist land loss was widespread. All leading rangatira of the North Island came to Ngāruawahia to support Potatau, symbolically laying their lands and service at his feet. Mountains, headlands and other prominent physical features were declared by rangatira present as the pou, or boundary markers, that, in spirit at least, delineated the Māori King's dominion.

Māori now seemed united behind Potatau Te Wherowhero and his promise to veto the sale of land. But war at Waitara in 1860 – over a disputed land sale – tested Māori resolve. During the war, 3000 Māori gathered at Ngāruawahia to affirm the resistance of the Kīngitanga to land sales.

After the war, in September 1861, a new Governor, George Grey, proposed a series of self-sufficient tribal rūnanga that might exercise a range of local powers of governance, including aspects of land titling and regulated land disposal. But though proposed on a grand scale, Māori chose not to participate, instead adhering to the Māori King's resistance. The outcome, once again, was war, fought over eighteen months throughout the Waikato. By war's end in 1864, the Kīngitanga was in full retreat, with land confiscations now enacted, aimed at the 'rebels' of Taranaki, Waikato and, thereafter, other

Te Pou Tutaki/FitzRoy's Pole, erected north of New Plymouth by Māori in 1847, marking the place beyond which land sales would not be permitted

(Danny Keenan)

iwi including Tauranga, Hauraki, Whakatohea, Ngāti Awa and Tūhoe.

Tightening the Legislative Hold on Land

During the wars, the government had taken advantage of its military successes against Māori by enacting a range of punitive measures. The land confiscations had arisen from such a measure – the New Zealand Settlements Act 1863. By this Act, huge tracts of Taranaki, the Waikato, Tauranga and the Bay of Plenty would be confiscated as punishment for 'the evil-disposed persons of the Native Race' who had 'taken up arms with the object of attempting [the] extermination or expulsion of the European settlers'.[284]

As a consequence, 1.2 million acres was confiscated in the Waikato with 1.2 million acres also taken by legislative compulsion in Taranaki. In Tauranga, 214,000 acres were taken, with 440,000 acres confiscated across the Bay of Plenty.[285]

Another example of punitive legislation was the Native Lands Act 1862 which, in effect, was a resurrection of the failed 1858 Native Territorial Rights Bill.[286] By this new Native Lands measure, the government was able to finally institute its policy of extinguishing Māori communal titles, replacing them with English individual land titles.

The 1862 Act also abolished Crown pre-emption. In 1840, when the Treaty of Waitangi was signed, pre-emption had been included to protect Māori from fraudulent sales with settlers. Land could then only be sold to the Crown, thereby guaranteeing fair practices and prices – at least, that was the theory.

In 1862, all this changed. Māori lands would now be opened up to Pākehā settlers who were empowered to negotiate sales directly with Māori individuals who

AHUWHENUA

The Weeping Woman. Moutoa Gardens/Pākaitore, Whanganui. New Zealand's first Land Wars memorial, erected in 1865
(Danny Keenan)

might be willing to sell, without Crown involvement or oversight – a controversial policy that was called 'free trade'.

By 1862, Pākehā settlers had succeeded in pressuring the government to disengage from the business of land acquisition altogether. To further this end, a Native Land Court was also established in 1862. The court's role would be to manage the process of title extinguishing, and the issuing of new individual titles to Māori who could then, if they wished, sell their individual holdings directly to Pākehā.

The new court would be made up of rangatira who would investigate land titles and issue deeds to Māori deemed to be the 'true owners', thereby guaranteeing an important element of customary oversight. A Resident Magistrate would convene the gatherings of rangatira, lending his legal authority to the process. In reality, tribal committees that were already established were utilised, where possible, with such committees ensuring that customary tribal protocols would be adhered to.[287]

Finally, the Māori 'right' to sell land as an individual – though not recognised in customary law – was confirmed to Māori by the government via the Native Rights Act 1865.[288] By this measure, Māori were conferred with the same 'freedoms' as Pākehā, which of course included the freedom to sell land, once individual titles had been granted (but it did not confer the freedom to vote).

However, in 1865, an amended version of the Native Lands Act was passed through Parliament. By this reforming measure, the exisiting local runanga-style Land Court process was abolished because it was seen to be too slow.

In its place, a 'regular' Native Land Court was set up that, consistent with other Pākehā courts, would use proper legal protocols, have sole judges sitting on a bench, with legal counsel, witnesses and applying legal precedent. Kaumātua could also be appointed as Court Native Assessors in order to assist with land title determinations. But they had no power to set final determinations; this authority lay soley with the judges.[289]

By this series of legislative measures enacted in the early 1860s, the government was able to embark upon its long process of wholesale Māori land dispossession.

Politics as Resistance

In 1867, twenty-seven years after the signing of the Treaty, Māori were finally granted the vote.

On 10 October 1867, the Māori Representation Act was introduced into Parliament, establishing four special electoral seats for Māori – Northern, Southern, Eastern and Western Māori.[290] Now enfranchised, Māori increasingly turned to politics to resist wholesale land dispossession and especially the depradations of the Native Land Court.

In 1868, for the first time, four Māori took their place as Members of the House of Representatives. The first to speak was Tareha Te Moananui of Eastern Māori who, on 4 August 1868, criticised the Native Land Court's contrary rulings. Māori owners 'entitled to the land' were not being included on Crown titles, whereas others 'not having a title' were represented on the legal grants being issued.

As a consequence, 'the lands of one person are sold by another'. Te Moananui also spoke of the turmoil amongst Māori, not least on the East Coast, over the unfair application of Native Lands legislation. 'That Act laid down by you is not working satisfactorily', he said, urging Parliament to take remedial action because the Act was instead 'working in a troublesome way'.[291]

The next Māori member to speak was Mete Kingi Paetahi of Western Māori, who also alluded to the inequities of the court's deliberations. Lands belonging to Whanganui Māori at Waitotara had been 'taken by the blade' – confiscated – but never returned, as promised. Whanganui Māori had been assured 'that the land of the peaceable Natives would be left to them'. Yet, when Waitotara had made application to the court for the lands to be returned, the court had ruled otherwise, granting the lands instead to others with 'fictitious claims'.[292]

Meanwhile, in 1869, the Kīngitanga and allies gathered at Tokangamutu, with Tukaroto Matutaera Te Wherowhero Tawhiao now installed as a second Māori King. A delegation of rangatira from Auckland, including Paora Tuhaere of Ngāti Whātua and Hone Mohi Tawhai of Ngā Puhi, also attended, encouraging the Māori King to engage with the Crown as a strategy to limit the continuing land losses. But the Māori King Tawhiao, now living in refuge after the wars amongst Ngāti Maniapoto, refused.

At the time, significant protests against land dispossession had also arisen in Hawke's Bay.

Since the 1840s, pastoralism had expanded rapidly into the region with large flocks of sheep and cattle imported from Australia. Vast land blocks had been leased from Māori for sheep grazing, though such pastoral leases were strictly unlawful. As a result, Ngāti Kahungunu became vulnerable, especially after 1865 when the government ceased land purchasing, leaving the field open to Pākehā capital.

Pastoralists were then able to register their 'unlawfully' leased blocks with the Native Land Court, forcing rapid sales without proper court deliberations as to the rights of true customary owners. Pastoralists were also permitted to force through foreclosures as a means to retrieve debt.

About 700,000 acres of Ngāti Kahungunu land was lost in this way, acquired by Pākehā pastoralists during the early years of the Native Land Court operations. A Commission of Enquiry established to investigate Māori protests presided in Napier from 3 February to 12 April 1873.[293] The commission's findings were disappointing for Māori, with commissioners ruling that no evidence of widespread fraud had been demonstrated.

Sentry Hill Redoubt, scene of the last encounter between British troops and Māori in Taranaki on 30 April 1864
(Danny Keenan)

By the early 1870s, the ultimate erosion of entire tribal estates was now threatened with significant acreages of customary land continuing to pass into Pākehā hands. From 1 July 1876 to 30 June 1877, the acreage of land to which titles were ordered by the Native Land Court equaled 488,997 acres. Of these land holdings, 190,792 acres – or 39 percent – came from Hawke's Bay alone, the second largest total in the country behind Auckland.

Losses through Reserves and Leases

Another means by which vast acreages of land were lost to Māori was through the creation of native reserves which, once acquired as rentals by Pākehā farmers, eventually became perpetual leases.

This complex and troublesome process mostly affected Māori who had suffered wholesale land confiscations after 1863. Where those confiscations were seen to have been enforced in error, the Compensation Court (a division of the Native Land Court) on occasions returned the land to Māori. But the lands were invariably returned as native reserves, with their communal titles now converted into any number of individual titles. This conversion of land titles – from communal to individual – rendered the lands imminently available for alienation. Large tribal estates therefore were soon acquired by Pākehā farmers, either by outright sale, or through medium term leases, later renewed as perpetual leases.

This occurred in many parts of New Zealand, perhaps most egregiously in Taranaki. By the late 1870s, as Taranaki Māori were returning to their pā and papa kāinga after a decade of warfare, confusion reigned as to the status of the land confiscations and, therefore, the security of customary holdings. This was particularly the case in South Taranaki, where sporadic tribal protests began to occur. Tribal insecurity was exacerbated by surveying, road-making and other aggressive government actions, like the clearing of forests and disruptions to grazing livestock and cultivations.

A Commission of Enquiry appointed to investigate the Taranaki situation reported its findings in 1880, leading to the passing of the West Coast Settlement (North Island) Act 1880. By this measure, a significant number of native reserves were to be finally set aside for Taranaki Māori, ostensibly securing their tribal footholds.

But the legislation was deeply flawed. The reserves spanned approximately 200,000 acres of land in total, with about 5000 Māori identified as possessing interests in the new reserves. This roughly equated to about 40 acres per person. At the time, had Māori been willing to lease the entire 200,000 acres, an income of about £10,000 per year could have been expected, less the costs of surveying and administration.

Matters were complicated however by the Public Trustee, who exercised almost total control over the reserves. When the Trustee sought to negotiate lease arrangements for Pākehā farmers with Māori owners, many Māori refused to cooperate, instead insisting that the lands be returned.[294]

As a consequence, the government enacted the West Coast Settlement Reserves Act 1881 which, over the next 100 years, would go through no fewer than fifteen major amendments and be investigated by a succession of commissions and committees of enquiry, not least the Waitangi Tribunal commencing in 1990.

Because this 1881 legislation awarded the Public Trustee full powers of reserve administration and disposal, the economic interests of Pākehā farmers quickly overrode the communal interests of Māori.

Against Māori wishes, leases were arranged to Pākehā farmers for periods of up to thirty years with rentals set at 'peppercorn' levels, a disingenuous arrangement, which effectively delivered the land as 'rent-free'. Farmers could also recover their costs for improvements, against their rents, despite the cheap lease arrangements.

Once awarded, Pākehā farmers fought hard to safeguard their leases from expiring, whereas Māori owners refused to renew the leases, instead wishing to retain and develop their lands. However, an 1883 amendment enabled leases to be renewed for a further thirty years; and a further 1885 amendment enabled Pākehā farmers to charge their survey costs against the reserves.

Thereafter, the Māori Trustee's unwillingness to fairly represent Māori interests had irreparable implications for Māori communal and farming development, especially given the vast sums of money earned from the lucrative leases in operation, as opposed to the paltry rentals actually passed on to Māori. And, in 1887, legislation was passed to indemnify the Public Trustee for immeasurable losses incurred by the Trust Office through derelict decision making over Māori assets.

By 1892, most of the leases were due to expire. Taranaki Māori who wished to commence farming themselves were adamant that the lands should now be returned, as legally required. However, in 1892, Premier John Ballance introduced the West

Coast Settlement Reserves Act, which introduced the concept of 'perpetual lease'. Under section 9 of the Act, when leases expired, Pākehā lessees were to be offered a 'perpetual lease' – effectively, a lease without end. This provision dealt a severe blow to Māori who still hoped that, one day, their lands might be reoccupied and an economic base developed.[295]

Thereafter, Taranaki Māori made numerous representations against both the perpetual leases and the arbitrary powers vested in the Public Trustee. In 1920, the Native Trustee took control from the Public Trustee, with an unintended consequence; large numbers of beneficial owners alienated their interests to Pākehā lessees. Fifteen years later, under the Native Purposes Act 1935, the Native Trustee was prevented from significantly raising the rentals paid by Pākehā lessees. In 1955, some 43 statutes governing the leasing of Māori reserves in Taranaki, Palmerston North, Wellington, Nelson and Westland were consolidated into a single Reserved Land Act. Provisions, however, continued to favour lessees, like the continuance of perpetual leases.

In 1963, all of the leased reserves then under the jurisdiction of West Coast Settlement legislation were united into a single reserve, named the Parininihi ki Waitotara Reserve. A corporation to manage the substantial interests of this new reserve was established in 1976, the Parininihi-ki-Waitotara Incorporation (PKW).

In 1997, the Maori Reserved Lands Amendment Act was passed, yet another measure enacted in the long history of legislation governing the leasing of the native reserves, originally set aside in the 1880s, supposedly for the benefit and security of Māori communities. 'While the [1997] Act did not terminate perpetual leasing, it did at least provide some mechanism for the PKW Incorporation to purchase back leases as they became available'.[296]

Thereafter, the PKW Incorporation went from strength to strength. In 1998, statutory rents were abolished with market rentals introduced. In 2000, PKW began purchasing lessee interests as they became available. By 2005, seventeen such interests had been acquired.

In 2006, now operating fifteen farms, PKW were awarded the Ahuwhenua Trophy for dairying. Judges described the incorporation as 'a very polished enterprise that had made significant progress', reflecting what could be achieved with corporate discipline whilst delivering to its Māori owners in a social and beneficial way.[297]

Māori Seek to Control the Losses

Such issues as protecting Māori lands weighed heavily upon early Māori parliamentarians. When Wiremu Pere took his seat in Parliament in 1884, his intent was clear: the protection of Māori land. Pere criticised the government for not granting legislative protections that would enable Māori to manage their own holdings. Without such protections, said Pere, Māori could not control, much less prevent, the continuing appropriation of land.[298]

One form of protection, advocated by Pere, would be the state-sanctioned tribal committee with powers to adjudicate on matters of land titles. In 1883, the government had enacted the Native Committees Empowering Act, which established a series of regional Māori committees. Their role was to advise the Native Land Court on a variety of issues, especially ownership disputes. In theory at least, this placed a vestige of control into tribal hands.

However, Māori resisted regional pan-tribal committees, instead much preferring committees at hapū or iwi level, thereby rendering this otherwise grand scheme unworkable. Complicating matters were the Land Court judges who, defensive of their own court domains, refused to work with the regional committees. Thereafter, lacking widespread support, not least from Māori, the regional committees soon lapsed.[299]

The next attempt to establish tribal self-sufficient committees was made by Native Minister John Ballance who, after some extensive deliberations with Māori, introduced into Parliament his Native Lands Administration Bill on 8 June 1886. Under this measure, tribal committees would be formed to manage their holdings, as well as determining which lands might be put up for lease or sale. As before, the idea was to put Māori in charge of the sale and purchase process, though not entirely.

Once these committees were up and running, Pere hoped that further powers would be conferred, such as the issuing of land titles which was then the sole responsibility of the Native Land Court. But once again, this new system of tribal committees failed to take hold, not least because Māori owners were still reluctant to divest their lands into the control of such committees, even though they were managed by their own.

Eventually, on 30 August 1888, Ballance's Act, setting up tribal committees, was repealed and replaced by the Native Land Act 1888, which removed all constraints – and committees – thereby restoring 'free trade' in Māori land, granting Pākehā the unfettered right to negotiate land purchases directly with Māori owners.

By this measure, the government removed all restrictions on Māori disposal of land, something that Ballance had rejected as being contrary to Māori interests. But, according to the government, Māori now had the freedom to dispose of their land as they wished, or not, a freedom that was popular with Māori.

In 24 January 1891, however, John Ballance would return to Parliament as the premier of the new Liberal government. By that stage, Māori land holdings had reduced to about 40 percent of the North Island, down from 80 percent in 1858. During the 1860s, the government had confiscated about 4 million acres, with a further 8 million passing to Pākehā ownership through the Native Land Court. Lands now remaining in Māori hands comprised approximately 11.6 million acres (or 4.7 million hectares).

By 1890, the Māori of Auckland (including the Far North) still possessed about 1 million acres, or 24 percent of their original customary estate, a substantial reduction from the 58 percent still possessed in 1860. The tribes of Hauraki still possessed about 300,000 acres, or 37 percent of their original holdings, a significant reduction from the 90 percent still possessed in 1860. In the Waikato, the tribes of Tainui still possessed about 400,000 acres, or 17 percent of their original holdings, a significant reduction from the 91 percent still retained in 1860. And in Taranaki, Māori now possessed about 540,000 acres, or 28 percent of their original holdings, a significant reduction from the 96 percent still retained in 1860.[300]

Liberal Land Policy

Māori land acquisition lay at the heart of Liberal policy, with Pākehā constituents clamouring for lands of their own. How this could be achieved – acquiring Māori land with secure titles – became the focus of a 1891 Royal Commission charged with investigating the state of Māori land laws.

The Commission travelled extensively throughout the country in 1891, interviewing witnesses and meeting with rangatira. A final report was filed in Parliament on 23 May 1891, ending an exhaustive consultation exercise achieved within a four-month parliamentary recess.

The commission's report was severely critical of the state of Native land laws, which had 'drifted from bad to worse'. The commission also criticised the Native Land Court, condemning its egregious impacts upon Māori communities by enabling 'private and individual dealings', opening the door for secrecy and fraud. The report paid particular attention to the 'chaos that the Native Land Court had created' by its continuing failure to recognise native tenure. A review of the Native Land Acts by the Commission since 1862 disclosed decades of legal attempts to force individual titles upon Māori, culminating in laws that were confusing and contradictory.

Land Purchases Continue

In December 1892, the Department of Native Affairs was abolished, with its Native Land Purchasing Branch transferred to the Department of Crown Lands in June 1892. Oversight of the Native Land Court passed to the Justice Department. The abolition of Native Affairs removed a significant impediment to the government's proposed land purchase programme, especially with the repeal of the Native Districts Regulations Act 1858 and the Native Circuit Court Act 1858.

Thereafter, the Liberals embarked on a programme of sustained land purchase, mainly achieved by the reintroduction of pre-emption. Once again, Māori land could only be sold to the Crown, not to private buyers. Between 1891 and 1911, the Liberal government purchased 3.1 million acres of Māori land. Over the same period, half a million acres of Māori land was also sold on the open market to private buyers who were granted exemptions from government pre-emption restrictions.[301]

New legislation quickly followed, such as the Native Land Purchase Act 1892. The purpose of this measure was to provide monies to facilitate the sale of Māori land 'for settlement purposes'. Another example was the Native Lands Validation Act 1892, which increased the powers of the Native Land Court to address the vexed area of disputed titles seen to be impeding the development of about 1 million acres of land.

Within a year, however, this Validation Act had been repealed and replaced with a revised Native Land (Validation of Titles) Act 1893, which established an entirely new 'Court of Record', to be called the Validation Court, with jurisdiction over lands situated within any Native Land Court District. The new Validation Court would investigate contested titles as an entirely separate operation from the Native Land Court.

In 1893, John Ballance died and was replaced as premier by Richard Seddon who, upon taking up office, declared that the continuing unavailability of Māori land was impeding development. Pākehā settlement could not proceed until this vexed issue was rectified, he said. Action was therefore needed 'to remove the great difficulties in the way of settlement owing to the unfortunate position of affairs as respecting the Native race and their lands'.[302]

When the Liberals had taken office in 1891, 11 million acres of land had remained in the hands of North Island Māori. Since then, the Liberals had focused on addressing other issues like land tax. Now, a change of land policy was necessary. The time had come, said Seddon, for Parliament to 'grapple with this all-important question' of removing the difficulties impeding settlement.

Moving Land Purchases Forward

Further Acts soon followed to bolster the government's land purchase programme. On 31 August 1893, the Native Land Purchase and Acquisition Act 1893 confirmed pre-emption as government policy, assigning to the government sole rights over the purchase of Māori land.

With 7 million acres of land still owned by Māori, the Act's preamble stated that such land was deemed to be 'lying waste and unproductive'. In the interests of Māori and 'Her majesty's other subjects', and for the extension of settlement, it was necessary that such land should be made available for disposal through lawful sale. The existing law, said the preamble, had failed to supply enough lands to meet the increasing demands of settlers. Colonisation was being impeded. It was therefore necessary to provide further means by which lands owned by Māori could be acquired.

A Native Land Purchase Board would be set up comprising five senior officials. Native Minister Seddon promised to bring justice to Māori by safeguarding their access to fair land prices. He was concerned about the impoverished state of some Māori, promising that capital would flow from sales, providing opportunities to develop remaining Māori land.

But Māori Members of Parliament staunchly opposed the Bill. Many petitions had been received from Māori, they pointed out, protesting against the measure. If this type of measure was applied to Pākehā, said one Māori member, Pākehā would strenuously object. If the government had wished to assist Māori, they should have introduced a measure to encourage Māori to retain the land and develop it themselves.

Tightening the Legislation

But government legal measures assembled to acquire Māori lands continued to be expanded and consolidated. Under Seddon's watch as Native Minister, between March 1894 and March 1899, 1,600,000 acres of land were purchased.

For example, on 28 September 1894, legislation was introduced to overhaul the Native Land Court. The move was necessary because it was 'impossible to understand the Native laws, or the laws relating to the Native Land Court'. A new Native Appellate Court would now therefore help to streamline Land Court proceedings.[303]

But the government was now forced to concede that Māori were almost totally landless. During a recent tour of the King Country, officials had met with supporters of the Māori King, who were now entirely without lands to live on. 'They told us tales of woe about their landless condition. They asked for land to be given to them so that they might settle upon it.' But others in the King Country had prospered. They had done so by separating themselves from the King movement 'because it was to their advantage to do so'.[304]

Shortly afterwards, a further Native Land Court Bill was introduced providing funds to underwrite loans to Pākehā settlers so that land purchases could continue, as well as attendant infrastructure projects. Other measures quickly followed, tightening the government's pre-emptive grip on Māori land sales, such as a Lands Improvement and Native Lands Acquisition Act 1894 and a new Public Works Act 1894, all ensuring that a 'virtual Crown monopoly over Maori land purchases' was firmly established.[305]

One deviation from this tightening process occurred in 1896. After many years of Tūhoe protest, the Urewera District Native Reserve Act was passed. By this measure, a 650,000 acre reserve was established within the Urewera district. The area would be surveyed with land blocks allocated to hapū, not individuals, as a protective measure against uncontrolled dispossession.[306]

Māori Oppose Further Legislation

By the late 1890s, however, for a brief few years, Māori finally succeeded in stalling the government's legislative measures aimed at land dispossession.

When a new Native Lands Administration Bill was introduced in 1898, it ran into immediate Māori opposition because it clearly lacked sufficient protections over prospective land losses. The Native Affairs Committee received major petitions from rangatira like Tamahau Mahupuku (with 3367 others); Tawake Pine (with nine others); Rauri Kepa (with twenty-nine others) and Matua Enoka (with seventy-one others). Detailed submissions were also received from Ngāti Tūwharetoa, voicing strong objections.[307]

As a consequence, Native Minister Seddon embarked on an extensive consultation tour of native districts, including Waitangi, Auckland, Waikato, Otorohanga, Rotorua, Hamua, Putiki, Papawai, Bay of Plenty and the East Coast.

One leading opponent, once again, was the indefatigable Wiremu Pere of Eastern Māori who spoke of the 'evil that [had] fallen on this Island' because of the land selling by Māori. Pākehā had acquired vast acreages thus far, he said, leaving Māori with little to live on. Pere asked that all land sales 'entirely be stopped'. Māori also wanted to borrow funds 'to enable us to improve the lands remaining to us'. The lands that remained in Māori hands should remain inalienable, he said. Pere also asked the government to restrain land purchase officers who were coming amongst the people, 'throwing corn for the fowls to eat'.[308]

Further evidence and petitions were filed before the Native Affairs Select Committee on 3 October 1899, with a substantial debate ensuing. Taking account of objections, an amendment was introduced on 19 October 1899. However, with opposition mounting, and with the clock ticking on the parliamentary session, the Bill lapsed.[309]

Partly as a consequence of the Bill's defeat, the portfolio of Native Affairs was handed to James Carroll, a senior Māori parliamentarian who represented Waiapu. Perhaps Carroll might succeed in passing appropriate Māori land legislation, where the government had failed.

New Māori Lands Legislation

During the 1890s, the Liberals had acquired 2.7 million acres through pre-emptive purchases, with a further 400,000 acres sold through private arrangements.[310] However, during the late 1890s, the rates of Māori land alienation had slowed because of continuing Māori opposition, especially to the government's much-vaunted Native Lands Administration Bill of 1898.

Now, as Minister of Native Affairs, Carroll proposed a revised Māori Land Administration Act 1900. Under this Act, six Māori land districts would be established with a Māori Land Council set up in each district. Each council would have Māori majority membership. Councils would be assigned powers to determine the 'ownership, partition, succession [and] the definition of relative interests' in their lands. This would involve conducting initial investigations of titles and ascertaining the customary owners. Tribal and hapū block committees would then be established, and with the participation of owners, Māori lands would be brought into economic production.

The James Carroll Walkway, Wairoa, which leads to the Ti Tree under which Wairoa's most famous son, Sir James Carroll, was born on 20 April 1827, and where his whenua was buried

(Danny Keenan)

Māori had long sought powers such as these. But they remained ambivalent at best about the primary function of the councils, which was to acquire blocks of land from Māori (voluntarily) for on-leasing or on-selling to Pākehā farmers. The Bill conferred a significant measure of autonomy to the councils. However, legal control and overall supervision remained a government responsibility, with the Native Land Court to be retained. As it turned out, the government was not prepared to transfer the court's functions to a tribal system of councils as yet untried.[311]

After 1900, as the Māori District Land Councils were being set up, Native Minister Carroll travelled extensively through the districts, meeting with tribal owners and observing progress on the ground. Though severely criticised by Pākehā, he urged Māori to at least offer their lands for leasehold in preference to outright sale.

By so doing, Māori could acquire much-needed rental monies to support land and community development. Carroll regretted that tribal owners could not obtain alternative finance from the government, largely on the basis of their collective titles which, for finance purposes, were seen as insecure.

However, it soon became obvious that Māori Land Councils were struggling to acquire lands for on-leasing or on-selling to Pākehā. Once title investigations were finished, Māori owners often preferred to retain their lands for their own use, rather than pass them on to Pākehā for leasing and development. Land Councils were therefore showing themselves to be disinclined to fast-track lands to Pākehā farmers, as provided in the legislation.

Though some councils did record some successes, the government soon realised that the Māori Lands Councils experiment was not functioning as expected. Pākehā parliamentarians were also dissatisfied with the slow rates at which lands were being 'opened up', often deriding James Carroll for his preference for Māori leasing. It now seemed that direct Crown purchasing would need to be resumed.

Amending the Legislation
To this end, amending legislation was soon being introduced. On 8 November 1901 a Māori Lands Administration Amendment Act 1901 was passed, making minor changes to the original 1900 Act dealing with issues like alienation proceedings. A further amendment Act followed on 3 October 1902, granting additional powers to councils to complete their title investigations. Most critically, councils could now compulsorily acquire Māori land.

The government had once opposed granting such compulsory powers to councils because compulsion in any form was universally opposed by Māori. Now, this was being reversed. In 1903, a further amendment dealt with such issues as quorum numbers, procedures where councils failed to report, tenures and vacancies affecting all councils, papa kāinga certificates and the registration of deeds executed by councils. The governor could now, on the recommendation of the council, remove any restrictions on the alienation of lands owned by Māori.

Incremental amending legislation continued with the Native Land Rating Act, which required Māori to now pay full rates on all lands still owned, even blocks in remote locations. Once again, an element of compulsion against Māori land appeared in legislation, with the Native Minister now empowered to vest lands in District Land Councils as a punitive measure for the non-payment of rates. Further amendments followed in 1904, with the passage of the Māori Land Claims Adjustment and Laws Amendment Act 1904. Under section 3 of that Act, certain lands over which the Minister of Native Affairs had discharged mortgages could now be vested in the District Land Councils, instead of being returned to owners. Once vested in the councils, the lands could be on-leased or sold to Pākehā for farming and development.

The most effective compulsory measure of all was the Māori Land Settlement Act 1905, enacted on 8 November 1905. Existing Māori Land Councils would now become Māori Land Boards with membership reducing to three, only one of whom would be Māori.

According to the government, given the large remaining areas of 'waste Native lands' still in the possession of Māori who were not able to undertake developments, 'we suggest a ready and quick method – namely, that they be directly invested in a Board, which can deal straight away with them for the purposes of settlement'. By such vesting of 'waste Native lands', which Māori clearly lacked the resources to farm themselves, leases would be arranged, contributing to a much-needed economic base for Māori – at least, this was the theory.

To date, the councils had cost the government £8289, yet only 6773 acres of land had been acquired. This was seen to be too slow. The whole council system was seen to be too cumbersome. Involving Māori in the management of their own lands, including lands for disposal, had in fact slowed down the rate of land coming available for development.

But Māori had seen some positives. Some 30,710 Māori owners had obtained titles through the system for 347,711 acres, without protracted sittings of the Native Land Court being necessary. As Māori saw it, the government had brought about an extensive Land Council system in 1900, which had given rangatira some powers over their lands, a system that was responsive to their needs, giving Māori leaders an important role to play.[312]

However, the government saw it differently, concluding that Carroll's Māori Lands Administration Act 1900 had not been a success. Only in Whanganui and the East Coast did Māori vest their lands with their local Land Councils. Elsewhere, isolation and distances contributed to the poor responses registered with some District Councils. Antagonisms between various tribes did not help, exacerbated by arbitrary boundaries, with some Māori refusing to permit councils to manage their lands. As a consequence, very little land was opened up for Pākehā settlement.[313]

Reviewing Māori Land

In 1907, the government determined that it needed an up-to-date inventory of all remaining Māori land holdings. The government also needed to know which of these lands should be retained by Māori and which lands could be divested to Pākehā farmers for development. Prior to 1907, no accounting of total Māori land holdings throughout New Zealand had ever been attempted.

A Royal Commission, which included Ngata as a member, would therefore meet with Māori in all parts of the country, collecting details as to lands still in tribal hands. Of those lands still retained, the commission would determine which portions were essential for family and papa kāinga support; which lands were needed for farming and economic development; and which lands could be vested with Pākehā and opened up for settlement.

The commission initially deliberated from March to July 1907, with a first report filed on 19 March 1907 covering Hawke's Bay. Legislation soon followed, with changes made to vesting orders and processes of land disposal, whether by lease or sale. Increased provision was made for lease transfers and subleases, and aspects of title registration were clarified.

During the 1907 summer recess, the commission continued its investigations, visiting Gisborne, Tolaga Bay, Tokomaru, Waipiro Bay, Waiomatatini and Te Araroa. The commission then visited seven centres in the Poverty Bay area. Other areas visited were the King Country and Whanganui, before moving north visiting at least sixteen centres near to, and north of, Auckland.

A final report was issued on 12 August 1908, identifying Māori lands available for Pākehā accession. One year later, a new Native Land Act was passed, setting out in legislation how those remaining Māori lands might be acquired. The Act was by far the largest piece of Māori legislation seen to date, comprising 441 sections. Divided into twenty-four parts, the 1909 Native Land Act repealed entirely, or partially, forty-nine Public Works Acts, eighteen Local Acts and two Private Acts.

A critical feature of the Act, as detailed in Part I and Part III, was the reform of the Native Land Court, with its functions substantially increased. The Māori Land Boards were also downgraded, losing their authority to define titles; they would now operate under the direct supervision of the Native Land Court. Māori Appellate Court powers were also increased in Part II of the Act.

To many Māori, the 1909 Native Land Act represented a dramatic reversal of the sentiments enshrined in the 1900 legislation, when issues of Māori political autonomy had been incorporated into a system of Land Councils offering Māori a vestige of independence in order to protect their land interests.

Māori Land Losses after 1910

By 1910, Māori held about 27 percent of the land in the North Island, or approximately 7.7 million acres (or 3.1 million hectares). In some areas of the North Island, Māori now possessed very little of their land, especially Auckland, Hauraki, Waikato, Taranaki, Poverty Bay, Hawke's Bay, Wairarapa and Wellington.

In 1910, the tribes of Auckland and the Far North retained 760,000 acres, or 18 percent of their original estates, a further reduction from 24 percent in 1890. In Hauraki, Māori still possessed 95,000 acres, or 12 percent of their original lands, a reduction from 37 percent in 1890. In the Waikato, the iwi of Tainui had retained 289,000 acres, comprising 12 percent of their original holdings, a reduction from 17 percent in 1890. And in Taranaki, the tribes still retained 272,000 acres, or 14 percent of their original estates, reduced from 28 percent in 1890.[314]

Following the election of a new Reform Government under William Massey in 1912, Crown acquisition of Māori lands continued but tapered off significantly.

About 3.5 million acres were sold between 1910 and 1939. Of these lands, some 2.3 million passed through the Māori Land Boards, which had acquired a range of compulsory vesting powers in 1905. Other Māori lands were onsold by the Native Trustee and other agencies that possessed various powers over Māori land holdings, not least the Native Land Court. These figures do not include Māori lands transferred to Pākehā as perpetual leases.

By 1939, the tribes of Auckland and the Far North retained some 218,000 acres, or 5 percent of their original estate, representing a further reduction from 18 percent in 1910. The tribes of Hauraki now possessed 7000 acres, or 1 percent of their original holdings, a further reduction from 12 percent possessed in 1910. In the Urewera, Tūhoe had retained 116,000 acres, or 11 percent of their original holdings, a dramatic reduction from 72 percent in 1910. In the Waikato, Māori had by now retained 32,000 acres, or 1 percent of their original lands, a reduction from 12 percent held in 1910. And in Taranaki, the eight tribes had retained 20,000 acres, or 1 percent of their original cumstomary estate, a reduction from 14 percent in 1910.[315]

Dispossession in Summary

As we have seen, by 1869, the government had already acquired about 7 million acres of the North Island's 28.5 million acres, as well as almost all of the South Island. Over the next 40 years to 1910, the government would acquire a further 8 million acres of the North Island, according to the government's own calculations though, as historians like Richard Boast have pointed out, these figures certainly understate the acreages acquired by 'hundreds and thousands of acres'.[316]

Be that as it may, the 8 million acres acquired by the government between 1869 and 1910 were purchased for just over £2 million. Thereafter, up until 1928, the government acquired a further 1.4 million acres at a much higher total cost of £3.4 million. Therefore, between 1869 and 1928, the government spent some £5.4 million acquiring 9.4 million acres of the North Island.

Given these figures, it is clear that, after 1910, when the government spent £3.4 million to acquire 1.4 million acres, the per-acre expenditure 'increased very substantially'.[317]

By 1920, on the eve of the land development schemes being launched, Māori had retained 4.7 million acres in tribal hands. Of these lands still possessed in 1920, about 2.8 million acres had been on-leased to Pākehā farmers. The East Coast Commissioner and Public Trustee were also leasing about 300,000 acres.

Of the 4.7 million acres owned by Māori in 1920, then, only about 380,000 acres were actually occupied by tribal communities. And in 1920, about 1.2 million acres of Māori land were unoccupied, vested variably in Māori Land Boards or under the control of the Native Land Court and other agencies, like the East Coast Commissioner. Finally, in 1920, only about 15,000 acres of Māori land remained in the original, customary tenure. Everything else had been investigated, extinguished and retitled by the Native Land Court, a process that had commenced in 1858.

Charting a Way Forward

On 10 December 1928, possibly much to his surprise, Sir Apirana Ngata had been awarded the coveted portfolio of Minister of Native Affairs. This came about because the United Party, of which he was a member, had won the election of 1928, against most predictions. Ngata, now aged fifty-four, would be ranked third in the new United cabinet and would occasionally act as Deputy Prime Minister. 'Always a hard worker, frequently far into the night, Ngata could get by with very little sleep, often mere catnaps. He entered into his new ministry with a ferocious energy that left men half his age, especially his hapless staff, trailing in his wake.'[318]

Since the advent of colonial settlement, Māori had lost prodigous amounts of customary land, devastating tribal economies. By the 1920s, not least because of the influence of Ngata, government policy towards Māori began to change, if slowly, reflecting

Relative Costs of Māori Land Purchases 1869–1928

	Acres acquired	**Pounds expended**	**Price per acre**
1869–1910	8 million	£2 million	6 shillings
1910–1928	1.4 million	£3.4 million	£2.3 shillings

'a strengthening commitment to Maori physical wellbeing'. Such a strengthening of government resolve was now essential if Māori were ever to emerge from the impacts of such large-scale losses inflicted since the 1860s.[319]

Once appointed, Minister Ngata set about rectifying those losses, as best he could, focusing on the land. His land development schemes after 1929 attracted little national attention. But each development scheme represented a small yet significant gain for Māori who were still as determined as ever, as stated by James Carroll in 1891, 'to become useful settlers and contribute to the wealth of the country'. Māori had once thrived as agricultural producers, and would do so again.[320]

Postscript: Māori Land Today

The Māori land acreage held by Māori today is about 1.5 million hectares, constituting 5.6 percent of New Zealand's total 26.9 million hectares. Like all citizens of New Zealand, Māori have of course the right to purchase and own 'general land', that is, land that is freely available on the market. Māori can also retain interests in ancestral estates, or 'Māori land'.

The largest concentration of Māori land in tribal hands is in Waiariki Māori Land Court district, with 426,595 hectares, comprising about 28 percent of all lands held by Māori. The Aotea district Māori have 334,207 hectares in tribal ownership, equalling some 22 percent of the total Māori holdings. A close third is the Tairawhiti district, where 310,631 hectares are held in tribal hands, comprising about 20 percent of all lands owned by Māori. Together, Māori within these three Māori Land Court districts possess about 70 percent of the current Māori land holdings.

Other areas possess smaller, but no less significant, acreages of Māori land. Taitokerau possesses about 140,000 hectares, or 9 percent of the overall Māori land holdings. Waikato-Maniapoto holds about 143,338 hectares, which also comprises 9 percent of the overall Māori land holdings. Tākitimu possesses some 88,000 hectares, or 6 percent of Māori land holdings. Finally, Te Waipounamu owns about 72,000 hectares, which equates to 4.7 percent of the overall holdings held by Māori.

According to a recent report issued by Massey University, 'Māori land today is generally of poorer quality than general land, largely because the most fertile and best suited land for agricultural production was sold or confiscated from its Māori owners in the 1800s or early 1900s'.

Accordingly, large areas of Māori land are utilised for forestry and conservation. A significant proportion of Māori land occurs in fragile natural environments such as wetlands or bordering lakes and rivers, thereby possessing marginal productive value at best. About 30 percent of Māori land is also landlocked, contributing to development difficulties. 'Owners of such land are [often] unable to access the land and have no option but to lease it to neighbours who have access.'[321]

In the end, such issues as exist today were certainly evident in abundance in the late 1920s and early 1930s. The instigation of Sir Apirana Ngata's land development schemes in 1929 and the launching of the Ahuwhenua Māori farming competition in 1933 both represented significant undertakings aimed at rapidly increasing and expanding the economic base of Māoridom, centred on the land.

From 1933, the Ahuwhenua Trophy competition in no small measure contributed to the strengthening and immeasurable transformation of that well-grounded Māori economy.

The Minister of Treaty Settlements, Hon Chris Finlayson, and Dame Tariana Turia, Member of Parliament for Te Tai Hauāuru, sign the Te Ātiawa Heads of Agreement at Rangiatea, New Plymouth, on 9 August 2014.

(Danny Keenan)

Te Awahohonu Forest Trust – Gwavas Station
(John Cowpland – Alphapix)

Chapter Ten

Revisiting the Early Ahuwhenua Farmers – 1933–1963

A New Way Forward

With the introduction of Sir Apirana Ngata's land development schemes in 1929, Māori were once again presented with an opportunity to work towards economic recovery and expansion. The land development schemes, and the Ahuwhenua Trophy competition, offered a way forward, which Māori farmers were determined to utilise.

Early Ahuwhenua Farmers

Following its inception in 1933, the Ahuwhenua Trophy competition attracted the interest of Māori farmers and competitors located in all parts of New Zealand. Detailed reports prepared by the Ahuwhenua judges, who travelled huge distances to visit competitors on their isolated farms, tell the stories of these pioneer Ahuwhenua farmers.

This chapter will briefly revisit some of these farmers and their families, highlighting the challenging conditions under which many were working. The chapter acknowledges these selfless endeavours, as described by the judges travelling across the development blocks, assessing farming practices, walking the land, speaking with the families. Their inspection reports necessarily focused on farming issues – things like animal husbandry, record keeping, cleanliness, stock quality, fencing, winter feed, shelter belts, fertiliser, noxious weeds and, not least, pasture quality. But occasionally, a judge would offer a personal observation or a background story, throwing compelling light on the lives of these early Māori farmers who contributed so much to the modern, thriving Ahuwhenua Trophy competition.

For every Māori farmer visited and assessed, however, there would have been innumerable other farmers, seldom noticed but equally working across the land development schemes, struggling to convert difficult landscapes into viable, economic entities. The earliest years were arduous because of isolation, inexperience and immeasurably strenuous environmental conditions – weather, soils, gorse, water, unbroken terrain – which demanded hard work and sacrifice from all of these earliest guardians of the soil.

Launching the Ahuwhenua Trophy Competition

When launching the Ahuwhenua Trophy competition at Ohinemutu Marae, Rotorua in 1933, Governor-General Lord Bledisloe complimented the seventy-one Māori entrants for their enthusiasm in supporting, and competing for, the new Māori farming trophy.

All entrants were divided into seven groups, with each group representing an area within Waiariki and the Bay of Plenty, where land development schemes had been launched. Ngata's earlier Native Lands legislation of 1929 had enabled Māori owners to take control of their remaining land holdings. For its part, the government had offered legal support, financial assistance and supervisory support.

In early 1931, Lord Bledisloe had visited a number of new land blocks being developed in Rotorua and the Bay of Plenty. Bledisloe was particularly impressed by the hard-working Māori farmers from areas like Te Kaha, Tōrere, Whakatōhea, Horohoro, Mourea and Maketū. Initially focused on Waiariki and the Bay of Plenty, by 1936, the Ahuwhenua Trophy competition had extended across all farming districts of Aotearoa.

Nominating and Judging

Competitors for the Ahuwhenua Trophy were in the main nominated by farming supervisors working within the Department of Native Affairs or, after 1945, the Department of Māori Affairs. In nominating potential competitors, supervisors provided a substantial amount of information, including name, tribal affiliation, location of farm, acreage of the property, buildings and farm implements, plus a statement as to the condition of the property before developments had commenced.

Details of stock being carried were also provided, with added details like butterfat quanities over the previous two or three years. This information was preceded by a general statement highlighting positive elements of the competitor's work record and achievements. The nomination forms were eventually passed to the appointed judges who, during these early years of the competition, were required to travel some distance in order to meet with widely dispersed Māori farmers and inspect their farms.

Ahuwhenua judges were appointed for varying periods of time. Throughout the early years, judges seem to have travelled about the Māori farms largely on their own. However, in 1945, the Department of Māori Affairs was revamped with an ethos to expedite the economic and social advancement of Māori who were, at that time, still largely a rural population.

Accordingly, departmental land supervisors now accompanied the judges in an effort to raise their supervisory profiles, offering advice and assistance to Māori farmers as they sought to progress their land developments. Newly appointed welfare officers also accompanied the Ahuwhenua judges, representing a heightened concern for the wellbeing of remote Māori families.

In 1948, Judge G A Blake first reported that he had visited thirteen competitors 'in company with either the Maori Affairs Supervisor, or the Welfare Officer'. Judge Blake was an instructor in agriculture from Matamata. One year later, he again reported that, in visiting Māori farms, he had been accompanied by a Maori Affairs Supervisor. Thereafter, supervisors regularly accompanied the judges who, in turn, appreciated their assistance and support.[322]

Visiting Māori

In 1933, the inaugural Judge, W D Dempster, who was a dairy instructor from Hamilton, visited at least an astonishing eighty-one farms spread across the central North Island. This did not include the farm of Hirinia Waititi of Cape Runaway because Waititi 'was not ready but he hoped to win first prize next year'. Unfortunately for Waititi, the Ahuwhenua competition lapsed between 1934 and 1935. However, in 1936, he did enter and was placed a creditable seventh with an overall mark of 72 percent.[323]

Two years later, the trophy was shared by two farmers who could not be readily distinguished.

In 1945, the winner of the trophy was Joe Wharekura from Horohoro. The win set a new record; previously, the trophy had been won twice by Horohoro farmers, and it would be won by Horohoro farmers again. Mr Wharekura had been placed third in the previous year's competition.

Daughters of the Soil

Throughout the early years, Māori women competed for the trophy with a creditable measure of success.

In 1938, Mrs Huinga Nepia of Tikitiki was placed fourth, with Mrs C R Beazley of Kaiaua, Pokeno, placed seventh. According to the judge, good work had been done on Mrs Beazley's farm by her husband in clearing and developing this 'small and difficult piece of land'. Mrs Beazley had also just commenced farming pigs but thus far 'there was no pig layout – only a fattening pen'.[324]

Ever competitive, like her husband, Huinga Nepia entered the competition again in 1939, and was placed sixth.[325] One year later, in 1940, the winner of the trophy was Mrs Tatai Hall of Te Teko who had been placed second the year earlier. Another Māori 'daughter of the soil' was Mrs Ngarangi Kohere of Rangitukia who was placed third in 1940, having also attained third place the previous year.[326]

Success for Māori women continued in 1941 when Hiria Wi Hongi was placed fifth. Unfortunately, like many other Māori women competing, no further information was provided in the judge's otherwise quite long report, except that her farm was placed into a B judging category. Farms in this category were described as being of a 'ploughable nature but which [had] not reached a high level of development'. Further, Hiria's farm was only partially coverted to permanent grass, with the rest covered in native vegetation.

Seven years later in 1948, Mrs Elizabeth F Clemens of Tauranga was offered 'special praise' for the way she had developed her farm out of a heavily gorse-infected area. Mrs Clemens was also commended for the winter shelter she had provided for her stock, a necessary safeguard that many Māori farmers continued to overlook. However, her practice of cutting the same field for hay each year, without rotation, was causing deterioration.[327]

Reward for Consistent Endeavour

Having been placed seventh in 1938, Mrs Beazley of Kaiaua returned in 1952, this time to win the competition. Mrs Beazley was to be 'heartily congratulated' wrote Judge Taylor who had by then acted as judge on five occasions. Mrs Beazley's win he said was 'richly deserved for consistent endeavour, often against tremendous difficulties in the early stages of development, and painstaking attention to those details of farm management that so frequently make all the difference between success and failure'.

Mrs Beazley's success over fifteen years of farming would constitute a source of personal satisfaction. But it should also serve, wrote Judge Taylor, 'as an inspiration to those of her people with whom she comes into contact'. Mrs Beazley was now undoubtedly 'a practical and enlightened leader of

Ahuwhenua judges were often impressed by Māori commitment to developing their often rugged and difficult farms.
(Danny Keenan)

her Race in the field of Agriculture', which was of course New Zealand's most important industry, added Judge Taylor.

One remarkable aspect of Mrs Beazley's success in 1952 also lay in the fact that, in that year, four Māori women had featured amongst the final eleven Ahuwhenua Trophy competitors: Mrs M Reid from Mangonui attained third place, with Mrs Elizabeth Clements of Omokoroa returning, this time to earn sixth place, with Winnie Shedlock of Towai placed ninth.

Third-placed Mrs M Reid of Mangonui was said to be farming the most difficult country 'and few would relish the prospect of breaking in the steep hillsides covered in gorse and tall manuka'. But in fact, this is what she 'and her resourceful family' had done for seventeen years. Mrs Reid had also been unwell for some considerable time, rendering her achievements all the more astonishing.

Following her positive showing in the 1948 competition, Mrs Clements of Omokoroa gained sixth place in 1952. According to Judge Taylor, Mrs Clements 'had worked hard and well with her husband, and later her son, to convert a one-time gorse and ragwort infested property into an excellent small dairy farm, producing 226 lb of butterfat per cow, or 149 lb per acre'. This achievement 'was highly creditable and must be a source of pride and satisfaction to those responsible'.

Winnie Shedlock's farm, located at Towai, had not been of a standard to attain a place in the competition. 'Nevertheless, the entry of this competitor does serve to show what a capable woman can achieve through perseverence, determination and the will to succeed, often against great difficulties.'[328]

An Example to Farming Neighbours

In 1954, following the division of the competition into two sections (sheep and cattle, and dairying), the winner of the dairying section was Mrs Mihi Stevens from Rangiāhua. 'In securing first place,' wrote Judge Taylor, 'the hard and often irksome toil of 22 years in a locality bad of access and far removed from sources of material aid has at last gained for Mihi Stevens due recognition and just reward'. Stevens had farmed her 89 acres for two long decades, from the days when it was 'just a Wiwi swamp cluttered up with Puriri logs and stumps'. Developments had been tedious but they had borne results.

Judge Taylor was obviously impressed by Stevens' achievements and success: 'In congratulating this competitor and her husband on their well-earned victory, I would like to think that the excellent example that they have for so long set their neighbouring farmers by their industry, knowledge of farming, initiative, steadiness and honesty of purpose will not only serve to inspire them to emulate their good work but will also point the way to achievement to their people as a whole.'

Reflecting further on Mihi's efforts, Judge Taylor applauded Māori farmers who strove to attain such high awards: 'For several years now I have judged this competition and in doing so my eyes have been opened to the high level of attainment many Māori farmers reach when they have the will and the determination to succeed.' That so many did succeed was clear evidence that 'the high standard often reached by European farmers is not beyond the capacity of Māori farmers to attain'. It was commendable that such Māori farmers sought 'to make a real success of their lives, as have the various competitors in this and past competitions'.[329]

Drought

When reporting on the creditable efforts of Māori farmers to convert and utilise their difficult landscapes, judges of the Ahuwhenua Trophy competition often remarked on 'the most difficult country' that 'few would relish the prospect of breaking in', as Judge Taylor had observed of the efforts of Mrs Reid in 1954. The task of creating viable, economic units out of the land development blocks was undoutedly difficult, demanding much from Māori farmers.

One of the most consistent problems faced by Ahuwhenua Trophy competitors, and therefore all Māori farmers, was drought. In 1944, the competition had attracted entries from nine Māori farming units. But with the exception of two from Northland, all others had suffered from egregious drought conditions, a fact which Ahuwhenua judging had needed to take account of.

In 1946, it was observed that Tupu Erueti of Tautoro was working 'on an exceptionally hard place to farm', especially during the dry weather. 'During this season, the water supply failed and the livestock had to be driven to water every day during the dry spell.' Mr Erueti was said to be 'a hard working reliable type of man and it must be admitted that he has done very well under the conditions he has been up against'.[330]

Drought conditions were again a factor in the judging in 1950. The standard of Māori farming was generally high, observed Judge K M Montgomery, who was an instructor in agriculture from Te Kuiti. However, the points to be awarded needed to take account of the continuing 'adverse climatic conditions which have prevailed during the past 4 months', not the least of which was an unseasonal lack of water.

That same year, P Toroa of Ruatōria 'made an excellent showing and thus far [reflected] his ability to get things done'. But rainfall had been scarce, detrimentally affecting his farming operations. However, the rain that fell a few days prior to inspection had revived his pastures and it was evident from the growth made that these would be very productive.[331]

One year later in 1951, the Ahuwhenua winner was Kopua Waihi, also of Ruatōria who had worked 'conscientiously over a very long period, his industry being reflected in the fine farm he [had] developed'. But severe climatic conditions (droughts and floods) 'have often been against him'.[332]

The competition judging of 1954 was also qualified by 'dry, droughty weather which was experienced throughout the dominion during January and February of 1954, having an adverse effect on the farming operations of most competitors'. In all cases, butterfat figures were much reduced, with stock also affected. 'Due note was taken of this unfortunate happening and in the allocation of points to each competitor, thus compensating them, based nonetheless on the merits of each case.'[333]

Particularly Difficult Properties

Judges often commented at length on how difficult many of the properties being developed actually were, often concluding that efforts underway seemed destined for disappointment. At the same time, Māori endeavours were praised.

Tiaki Tamaki of Pirongia had been observed in 1949 working a property badly affected by ragwort, which had not been brought under control by the time of the judge's visit. In fact, there was the possibility that Mr Tamaki might be forced out of dairying and into sheep, either wholly or to a larger extent than at present, should his issue with ragwort and other similar plants continue. Two years earlier in 1947, Mr Tamaki had been observed farming on soil that was much poorer than that of the other competitors. It was reported, 'The property only awaits unrationed fertiliser supplies to show a big increase in production'.[334] Despite these problems, Mr Tamaki worked resolutely; and, in particular, he 'outclassed all other competitors in the amount of supplementary feed he had provided for the winter'.

Also in 1949, Ahuwhenua Trophy competitors Waerata Harris of Inglewood and Irihapeti Tuhakaraina of Tauranga were both complimented for their remarkable work in bringing into production areas that had 'formerly been completely abandoned to weeds'. Both farms now reflected great credit to their occupiers.[335]

A competitor in 1950, Tapuae Rogers, was observed farming a leasehold property with his two sons 'in a most efficient and creditable way'. This effort was all the more remarkable, noted the judge, because access to the property was a major and continuing issue: 'it was so bad that even a jeep could barely negotiate the hazardous roads in winter'.

Another competitor in 1950, Kuratau Hau, was away from his farm when the judge called. His stock had also been turned out into the bush on another property. Accordingly, Judge Montgomery could not speak with Kuratau, nor assess his stock.

However, an inspection of his farm did proceed. Kuratau and his family had worked very hard over the previous sixteen years, observed the judge. The land had been exceptionally difficult to break in, being poor clay country covered by mānuka and gorse. 'And to make matters worse, access was limited to a poorly-formed sledge track which became practically impassable during the winter.'[336]

One year later in 1951, competitor A T Hetet had hosted a farm inspection. Mr Hetet had taken over his farm, wrote Judge Taylor, three years earlier following an extended period of leasehold where the farm had been allowed to deteriorate, especially with broom, ragwort and blackberry. Mr Hetet was working hard to address the problem; 'the effort taken to convert and develop this land', observed the judge, 'could only be the work of a keen industrious and resourceful man'.[337]

Providing Shelter

A decade later in Taranaki, E Tamati of Bell Block had won the 1965 dairying category of the Ahuwhenua competition. Mr Tamati had been praised for the quality of his farm, though he did face an issue perhaps unique to Taranaki; that of fragmented titles, which had meant that his property was divided up into several scattered units. The judge reported that Mr Tamati had won the competition, despite the awkward nature of his fragmented farm: 'He has shown what can be done with bits and pieces of land.'

A short distance away, third-place Mr T Manu of Opunake had been observed struggling with high boxthorn hedges that were proving difficult to eradicate, compromising his farming operations. However, the judge did commend Mr Manu on his ingenuity in building grandstand accommodation, overlooking his dairy shed, for his young children while milking was in progress.[338]

One year later in 1966, Mr J Niwa of Puniho, also in Taranaki, had been adjudged as second in the dairy category because of his high rate of production per acre, which had been commendable. However, an urgent problem facing Mr Niwa was the need for shelter. 'An impression gained at the time of the visit was that shelter was badly needed, there being a very cold wind blowing in from the sea, apparently a very common occurrence by the look of the few trees.'[339]

Isolation

Judges often remarked on how remote some Māori farms were, posing particular difficulties for Māori farmers who were so isolated.

In 1949, one judge reported that John Tiwha of Marunui had a large and difficult property and was 'making an excellent job of it with a considerable handicap in that he has to do most of the work himself without a great deal of outside assistance'.[340] The winner of the competition in 1950 was G Thompson of Otorohanga, whose win was remarkable because 'he has had practically no supervision and has had to rely on his own initiative'.[341]

Of all the farms visited in 1951, reported the judge of that year, the property of John Savage 'was the most illustrative of what a Māori farmer could achieve'. Mr Savage had lived on his farm for over fifty years, during which time 'he had battled along unaided in a most isolated place'. Mr Savage had taken on other labouring work in order to acquire monies to develop his farm.

'Progress was necessarily slow and many times he must have felt disheartened as a consequence.' Nevertheless, 'he stuck to his guns and refused to surrender'. Assistance did eventually come in 1939 in the way of a small loan from Māori Affairs, along with the assistance of good supervisors.[342] Eventually, in 1947, Mr Savage earned second place in the competition. At that time, he was observed to be farming an 'isolated locality seldom met with in dairying' and performing 'a very good job'. His homestead surroundings, orchard and garden 'are to be particularly commended'.[343]

Illness

In 1944, Judge J F Shepherd, who was a fields instructor from Hamilton, observed that one competitor, Mr Taitoko Bailey of Opunake, had died shortly after his inspection. In a letter to the Native Department, Judge Shepherd expressed his remorse and asked for the address of Mr Bailey's next of kin.[344]

Katene Huriwai of Tikitere had also toiled assiduously on his otherwise difficult property, being too ill to meet with Judge Shepherd when visited.[345]

In 1948, Judge G A Blake from Matamata complimented Patuwahine Albert of Omaio especially because of his advanced age. In his report, the judge made reference to the fact that 'this old couple have

Isolation, difficult terrain and lack of adequate transport severely taxed the resolve of the early Ahuwhenua farmers.

(Danny Keenan)

done remarkably well and now have a well subdivided farm where clumps of native bush have been reserved for shelter'. Milking, however, was laboured because it was still being done by hand; and it was recommended that 'the milking shed should now have milking machines installed'.[346]

Three years later in 1951, Judge Taylor observed Mr K A McGregor of Waihi developing a property that McGregor had once worked on as an employee of the Māori Affairs Department. 'Unfortunately he has been a sick man for some considerable time and has had to depend largely upon his wife and daughters of school age to carry on the most essential work on the farm', Judge Taylor reported. Also in 1951, Manu Stainton had inherited a family property bedevilled by constant flooding.[347]

In 1966, the winner of the dairy category was Mr A K Mauriohooho of Owairaka, Waikato. Despite his ill-health, he had managed to continue working with the assistance of his daughter who took a close interest in farming. A year earlier, Mr Mauriohooho had been placed second. As observed by the judge in 1966, his achievement in winning was all the more remarkable because of his continuing ill health. 'His daughter must be commended for the way that she has taken part in helping with the farm management.' Nonetheless, Mr Mauriohooho had 'gamely come forward into the competition, as had so many other Māori farmers who were similarly afflicted'.[348]

Pride in the Family Home

Ahuwhenua competitors were required to satisfy a number of criteria in order to be awarded points by the judges. One of these criteria related to the cleanliness of the farm, buildings, implements and the family home. Many Māori competitors scored particularly well in this category.

In 1946, Judge T E Rodda described the eventual winner, Henare Paraone of Clevedon, as a 'good all-round farmer with knowledge of dairying, fat land raising, bacon production and beef production'. His houses and buildings were also tidy. One particular feature noticed by Judge Rodda was the number of orante carvings placed about the farm: 'Māori carvings adorn many of his farm buildings.'[349]

The property of competitor Eruera Hoera of Kaitaia in 1946 was said to be 'too hilly for the breed of cow' he was running. A more 'productive type of cow' was recommended by Judge Rodda. 'Perhaps a Jersey-Shorthorn cross would be more suitable for butterfat production.' One year later, in 1947, the judge again recommended a change to Mr Hoera's herd. Perhaps the purchase of ten to twelve 'good jersey in-calf heifers and a good jersey bull' was the answer. However, turning his attention to Mr Hoera's farm and buildings, Judge Rodda commented that great pride was taken by this family in their home and garden 'which are exceptionally well kept'.[350]

Mr Hoera evetually won the competition in 1949. The judge observed that Mr Hoera had done an exceptionally good job in breaking in a farm from standing bush. The financial assistance he had received from Māori Affairs 'has been comparatively little', yet all of his debts were repaid, while 'he has a substantial sum to his credit'. The hard work had mostly been done single-handed 'and the farm today would be regarded with pride by any farmer, Maori or Pakeha'. It was easily the most difficult country farmed by any of the competitors. 'This farmer can truly be said to be a "son of the soil".'[351]

In 1948, Judge Blake commented that the outstanding feature of Bunny M Otene's farm, near Hastings, was 'the homestead and garden with its beautifully furnished home kept spotlessly clean'.[352] Mr Herewini Rewa's spacious grounds were also neat and tidy, the result of much hard work when farm duties were finished.[353]

Families Farming Together

On occasion, judges noted that particular farming endeavours were the product of family members working together.

A good example was the farm managed by the Potaka brothers at Maketu in 1951. The brothers had taken over their father's farm, following his death in 1948. The older brother had already done a substantial amount of the work during its development phase, when the farm had been covered in blackberry and ragwort. The younger brother had not long returned to the farm; the two brothers were now working well together. Two full-time male workers were also employed on the farm, providing an advantage in manpower not enjoyed by other competitors.

Also in 1951, one competitor named Eru Mako Pou worked as an officer in the Maori Affairs Department in Kaikohe. Because of this full-time position, he could only devote his spare time to the development of the farm. Despite this, he and his sons (school or college age) had managed to convert a one-time gorse and blackberry infested area to its present state of advanced development. Another competitor that year was Noti Tiopira who had taken over his father's farm about three years prior, after his father had fallen ill and the farm had been allowed to deteriorate. Through Mr Tiopira's effort, the farm was now undergoing a dramatic transformation.[354]

Joseph William Hedley was aged sixty with a grown family when he won the Ahuwhenua Trophy in 1955, and he earned a place in 1959, 1960 and 1961. Hedley was described as 'a man of outstanding character, highly regarded in this district'. Both he and his wife had more recently taken a keen interest in the welfare of local Māori, with much of the farm management being taken over by their son.[355]

One of the competitors in 1962 also had close family connections. The winner of the dairy category, Mr Wiremu Maki, alias Ngawati Maki, was working a substantial property that he had originally broken in with his brother. However, both now had grown families. Because the land being farmed could not support both families, the brothers had decided to acquire a second property, which was now being managed by the second brother.[356]

Returned Soldiers

In the post-war period, a number of former soldiers were rewarded with Ahuwhenua recognition. One of these was Joe Wharekura of Ngāti Kahungunu who farmed a part of the Rongomaipapa Development Scheme, Rotorua. Wharekura, who was declared the winner in 1945, had worked on his particular farm since the early 1930s. During his war-time absence, his wife had managed the farm.[357]

Another was Tikirau Callaghan of Whānau a Apanui descent. Mr Callaghan won the award in 1948. After service during the war, Callaghan had purchased land at Ōrete, near Waihau Bay. The judge in 1948 described him as 'an exceptionally good dairy farmer and his management stands out as an example to many pakeha farmers'. He was presented with his award on 20 October 1948 by his former commanding officer, now Governor-General Sir Bernard Freyberg.[358]

A competitor in 1951 named Hohaia Puriri had occupied his farm for seven years, less two years spent on active service in Korea.[359] In 1954, five former servicemen gained minor placings in the Ahuwhenua competition of that year. Mr E Kingi had occupied a farm of 423 acres for about two years, following his army discharge. Mr E Pohio had managed his 445 acres for about the same length of time. Two brothers, Mr D Royal and Mr T Royal were also occupying adjoining properties, and had done so for three years. The final ex-serviceman to occupy a farm, and coming to the notice of the Ahuwhenua judges, was Mr John Joseph Reid. Aged forty-four and single, Mr Reid nonetheless 'comes from a family of high standing', the judge reported.

All five ex-servicemen had acquired their farms with the assistance of the Department of Maori Affairs acting as agent for the Rehabilitation Board. Judging such men, wrote the judge, would be difficult because only time would tell if they were able to demonstrate such qualities as were demonstrated by other competitors with much longer experience.[360]

Judges Supportive of Māori Farmers

Generally speaking, the reports of the Ahuwhenua judges were quite long and very detailed, as befits the seriousness with which judges undertook their responsibilities. Not surprisingly, the reports focus on the minutiae of farming practice, given that their purpose was to nominate winners and place-getters each year for the awarding of the prestigious Ahuwhenua Trophy.

In the course of their duties, many judges took time out to comment on how they saw Māori farmers performing in the management of their development properties. Though occasionally couched in cautious tones, most observations of Māori made by the judges were in fact very complimentary, recognising the effort and hard work being applied by Māori farmers as they progressed their land development aspirations.

In 1949, Judge A V Allo stated in his report that he wished to give full credit to each competitor that year 'for the effort he has put into the development and improvement of his farm since occupation, taking into consideration the improvement in production and pastures', all of which he regarded as exceptional. Judge Allo added that, given the great difficulties that Māori farmers had to contend with, such as difficult topography, poorly drained areas of swamp or very weedy ground, all Māori farmers visited were demonstrating 'considerable ability in coping with these problems and making well farmed units of their areas'.[361]

Judges were also, on occasion, very complimentary of indivual Māori farmers. In 1950, Judge Montgomery praised Mr Sam Shelford of Whakapara for 'his industry, initiative and resourcefulness as a farmer – in fact, his efforts would do credit to a European farmer of more than average ability'. That same year, the farm of F Hei from Te Kaha reflected an air of cleanliness, tidiness and attention to the smallest detail.[362]

In 1953, nine farms were entered into the competition. According to Judge Taylor, the winner, Rohe Takiari, was 'opposed by a group of eight rivals providing the highest standard of farming efficiency met with in four years of experience judging this competition'. For this reason, it surely was an honour to be placed at all, but to gain highest ranking, 'as Mr Takiari has done, is something to be proud of indeed'.

Judge Taylor then expressed the hope, perhaps reflecting the view of all of the judges, that Mr Takiari's achievement would 'inspire him to still greater effort in the future'. In the years to come, he would be able to look back with great satisfaction on a job well done – a 'monument to his own industry and resourcefulness, and an example to others of his Race'.[363]

Quentin Bedwell

Tataiwhetu Trust
(John Cowpland – Alphapix)

Chapter Eleven

Ahuwhenua Trophy Competition Continues 2014–2021

2014 Dairy

Showcasing Ahuwhenua Farmers

When the three finalists for the Ahuwhenua Trophy Bank of New Zealand (BNZ) Māori Excellence in Farming Award were announced in 2014, the Minister of Māori Development, Hon Dr Pita Sharples hosted a special function at Parliament. It was the first time that a function showcasing the awards had been held in Wellington. Leaders of farming, Te Ao Māori and agribusiness attended, as did represenatives and whānau of all of the finalists. The venue was appreciated by the Ahuwhenua Trophy Management Committee and finalists, who welcomed the opportunity to showcase the success of Ahuwhenua before such a diverse and influential audience. Dr Sharples thanked the Ahuwhenua Trophy Management Committee for their continuing achievement in safeguarding and promoting this prestigious award, which had arisen in 1933 from the foresight of rangatira like Sir Apirana Ngata, along with Governor-General Lord Bledisloe.

The Chairman of the Ahuwhenua Trophy Manangement Committee, Mr Kingi Smiler, applauded the 2014 dairy finalists. Each one had attained the highest standards, not only in their farming operations but also 'in their commitment as kaitiaki of their lands'. All finalists had demonstrated outstanding innovation in organising and managing their farms. Māori were now significant players in the New Zealand economy, particularly in the dairy industry. 'Māori farms always aim to be highly sucessful and profitable operations, but in so doing they do not compromise their commitment to key values such as caring for the land and their people', he said.[364]

The Finalists and Field Days

Three finalists were selected for the 2014 Ahuwhenua Trophy. The first was the Putauaki Trust – Himiona Farm, a new entrant into dairying situated on the scenic Rangitaiki Plains between Te Teko and Edgecumbe. The Putauaki Trust hosted a field day on their property on 11 March 2014. A second finalist was Te Rua o Te Moko Ltd situated on the fertile South Taranaki Plains just north of Hāwera.

The 2014 finalists

Back row, left to right: Kiriwaitingi Rei, Putauaki Trust – Himiona Farms; Dion Maaka, Te Rua o Te Moko Ltd; and Enid Ratahi-Pryor, Ngāti Awa Farms (Rangitaiki) Joint Venture – Ngakauroa Farm

Front row, left to right: Hon Pita Sharples, Minister of Māori Affairs; and Hon Nathan Guy, Minister of Primary Industries

(John Cowpland – Alphapix)

The Te Rua o Te Moko Trust organised its field day on 14 March 2014. The third finalist was the Ngāti Awa Farms Ltd – Ngakauroa Farm, which was located in very good dairy farming country about 12 kilometres south of Whakatane. Ngāti Awa Farms hosted their field day on 19 March 2014.[365]

The Winner is Announced

The winner of the 2014 Ahuwhenua Trophy for dairying was the Te Rua o Te Moko Ltd, from South Taranaki. The award was announced by the Minister of Māori Affairs, Dr Pita Sharples, at a special awards evening held in Tauranga on 13 June 2014. The awards evening was attended by the Governor-General, Sir Jerry Mateparae, the Minister for Primary Industries, Hon Nathan Guy, finalists and their whānau, and other leading representatives of the agribusiness sector.

The Ahuwhenua Trophy was presented to the Chairman of Te Rua o Te Moko Ltd, Mr Dion Maaka, by Sir Jerry Mateparae. Mr Maaka was also presented

Dion Maaka, Chairman of Te Rua o Te Moko Ltd, winner of the 2014 Ahuwhenua Trophy for dairying

(John Cowpland – Alphapix)

with a replica of the trophy by Hon Nathan Guy and the winner's medal by the Chief Executive of the Bank of New Zealand, Mr Anthony Healy. Te Rua o Te Moko Ltd also received about $40,000 in prizes donated by the Ahuwhenua Trophy sponsors.

The Chairman of the Ahuwhenua Trophy Management Committee, Mr Kingi Smiler, commended Te Rua o Te Moko Ltd for its success, describing it as a 'shining example' of how Māori were collaborating by combining small trusts in order to create larger and more economically viable enterprises, which together served to enhance and strengthen the Māori and New Zealand economy.

An outstanding feature of the 2014 competition, said Mr Smiler, was that 'all finalists were an amalgam of smaller entities' that had put aside their individual interests for the greater good of a larger whānau. This approach had commenced eighty-one years ago, encouraged by Sir Apirana Ngata. 'Today, the collaborative approach is now gaining greater momentum'; and this was clearly evident 'resulting in some outstanding achievements'.[366]

Ahuwhenua Trophy Finalist Profiles

Te Rua o Te Moko Ltd

Located just north of Hāwera on Highway 45 near Manaia, Te Rua o Te Moko Ltd managed 500 Kiwi-cross cows on a 170-hectare effective milking platform. The farm was a highly successful operation, which by 2014, was in its fifth season, producing 190,000 kg of milk solids. What was so special about this operation was its status as an aggregation of four separate Māori Trusts which, over time, had combined their resources and whānau interests to create Te Rua o Te Moko Ltd. As a consequence, Te Rua o Te Moko was seen as an economically and environmentally sustainable dairy operation.

Combining many blocks into a single entity had marked the beginning of a new era for the 1100 Ngā Ruahine owners, presenting them with 'an exciting vision for the future'. Individually, the blocks had been too small to farm as viable units. However, thanks to the present combined entity, enhanced financial returns had been possible for whānau and owners. Beyond operating as a commercial dairy farm, Te Rua o Te Moko was also managing a training operation for Ngā Ruahine shareholders who could whakapapa to the land. However, other interested young people were also able to participate. The training operation, managed by Land Based Training, had seen eight young people graduate in 2013. All eight had since secured employment on dairy farms, with a further eight enrolling in the new year.[367]

When assessing Te Rua o Te Moko Ltd, the judges had been impressed by Te Rua o Te Moko's strong goverance experience in strategic, financial and technical areas. Their Land Based Training Joint Venture was seen as a postive strategy to engage and train local Māori, encouraging young people especially to become involved in agriculture on their own mana whenua. Financial analysis using DairyBase reflected a strong business model generating good profits with low debts. Production had dropped over 2012/2013 with changes to the herd and calving patterns, though this was short term, with milk production rising to 400 kg of milk solids per cow, up from 300 kg of milk solids per cow.[368]

Putauaki Trust – Himiona Farm

Putauaki Trust was a relatively new entrant into dairying, having acquired the Himiona Farm in 2006. The farm was located on the Rangitaiki Plains in the Bay of Plenty, near Te Teko. The original Himiona Farm purchase of 57 acres had been too small to constitute

Te Rua o Te Moko Ltd Field Day, Manaia, 2014
(John Cowpland – Alphapix)

an economic unit. However, the farm had stood opposite the Kokohinau Marae which rendered the land as spiritually significant.

Himiona Farm was also surrounded by a number of blocks owned by individual whānau who were all shareholders in the Putauaki Trust. Many of these blocks did not adjoin each other, creating problems for the trust. However, since 2006, the trust had managed to negotiate leases for many of these small blocks and, as a consequence, they were able to enlarge their milking platform to the current 177 hectares.

Himiona Farm was now running 570 Kiwi-cross cows, and in 2014 they were on target to produce 240,000 kg of milk solids. The farm was managed by a variable order sharemilker who also employed two full-time staff. The farm possessed a 40-aside herringbone dairy shed which was 'a far cry from the original walk through' seen on the property. A five-year plan was aimed at securing more leases with the intention of expanding the milking platform to 213 hectares, with 700 cows producing 280,000 kg of milk solids.[369]

When visiting the Himiona Farm, judges had been impressed by the high levels of professionalism evident within the trust with a clear set of values informing business activities. Trustees were also very centred on cultural practices, especially recognising the contribution of kaumātua. A solid financial performance had also been evident, with low cost structures and a strong equity position. Hard work had clearly gone into developing a challenging terrain into the current dairying unit, with sound environmental strategies in place.[370]

Ngāti Awa Farms Ltd – Ngakauroa Farm

Ngakauroa Farm was located on the fertile Rangitaiki Plains just south of Whakatane. The farm came under the Ngāti Awa umbrella when purchased by Ngāti Awa Group Holdings in 2008. Currently, Ngāti Awa Farms Trust comprise five trusts: Kiwinui, Putauaki, Omataroa, Te Ihukatia and M K Ratahi.

The Ngakauroa Farm was the first significant asset purchased by Ngāti Awa following the settlement of their Treaty claim in 2005. Currently, about 22,000 people are affiliated to Ngāti Awa.

Ngāti Awa Group Holdings owned 51 percent of the farm. The balance of 49 percent was owned by the various trusts affiliated to Ngāti Awa. Ngakauroa Farm managed 620 Friesian cows on the 186 hectare

Kokohinau Marae, Te Teko, 2014, Putauaki Trust – Himiona Farm Field Day
(John Cowpland – Alphapix)

Ngāti Awa Farms Ltd – Ngakauroa Farm Field Day, Whakatane, 2014
(John Cowpland – Alphapix)

The 2015 finalists and officials

Front row, left to right: Hon Te Ururoa Flavell, Minister for Māori Development; Marty and Janice Charteris, Marenga Station; Bart and Nukuhia Hadfield, Mangaroa Station; Chad Paraone and Josh Williamson, Paua Station Parengarenga Incorporation; Hon Nathan Guy, Minister of Primary Industries

Back row: Scott Campbell, Chief Executive Officer Beef & Lamb; Jamie Tuuta, Māori Trustee; Michelle Hippolite, Chief Executive Te Puni Kōkiri; Ben Dalton, Deputy Director General Ministry for Primary Industries; John Janssen, Head of Sectors Bank of New Zealand

(John Cowpland – Alphapix)

milking platform. The cows were all milked through a 40-aside herringbone shed. The aim of farm managers was to attain 240,000 kg of milk solids in 2014. The herd on the farm was owned by Ngāti Awa Group Holdings but was managed under a variable order sharemilker.

Environmental issues ranked very highly on the agenda of the Ngakauroa Farm. Managers were especially committed to reducing nitrogen leaching on the property, and to further reducing their environmental footprint.[371]

What most impressed the judges when visiting Ngakauroa Farm was the commitment of Ngāti Awa Farms Ltd to strong values and principles, with a clear vision of returning lands to the Ngāti Awa ancestral fold. Social, community and ngā tikanga values were seen to be a positive strength, as highlighted in Ngāti Awa's commendable field day presentation. DairyBase analysis also indicated a highly profitable operation with results well above benchmarks. Good levels of formal monitoring were also evident in relation to pasture covers, cow conditions, reproduction and moisture levels. An environmental specialist had also been engaged as part of the management team.[372]

2015 Sheep and Beef

Return to Parliament

Parliament's ornate and historic Grand Hall once again provided a fitting venue for the announcement of the three finalists selected to compete for the 2015 Ahuwhenua Trophy BNZ Māori Excellence in Farming Award for sheep and beef. Like the Grand Hall itself, the Ahuwhenua Trophy inaugurated by Sir Apirana Ngata and Governor-General Lord Bledisloe in 1933 was equally steeped in history.[373]

The announcement of the three finalists was made by the Minister of Primary Industries, Hon Nathan Guy, with the Minister for Māori Development, Hon Te Ururoa Flavell, who presented medals to the three finalists. About eighty guests from the agribusiness industry also attended the ceremony, which was designed to promote, not only the finalists in the 2015 competition, but also the wider achievements of Māori agribusiness and its outstanding continuing contribution to New Zealand's economy.

Minister Guy referred to this significant contribution of Māori agribusiness to the national economy. This included $3.5 billion earned by Māori forestry alone, he said. The Miraka Dairy Company had also won the Māori Excellence in Export Award at the 2015 New Zealand International Business Awards, an outstanding achievement. It was imperative that such successes be celebrated. This was why the staging of the 2015 finalist announcements in Parliament was 'hugely important, not only for Māori but also for the primary sector as a whole', added the minister. It was also important that agriculture continued to sell itself to the wider community, not least to the Māori community, in order to attract the best and brightest young people. The Ahuwhenua Trophy played an immense role in promoting such Māori achievements in agriculture.

Hon Te Ururoa Flavell applauded the Ahuwhenua Trophy competition for the significant profile it had attained over recent years. Witnessing Māori participation in the competition, and especially seeing the joy of successful whānau celebrating their achievement, 'was a delight to him'. Such celebrations at Ahuwhenua awards functions were important because 'they celebrated the successes of Māori and enabled these to be shared with others'.[374]

The 2015 winners, Bart (left) and Nukuhia Hadfield (right) with Kingi Smiler, Chairman of the Ahuwhenua Trophy Management Committee (centre)
(John Cowpland – Alphapix)

The Finalists and Field Days

Three finalists were announced for 2015. The first, Mangaroa Station, was located in the Ruakituri Valley in Hawke's Bay and had been managed by its current occupiers since 2002. Mangaroa Station hosted their field day on 22 April 2015.

The second finalist was Maranga Station, located near Gisborne; this was 'a typical East Coast hill country sheep and dairy property'. A field day for the Ahuwhenua judges and interested agribusiness leaders was arranged by Maranga Station for 7 May 2015. The third finalist, Paua Station, was based in the small, isolated community of Te Kao in the Far North. Paua Station hosted its field day on 30 April 2015.

The Winner is Announced

The winner of the 2015 Ahuwhenua Trophy for sheep and beef was Mangaroa Station, located in the Ruakituri Valley, owned by Bart and Nukuhia Hadfield. Mangaroa Station's prestigious achievement was announced by the Minister for Primary Industries, Hon Nathan Guy at a special function hosted in Whanganui with 700 dignitaries, families and friends in attendance. Included were the Minister for Māori Development, Hon Te Ururoa Flavell, and the Māori King, Tūheitia Potatau Te Wherowhero VII.

The Ahuwhenua Trophy was presented to Bart and Nukuhia Hadfield by Minister Flavell. The couple were also presented with a special medal by Shelley Ruha representing the Bank of New Zealand, the competition's major sponsor. The presentation of the trophy came at the end of an intensive four-month period, which involved assessing the candidates, selecting and announcing the finalists, and holding field days on their respective properties which, whilst being open to the public, were an essential part of the judging process.

Mr John Janssen, Head of Agribusiness for the Bank of New Zealand, described the standard of the 2015 finalists as 'outstanding'. There was no doubt, he said, that Māori agribusiness had come of age and that Māori farmers were featuring amongst the best practitioners in the country. The Bank of New Zealand's long term involvement with the Ahuwhenua Trophy had seen a significant 'raising of the bar' in the performance of trusts, incorporations and individuals entering the competition. Māori farmers everywhere were 'answering the call of Sir Apirana Ngata and Lord Bledisloe' to double their efforts for the economic benefit of themselves and the country.[375]

Finalist Profiles

Mangaroa Station

In 2001, Bart and Nukuhia Hadfield had joined with family to purchase the leases for two adjoining properties, one of which was Mangaroa Station, in the Ruakituri Valley. In time, it was hoped, whānau members could purchase their own farms. Bart and Nukuhia had earlier farmed in Taihape. In 2002, the whānau purchased Mangaroa Station. Equity was secured by managing a low-cost operation, greatly enhanced by long family experience of working most of their lives in the sheep industry.

In 2008, Bart and Nukuhia acquired Mangaroa outright. By then, they knew the property well and had the confidence to take out a substantial mortgage. By 2015, significant improvements had been attained in asset acquisition and farm productivity. New buildings, cropping and regrassing had also been achieved, with gorse and scrub cleared and new fencing installed. The Hadfields now planned on docking 10,000 lambs by 2016. They had once set a similar target, in 2014, which they had succeeded in attaining, thereafter being awarded the 2014 Best Flock Award in the Ewe Hogget Competition.[376]

When visiting Mangaroa Station, judges had been impressed with the very clear vision and core planning that underscored Mangaroa's extensive operations. Great community contributions were also evident through their involvement in shearing. Excellent performance on difficult hill country was also an outstanding feature, with a strong focus on improving soil fertility through specific exacting targets and regular monitoring. An excellent baseline production performance was also evident in the main sheep enterprise.[377]

Mangaroa Station Field Day, Ruakituri Valley, 2015

(John Cowpland – Alphapix)

Maranga Station

Maranga Station was located about 30 kilometres south-west of Gisborne. The area occupied by the station was of particular significance to local Māori, being criss-crossed by ancient walking tracks and trading routes linking coastal hapū with whānaunga living inland. Mutton birds had also once nested on the property, before the unfortunate advent of introduced predators.

Owners Janice and Marty Charteris had by 2015 transformed Maranga Station into a highly successful operation. Their first farm purchase had occurred in 1989, comprising a 285-hectare unit on which they had managed 3400 stock units. They had also leased a further 56 hectares before finally purchasing Maranga, at which point they consolidated all their holdings into Maranga Station.

Maranga Station comprised 850 hectares on which 6000 Romdale sheep and 1200 Angus cattle were managed, alongside 600 goats, which were mustered annually. The property comprised a mix of terrains, ranging from about 50 hectares of flat land to 500 hectares of moderate rolling country. The balance was steep hill country suitable for sheep.[378]

Many postive factors pertaining to Maranga Station caught the eye of the judges. The setting and achieving of goals underpinning farm expansion and development were commended, with excellent use of professional help as well as strong local support. The station mentored a number of young budding farmers as part of active volunteering programmes. A key focus on securing the land as a significant asset was also evident, as seen in infrastructure developments like fences, fertiliser and water. Feed production had been progressively improved with a particular focus on elements like water reticulation and a nutrient management plan.[379]

Paua Station

Paua Station was located about 80 kilometres north of Kaitaia, near the settlement of Te Kao and was owned by the Parengarenga Incorporation. The traditonal Parengarenga mana whenua of Te Aupouri extends from Te Kao in the south to Thoms Landing in the north, encompassing the scenic Parengarenga Harbour to the east.

In 1956, nine blocks of Māori land south of the Parengarenga Harbour were consolidated into a single block known as Parengarenga Toopu. In 1965, Parengarenga Toopu was divided into two blocks, one to be developed as forest and the other to be developed into two farming stations. These two new farm stations became known as Paua Development Scheme and Te Rangi Development Scheme. The Parengarenga Incorporation was established in 1965. Māori owners thereafter sought full control of their lands and this was achieved in 1988.

Paua Station comprised 2430 hectares of easy rolling coastal sand country on which 2800 Angus cattle and 7000 sheep were being managed, of which 6100 were ewes with 118 percent lambing percentage. Paua Station was operating mainly as a finishing property, a policy that dovetailed into the Te Rangi Station, also owned by Parengarenga Incorporation. Paua's policy was to increase stock numbers, better adapting to the hot dry summer conditions experienced in the Far North. For example, Paua Station was lambing in June or July, with all lambs off the farm by Christmas. This enabled premium prices to be attained whilst alleviating feed pressures over the summer. The station was also breeding its own Angus bulls.[380]

Maranga Station Field Day, Tiniroto, 2015
(John Cowpland – Alphapix)

Powhiri, Paua Station Field Day, Potahi Marae Te Kao, 2015
(John Cowpland – Alphapix)

Judges who visited Paua Station commended the Parengarenga Incorporation for their seconding of skilled advisors to their farm committee and for their impressive five-year strategic plans, which clearly set manageable financial objectives and strategies. Paua Station's emphasis in providing employment for local young people interested in farming was also applauded. Significant developments had been achieved from the farm's cashflow, with a good focus on improving infrastructure such as fencing, sub-divisions and water development. Extensive riparian fencing was also underway with all boundaries facing the Harbour fenced off to protect water quality.[381]

2016 Dairy

Competition Launched in Style

In November 2015, during the Federation of Māori Authorities conference in Wellington, the Ahuwhenua Trophy BNZ Excellence in Farming Award competition for 2016 was launched. The Ahuwhenua competition presented 'a great focus to Māori working in the agricultural sector', said the Minister for Māori Development, Hon Te Ururoa Flavell, during the launch. That such competitions also instigated 'friendly rivalry' among Māori farmers 'was a great thing because it forced eveyone to lift their standards'.[382]

In recent years, said Minister Flavell, many of the competition finalists had belonged to trusts that had collaborated and consolidated their blocks of land. This had been done to achieve greater scale 'and therefore better economic returns for their people'. However, extensive tracts of Māori land still remained unproductive. The minister was sure that the Ahuwhenua Trophy competition would continue to be a significant catalyst for change and progress in this regard. Māori agriculture also offered significant and wide-ranging career opportunities for young people, with many iwi offering impressive training programmes. The Ahuwhenua Young Māori Farmer of the Year Award, introduced in 2012, was also to be applauded for doing so much to encourage enterprising young Māori people into the agricultural sector.

The Chairman of the Ahuwhenua Trophy Management Committee, Mr Kingi Smiler, commented that the Ahuwhenua competition offered an irreplaceable opportunity for Māori farmers, trusts and incorporations to gain valuable professional

2016 finalists
From left: James Russell, Chairman, Rakaia Incorporation; Hon Te Ururoa Flavell,
Minister for Māori Development; Hon Nathan Guy, Minister of Primary Industries;
Tuhi Watkinson, Chair, Tewi Trust; and Andrew Priest, Chief Executive, Ngāi Tahu Farming Limited

(John Cowpland – Alphapix)

feedback on their farming operations. The judges of the competition were highly skilled professionals with many years of agribusiness experience. They also had significant access to data against which they could benchmark individual Māori operations. Previous competitors had gained feedback of enormous value to their ongoing farming operations.

For too long, said Mr Smiler, Māori had failed to tell their own people, and all New Zealanders, of their success stories. Māori farming was one of these success stories. Throughout the country, great things were being achieved on the land owned by Māori. All encouragement was therefore given to Māori to enter the competition, the rewards for which remained substantial.[383]

The Finalists and Field Days

In March 2016, the three finalists for the 2016 Ahuwhenua Trophy BNZ Māori Excellence in Farming Award for Dairy were announced at a special function in Parliament. The annoucement was made by the Minister for Māori Development, Hon Te Ururoa Flavell.

The function was attended by more than one hundred people including leaders of agribusiness, Members of Parliament, representitives of the Ahuwhenua Trophy sponsors and, not least, whānau and supporters of the finalists.

The first finalist was Ngāi Tahu Farming Limited, which owned farms near Oxford, in northern Canterbury. Ngāi Tahu Farming Limited hosted a field day on 8 March 2016. The second finalist was the Tewi Trust, which was situated near the small Waikato settlement of Okoroire, near Tirau, an area famous for its hot springs. The Tewi Trust hosted their field day on 11 March 2016. The third finalist was the Proprietors of Rakaia Incorporation's Tahu a Tao farm, located near Ashburton, a property that had a long history dating back to the 1880s. The Proprietors of Rakaia Incorporation hosted their field day on 16 March 2016.

At the announcement function, Minister Flavell congratulated the three finalists for their sheer hard work and for fulfilling the legacy left by Sir Apirana Ngata who, in introducing the competition in 1933, had inspired Māori farmers ever since to acquire proficiency and skill in their farming endeavours. For the first time in the recent history of the competition, two of the finalists' farms were in the South Island, with the third finalist in the Waikato. All three farms were 'dedicated to the economic, social and cultural welfare of Māori people', also sharing a commitment to good environmental practice and Māori values.[384]

The Winner is Announced

For the first time in the trophy's eighty-three-year history, a South Island farm was awarded the Ahuwhenua Trophy BNZ Māori Excellence in Farming Award. On 20 May 2016 at a special function in Hamilton attended by 850 people, including the Minister for Primary Industries, Hon Nathan Guy, and other dignitaries, the Proprietors of Rakaia Incorporation were announced as the Ahuwhenua Trophy 2016 winner.

The Ahuwhenua Trophy was presented to Rakaia Incorporation by the Minister for Māori Development, Hon Te Ururoa Flavell, along with a replica trophy. They were also presented with a special medal by Pierre Tohe representing the competition's principal sponsor, the Bank of New Zealand.

The 2016 Dairy winner, Rakaia Incorporation, Chairman James Russell

(John Cowpland – Alphapix)

Mr Kingi Smiler, Chairman of the Ahuwhenua Trophy Management Committee, commended the finalists, saying that their achievements had mirrored the founding aspirational vision of Sir Apirana Ngata and Lord Bledisloe. It was an amazing thing, said Mr Smiler, that every year, another group of outstanding finalists came forward, with, ultimately, a winner 'who joins the ever expanding alumni that personifies the success of Māori agribusiness'.

Mr John Janssen, Head of Agribusiness for the principal sponsor, the Bank of New Zealand, said that the standard of the 2016 finalists had been exceptional, and that 'BNZ is proud to be the principal sponsor of this competition and to be a part of the movement that is raising the standard and profile of Māori agribusiness'. All New Zealand farmers could learn much from their Māori counterparts. 'There was no doubt that the Ahuwhenua Trophy competition was the catalyst for much of the success of Māori farming,' he said.[385]

Finalist Profiles

The Proprietors of Rakaia Incorporation

Rakaia Incorporation's Tahu a Tao farm comprised 216 hectares and managed 830 Kiwi-cross cows, producing 371,294 kg of milk solids over the previous year. In addition to the cows in milk, a further 200 replacement calves were reared each year. The farm reflected a strong farm culture with particular emphasis on animal welfare.

Tahu a Tao farm was located 8 miles from the mouth of the Rakaia River and was once the scene of great difficulties for travellers, particularly crossing the rivers. Today the soils were free draining and ideal for dairying. Soil tests were taken annually and feed budgeting was a key component of farm management.

Tahu a Tao was converted to dairying in 1996, thereafter operating in conjunction with 50/50 sharemilkers. The resident sharemilkers had been there for nine years. When the farm was initially converted, old pastures were sprayed out and new ones planted. In more recent times, the incorporation had made concerted efforts to repasture with the latest cultivars of seed as a key means to boost production. With conversion came new fences, stock water and irrigation and a fifty-bail rotary dairy shed. Two bores supplied the four irrigators that watered the property. However, irrigation was at the foundation of the overall dairy system. It was tightly managed using water and soil measuring technologies.

Rakaia Incorporation Field Day, Rakaia, 2016
(John Cowpland – Alphapix)

Cultural and environmental factors also played a large role in determining strategies. An example was the recycling system utilised on the farm. All farm waste was stored in containers, with silage wrap and other detritus stored so that it could not become windblown. An effluent handling system also reduced the risk of nitrogen leaching.[386]

When visiting the Proprietors of Rakaia Incorporation's Tahu a Tao farm, the judges were impressed with the professional nature of the management team, with very clear financial, investment and farming strategies. The incorporation also encouraged participation in the Whenua Kura Māori training and leadership programme. Their business strategy was designed to facilitate the purchase of more land, when debt levels were sufficiently low. Production per cow had been well above benchmark over the previous three years, with a strong environmental focus also in place.[387]

Ngāi Tahu Farming Limited

Ngāi Tahu's dairy farming operation comprised seven dairy units, including Te Ahu Pātiki and Maungatere, which were located side by side near Oxford in northern Canterbury. The area was known as Whenua Hou and was originally an extensive pine forest. However, Ngāi Tahu decided to convert to dairying in order to maximise the economic prospects of the property.

The farm came on line during the 2012/2013 season, with full development supporting thirteen dairy farms and seven dairy support farms across the property. All of the land was flat and most of the milking platforms were well irrigated. The milking platform for Te Ahu was 355 hectares with 1251 Kiwi-cross cows that, over the previous season, produced 468,747 kg of milk solids. There were also a number of dairy supports owned by Ngāi Tahu where stock from these farms could be sent for the autumn.

The soils were quite sterile, but fertility would be increased by enhancing the organic components in the ground. Both farms had sixty-four section rotary sheds and incorporated modern technology to minimise the environmental footprint. Modern irrigation systems had been introduced with many innovative features, allowing for the efficient use of water. Five people were working on the two farms, though the overall Ngāi Tahu Farming operation employed seventy people.[388]

The visiting judges commended Ngāi Tahu Farming Limited for their clear vision for the future, with support and input from all levels of mana whenua.

Ngāi Tahu Farming Limited Field Day. Oxford, 2016
(John Cowpland – Alphapix)

Strong leadership and a wide range of skills were also evident at governance level. A clear set of Ngāi Tahu values was communicated and positioned at the core of all future planning. Individual reporting on each farm ensured clarity and internal benchmarking. Also commended was the desire to attain a predominantly pasture-based system to ensure farms were flexible in the current farming environment. Precision farming was also a definite focus, with a clear genetic selection policy in place.[389]

Tewi Trust

Tewi Trust was located near the small town of Okoroire, near Tirau, in eastern Waikato. The area was famous for its hot springs, where tūpuna of old could rest, and a beautiful hotel. The trust had fifty-three shareholders and was named after one of the original owners, Mr Tewi Hoera, who had passed the land on to his daughters. The daughters had in turn leased the property to a local farmer, with a lease renewal occurring in the early 1940s for thirty years.

When the lease expired, two parcels of the land were returned to the whānau, though they were separated by a private farm. During the 1970s, the milk shed was upgraded and other infrastructure improvements were also completed. A 50/50 share-milking arrangement was also agreed to. In 2000, the trust purchased the property that separated the two parcels of land.

The farm was operated by 50/50 sharemilkers and comprised a 138-hectare milking platform on which a 430 strong Friesian cross herd produced 174,405 kg of milk solids. Milking was performed in a 40-aside herringbone shed. The land itself was flat to gently rolling, with Tirau sandy loam soil that was ideal for dairying.

Up to 6 hectares of turnips and chicory were grown each year to protect the stock against the long, dry summers. Silage amounting to 90 tons was also produced per annum. Tewi Trust was a great example of how a small and determined group of whānau had, over years of challenge and adversity, achieved their aspirations whilst remaining focused on Māori values.[390]

The judges commended Tewi Trust for their admirable governance group, which maintained a strong relationship and trust with key advisors and sharemilkers. An ongoing determination to reduce debt and keep farm costs low had also paid dividends. Trustees were familiar with and receptive to their whānau. The business was also seen to be in a very stable financial position. The farm was well managed

Tewi Trust Field Day, Okoroire, 2016

(John Cowpland – Alphapix)

with production levels of 1186 kg of milk solids and 383 kg of milk solids per cow, with relatively low levels of supplement use. Health and safety issues were well recognised and the farm was fully compliant with Environment Waikato regulations.[391]

2017 Sheep and Beef

Big Turnout for Announcement of Finalists

More than one hundred people gathered in Parliament's historic Grand Hall for the announcement of the 2017 finalists for the Ahuwhenua Trophy BNZ Māori Excellence in Farming Award for sheep and beef. The finalists were announced by the Minister for Primary Industries, Hon Nathan Guy, in association with the Minister for Māori Development, Hon Te Ururoa Flavell.

Speaking before the gathering of parliamentarians, leaders of agribusiness and whānau of the finalists, Minister Flavell described the Ahuwhenua Trophy as possessing significant mana, given the high-powered assembly of participants and involved parties. Māori could do more to publicise their achievements, he added, because the Ahuwhenua Trophy was undoubtedly a prestigious and premier agricultural award, not just for Māori but for all of New Zealand.

Minister Flavell complimented the finalists, stating that they should be 'very proud of their achievement in making the finals'.[392]

Minister Guy commented that the farms of the finalists were 'shining examples of the commitment Māori farmers had to sustainability, developing their lands for future generations'. Minister Guy was proud to acknowledge the key role Māori played in New Zealand's primary industries. 'The asset base of the Māori economy was worth over $42 billion, most of which was strongly focused on the primary industries. Māori collectively owned 40 percent of forestry land and 38 percent of fishing quota, just to name two examples. Right across the country as a whole, Māori were successful economic players and many of their companies and entities are amongst the top performing commercial operations in New Zealand. To top this off was the Ahuwhenua Trophy which was showcasing excellence.'

The Chairman of the Ahuwhenua Trophy Management Committee, Mr Kingi Smiler, said it was gratifying to once again see three top-quality sheep and beef farms selected as finalists for the prestigious trophy. 'All are performing very well in some of the most challenging times the New Zealand primary sector has encountered, with

2017 finalists

Back row, left to right: Hon Te Ururoa Flavell, Minister for Māori Development; Weo Mang, Trustee, Pukepoto Farm Trust; Ronald and Justine King, Puketawa Station; and Hon Nathan Guy, Minister of Primary Industries

Front row, left to right: Aloma Shearer, Chair, Pukepoto Farm Trust; Raniera Tau, Chairman, Omapere Rangihamama Trust

(John Cowpland – Alphapix)

volatile global markets and low prices together with a need to adapt to climate change.' Māori who ran these operations nonetheless were positive and confident about their future.[393]

Finalists and Field Days

Three finalists were chosen for the 2017 Ahuwhenua Trophy BNZ Māori Excellence in Training Award for sheep and beef.

The first was Omapere Taraire E & Rangihamama X3 Ahu Whenua Trust, which was situated in the Far North, 2 kilometres northwest of Kaikohe, comprising properties regarded by shareholders as taonga tukuiho, given to them over time by their tūpuna. The Omapere Taraire E & Rangihamama X3 Ahu Whenua Trust hosted their field day on 20 April 2017.

Second finalist was R A & J G King Partnership, Puketawa Station, which was a sheep and beef breeding unit located at Tiraumea, about forty-five minutes east of Paihiatua in the Wairarapa. R A & J G King Partnership, Puketawa Station organised their field day for 27 April 2017.

The third finalist was Pukepoto Farm Trust, a small farm trust with about 1000 owners situated near the tiny but picturesque settlement of Ongarue, about twenty minutes north of Taumarunui. A field day hosted by the Pukepoto Farm Trust was held on 4 May 2017.

The Winner is Announced

Out of the three outstanding finalists, the Omapere Taraire E & Rangihamama X3 Ahu Whenua Trust was declared the winner at an awards evening in Whangarei. This event was attended by 700 people including Māori leaders, politicians, agribusiness professionals and whānau of the finalists. The announcement was made by the Prime Minister, The Rt Hon Bill English. The Chairman of the Omapere Taraire Trust, Mr Raniera (Sonny) Tau, was presented with the trophy, a special medal and more than $40,000 in prizes.

Omapere Rangihamama Trust whānau celebrating their win with the Prime Minister, The Rt Hon Bill English, and Minister for Māori Development, Hon Te Ururoa Flavell

(John Cowpland – Alphapix)

The Chairman of the Ahuwhenua Trophy Management Committee, Mr Kingi Smiler, congratulated Omapere Taraire E & Rangihamama X3 Ahu Whenua Trust, saying that, like all of the finalists, the trust possessed a 'strong strategic and practical commitment to improving the environment of the property' and that this was benefitting their whānau and all other people in their northern district. The trust was working hard to encourage its young people to make a career in farming by offering scholarships 'and this once again highlights their intergenerational strategic thinking'. Observers of the Omapere Taraire E & Rangihamama X3 Ahu Whenua Trust's operations could not but be impressed with the passion and commitment shown in 'making the best out of some challenging country'.

New Zealand was fortunate to have such Māori farmers, as represented in the 2017 finalists, said Mr Smiler, because 'it is in their DNA as kaitiaki to manage the fragile environment as an investment for future generations'. The spiritual closeness to the land was vital in a modern society where consumers not only wanted food, 'they also wanted assurances that the food was produced sustainably and ethically'.[394]

Finalist Profiles

Omapere Taraire E & Rangihamama X3 Ahu Whenua Trust

The farm managed by the Omapere Rangihamama Trust was located about 2 kilometres south-west of Kaikohe in the Far North. The land over which the farm stood had strong valued connections through whakapapa and tūpuna to present-day shareholders. Until the 1950s, the Omapere lands comprised a myriad of small titles utilised by whānau for near-subsistence food gardens and small dairy farms. In time, many of these small titles became uneconomic and fell into neglect.

In 1953, enabled by the Māori Affairs Act of that year, many of these titles were amalgamated into a single trust known as Omapere Taraire E & Rangihamama X3 Ahu Whenua Trust. During the subsequent process of consolidating titles, the Māori Trustee had assumed a significant holding, especially through the acquisition of a substantial number of shares. Thereafter, the property was managed as a sheep and beef unit under the supervision of the Department of Māori Affairs.

Powhiri, Omapere Rangihamama Field Day, Kohewhata Marae, Kaikohe, 2017
(John Cowpland – Alphapix)

With the arrival of new trust leadership in 2007, a concerted attempt was made by trustees to purchase the Māori Trustee's interests in the farm; and this continues as an ongoing strategy in the hope that the farm will one day be owned outright by the Omapere Taraire E & Rangihamama X3 Ahu Whenua Trust. The trust at present owns 54.86 percent of the shares.

In the past, the farm operated as a sheep and beef unit. A move away from this was driven by better returns for bull beef, as opposed to poorer returns for wool, sheep and lamb. At present, environmental concerns were of the utmost importance, especially given the land's historic value to Māori. Omapere Taraire E & Rangihamama X3 Ahu Whenua Trust saw itself as being ahead of legislation, in this regard.[395]

When the judges visited the Omapere Rangihamama Trust's Omapere Farm, they were impressed by the fact that each of the trustees and managers knew their roles and responsibilities in executing the overall trust strategy and that of the sheep and beef unit. Strong connections to Putahi Maunga were also evident, with strong support for papa kāinga housing and educational scholarships. Judges noted the successful implementation of a strategic decision to reduce the proportion of sheep leading to an economic farm surplus. Soil maps were being used for targeted improvements to feed production. A waterway fencing programme had also commenced, which was now nearing completion.[396]

R A & J G King Partnership, Puketawa Station

R A & J G King Partnership, Puketawa Station was a sheep and breeding unit located at Tiraumea, about forty-five minutes east of Paihiatua in the northern Wairarapa. The farm comprised 1108 hectares (900 effective) and was mainly medium to steep hill country with some rolling contours. Breeding stock on the farm comprised 850 ewe replacements and 148 stud ewes. Cattle numbers consisted of 144 mainly mixed Hereford cows plus replacements.

Puketawa Station was purchased by Ronald and Justine King in 2013 when the farm was in a 'tired' condition. Since then, enormous efforts had been made to improve pastures and infrastructure. Prior to owning Puketawa Station, Mr King had owned a 280-acre sheep and beef farm near Whangamomona. In 2001, Mr King joined family in purchasing Mangaroa

Puketawa Station Field Day, Tiraumea, 2017
(John Cowpland – Alphapix)

Pukepoto Field Day, Ongarue, 2017
(John Cowpland – Alphapix)

Station in the Ruakituri Valley, Hawke's Bay. A further family purchase had also been Ruakaka Station in the Tiniroto Valley. All siblings now owned their own farms. In fact, Mr King's sister, Nukuhia Hadfield and her husband Bart had won the Ahuwhenua Trophy in 2015 with Mangaroa Station.

At Puketawa Station, Mrs King's father lived on the farm, offering his advice and assistance in management. Other family members assisted on the farm from time to time. Mrs King had also established the Alfredton Riding Club, and her husband judged in the Golden Shears competitions. Long-term developments included gorse planting, retiring areas to address erosion, riparian planting and preserving the environment.[397]

The judges of the Ahuwhenua competition commended the Kings for their clear business vision for their farm, which largely focused on increasing stock units through greater productivity and repaying debt. A strong strategy was also evident for developing land and infrastructure with a complementary strategy on livestock productivity. The farm also reflected its strong and valued whānau connections, which worked for the benefit of the business and future generations. Clear financial targets had been set, with feed production enhanced by recent reviews of fertiliser strategies. Important environmental decisions included the maintenance of mānuka and tree cover on steep hillsides, alleviated by spot spraying to benefit production.[398]

Pukepoto Farm Trust

Pukepoto Farm Trust comprised just over 1000 owners and was located near Ongarue, north of Taumarunui. Though Pukepoto means 'small hill', the farm was in fact a typical hill country property, with high steep hills and gullies dropping down to feed the Ongarue and Ohura Rivers, which flowed into the Whanganui River.

Pukepoto Farm Trust's property comprised 1400 hectares of which 1000 was farmed. About 100 hectares were covenanted under the Ngā Whenua Rāhui scheme. There was also 62 hectares in plantation pine, and the remainder of any unfarmed was scrub, much of which was being retained to prevent erosion.

The property wintered a flock of 6000 Romney ewes and a herd of 300 mainly Angus cattle. The trust was established by the Māori Land Court in 1878 as Pukepoto Forest and Farm Trust. In 1994, it was renamed Pukepoto Farm Trust.

The land was regarded as sacred, especially since the oldest son of Rereahu, who was named Te Ihingarangi, was born on the whenua. In the old days, toa from Taranaki and Whanganui often passed through the area, rendering it unsafe for Māori to build papa kāinga in this area.

In recent years, Pukepoto Farm Trustees had worked hard to elevate the farm's performance, working closely with owners and other stakeholders. New staff had been engaged with improved stock policies adopted. Trustees had stressed the centrality of financial, social and environmental issues to their forward strategies, as they continued to deliver real benefits to whānau of Pukepoto.[399]

When the judges visited Pukepoto Farm Trust, they observed a collaborative approach to governance where important decision making was arrived at by consensus. Trustees used their considerable skills to participate in both governance and management. The trust was seen to be on track with its strategies and policies and clearly had the confidence of shareholders. Clear and simple key performance indicators were evident with strong returns noted. A regular regrassing policy was in place with regular soil testing, alongside a clear focus to improve calving and lambing percentages.[400]

2018 Dairy

Top Māori Dairy Farm Finalists Announced

As with previous years, great excitement accompanied the announcement of the 2018 finalists in the Ahuwhenua Trophy BNZ Māori Excellence in Farming Award for dairy. The announcement was made in Parliament by the Minister for Māori Development, Hon Nanaia Mahuta, before an audience of one hundred people with extensive connections to the agribusiness sector, as well as politicians, other business leaders and whānau of the finalists.

Speaking to the assembled audience, Minister Mahuta commented that the Ahuwhenua Trophy competition was close to her heart as she currently occupied both portfolios of Māori Development and Environment. She said it was a pleasure to announce the 2018 finalists and to acknowledge and celebrate the excellence and innovation on show in Māori farming, especially as exemplified by the prestigious Ahuwhenua Trophy. Minister Mahuta said that ongoing government programmes and pending legislative changes would continue to encourage such excellence and innovation amongst Māori across many economic sectors, of which agriculture was one of the most important.

The Minister of Agriculture, Damien O'Connor, said the Ahuwhenua Trophy was a wonderful event that celebrated the success of Māori and their significant contribution to the agribusiness sector. 'The land delivers us the opportunity to have a better future but it does not guarantee it,' he said. 'So, this competition was about safeguarding the business of the land', as well as safeguarding those that generated the wealth 'that could be transferred onto not just iwi and whānau but the whole country'.

The Chairman of the Ahuwhenua Trophy Management Committee, Mr Kingi Smiler, commented that it was gratifying once again to see the top Māori dairy farms selected as finalists for this prestigious trophy. 'We are a long way from our markets, yet through innovation and determination we are able to put a wide variety of quality products on supermarket shelves, restaurant menus and processing plants around the world,' he said. Māori had managed fragile environments for a long time, generating sustainable economies, 'and so it would continue'.[401]

The Finalists and Field Days

Two finalists were chosen for the 2018 Ahuwhenua Trophy BNZ Māori Excellence in Training Award for dairy farming.

The first finalist was the Proprietors of Mawhera Incorporation, which managed Mawhera Tuatahi Farm, located in the Arahura Valley just north of Hokitika. The Proprietors of Mawhera Incorporation hosted a field day on 5 April 2018.

Onuku Māori Lands Trust were the second finalist. This trust managed a dairy farm near Lake Rotomahana, about 30 miles to the south of Rotorua. A field day was organised by Onuku Māori Lands Trust for 12 April 2018.

Onuku Māori Lands Trust whānau celebrate their win with Governor-General The Rt Hon Dame Patsy Reddy (front, third from left), Hon Nanaia Mahuta, Minister for Māori Development (left), and Hon Meka Whaitiri, Associate Minister of Agriculture (second from left)

(John Cowpland – Alphapix)

The Winner is Announced

Out of the two exceptional 2018 finalists, the Onuku Māori Lands Trust was declared the winner of the Ahuwhenua Trophy for dairy. The announcement was made by the Governor-General The Rt Hon Dame Patsy Reddy at a special awards function held in Christchurch on 25 May 2018. The event was attended by 650 people including the Minister for Māori Development, Hon Nanaia Mahuta, the Associate Minister of Agriculture, Hon Meka Whaitiri, and other dignitaries representing agribusiness, central and local government and Māori communities.

Both nominated dairy farms set an exceptional standard in their farming and governance operations, said Mr Kingi Smiler, Chairman of the Ahuwhenua Trophy Management Committee. 'I congratulate the Onuku Māori Lands Trust for their great farming operation and their commitment to the values of Sir Apirana Ngata and Lord Bledisloe.' He said that, every year, the trophy attracted an astounding cohort of Māori farmers willing to showcase their successes in the agribusiness sector, 'and long may this continue'.[402]

Finalist Profiles

Onuku Māori Lands Trust

The Onuku Māori Lands Trust operated the Boundary Road Farm comprising a 72-hectare block near Lake Rotomahana, south of Rotorua. The trust milked 220 cows which produced about 90,000 kg of milk solids.

In 1882, the land now known as the Onuku Māori Lands Trust was partitioned and handed back to Ngāti Rangitihi. When Mount Tarawera erupted in June 1886, Rangitihi Pā, located on the shores of Lake Tararewa, was tragically wiped out. The surrounding land including Onuku was covered with Rotomahana mud and portions of Tararewa ash and gravel. These soils today make up the base of the Onuku Farms. The trust estate stretches from Mount Tararewa in the north to Timberlands in the south.

From 1963, the land was acquired and developed by the Department of Lands and Survey and handed back to Ngāti Rangitihi in 1982, when the Onuku Māori Lands Trust was established. Originally one large drystock farming operation, trustees were determined to diversify and so dairy was seen as the best option.

Alan Rondon, Farm Manager, Onuku Māori Lands Trust, Rotomahana
(John Cowpland – Alphapix)

By 2018, the trust had developed four farms, a drystock farm, forestry, a natural reserve and a mānuka plantation. Onuku had also developed outside the farm gates, establishing an exporting honey business known as Onuku Honey. A new property acquired in 2004 was now performing well, generating 1214 kg of milk solids per hectare.

Onuku Māori Lands Trust had a strong environmental focus with Onuku participating in Project Rerewhakaaitu, a local voluntary farming initiative helping to protect local waterways and lakes. Management also focused on reducing nitrogen and phosphate and achieving the highest possible animal welfare. Onuku had recently moved from sharemilking to herd ownership, and the trust now owned herds on Dairy Two and Boundary Road Farms. The current strategy also sought to train Onuku owners in farm management, thereby creating pathways to larger dairy farms on Onuku.[403]

When the judges visited Onuku Māori Lands Trust, they commended the trustees on their strong sense of responsibility, as demonstrated to their people, with a clear mission statement comprising vision and cultural values forming an essential part of their business strategy. Good linkages and relationships existed between trustees and governance. Grants and scholarships were available, exhibiting a strong sense of community and ngā tikanga Māori.

The Boundary Road Farm showed good levels of profitably, generated by strong levels of consistent production. Judges observed excellent pasture quality, with very strong awareness of environmental issues.[404]

The Proprietors of Mawhera Incorporation

Mawhera Tuatahi Farm consisted of 348 hectares in the Arahura Valley, north of Hokitika on the West Coast. The milking platform was 257 hectares and the 500 cows produced 190,000 kg of milk solids. The property's development dated from the 1800s when Ngāi Tahu gained occupation of the mana whenua of Te Tai Poutini – the West Coast. During the nineteenth century, the Crown purchased 3.1 million hectares of land but excluded 4139 hectares to be held in 54 reserves along the western coastline. Thirty-nine of these reserves had been assigned as individual allotments for use by their Ngāi Tahu owners. The rest had been leased to provide income for the Ngāi Tahu iwi.

The Arahura riverbed was exempt from the purchase. The land within the Arahura Māori Reserve was leased out to farmers via perpetual leases by various agents of the Crown. After the Māori Affairs Amendment Act was passed in 1967, Ngāi Tahu sought urgent Crown agreement to relinquish the perpetual leases. In 1973, a royal commission determined that twenty-four of the Poutini reserves should be vested in a Māori incorporation.

Field day at Mawhera Incorporation, Arahura Marae, Hokitika, 2019
(John Cowpland – Alphapix)

Consequently, The Proprietors of Mawhera Incorporation was established by Order In Council dated 17 May 1976, otherwise known as the Mawhera Incorporation Order. In 1993, the incorporation acquired the leases of nine properties, comprising 194 hectares on the south bank of the river. The incorporation further purchased European freehold land adjacent to the Arahura Reserve – some 62 hectares – thus increasing the milking platform to 210 hectares. Wellstar farm was acquired in 2009, comprising 12 hectares of Māori leasehold land and 80 hectares of European freehold land adjacent to the Arahura Reserve.

In recent years, extensive development had occurred on infrastructure in particular. The milking was 50/50 sharemilking, bringing experience, motivation and technical expertise to the business. Dairy now comprised the incorporation's major investment, focused on the two farms, Mawhera Tuatahi and Te Hewera, with a third dairy farm on the Karamea Reserve called Umere.[405]

Many aspects of The Proprietors of Mawhera Incorporation's operations impressed the judges when they visited. The Committee of Management reflected a diverse range of skills, backgrounds and experience. The farm was also used to advance local agricultural training, with sharemilkers assisting in many marae functions. The Mawhera Incorporation's farming system was historically low cost, which had delivered strong cash surpluses. Good use was made of winter crops to manage rotation length during quite severe winters. Given the three previous tough seasons, per cow production had remained consistent. A proposed river buffer zone was thought to be an excellent idea.[406]

2019 Sheep and Beef

Finalists at the Top of Their Game

The three finalists for the 2019 Ahuwhenua Trophy BNZ Excellence in Māori Farming Award for sheep and beef were announced at a special gathering at Parliament attended by politicians, agribusiness leaders, sponsors and Māori with strong links to the agricultural sector, including whānau of the finalists.

The announcement of the finalists was made by the Minister of Agriculture, Hon Damien O'Connor. Representatives of the finalists had flown to Wellington

2019 finalists

From left: Carl Read-Jones, Farm Operations Manager; Marty Van Heuckelum, Farm Manager, Te Awahohonu Forest Trust, Gwavas Station; Pania and Eugene King, Kiriroa Station; Richard Scholefield, General Manager, Whangara Farms

(John Cowpland – Alphapix)

the night before and, whilst some had been in Parliament before, for others it was clearly a new experience. Finalists were presented with medals by the Minister of Agriculture, who addressed the gathering with the Minister for Māori Development, Hon Nanaia Mahuta.

Minister O'Connor said that this year's finalists were at the 'top of their game' and provided the inspiration for others to contribute to the growing of the New Zealand economy. 'We have to encourage excellence right across agriculture and horticulture,' he said. Though many challenges lay ahead, the government was committed to providing enhanced opportunities for young people to get into agriculture. Acknowledging that significant progress had been made across the agribusiness sector, Minister O'Connor said it was 'imperative that we build a future that can be seen as bright and rewarding'.

The Minister for Māori Development, Hon Nanaia Mahuta, noted the history of the Ahuwhenua awards, saying that they had been part of a long tradition that had begun with Sir Apirana Ngata in 1933. The 2019 Ahuwhenua Trophy once again reminded the country of the significant share that beef and sheep farming made to the $50 billion earned by Māori enterprise.

The competition also emphasised – especially this year – the importance of Te Tairawhiti, with all three finalists coming from the East Coast.

The Chairman of the Ahuwhenua Trophy Management Committee, Mr Kingi Smiler, said that the high calibre of this year's finalists demonstrated the strength of the Māori agribusiness sector. Selecting three finalists from such an outstanding field of entrants had been a difficult task. 'This competition is prestigious and people actively seek to enter this event to showcase the quality of their farming enterprises.' What made sheep and beef so special was that, in most cases, properties were located in remote hill country areas which itself made farming challenging throughout the year but especially in times of adversity.[407]

The Finalists and Field Days

Three exceptional Māori farms were announced as the 2019 finalists.

The first finalist was Whangara Farms, which were situated about 35 kilometres north of Gisborne, comprising a partnership of three Māori incorporations. Whangara Farms hosted a field day on 4 April 2019.

Winners for 2019, Pania and Eugene King of Kiriroa Station, Matawai
(John Cowpland – Alphapix)

Kiriroa Station Field Day, Matawai, 2019
(John Cowpland – Alphapix)

Te Awahohonu Forest Trust – Gwavas Station was the second finalist, situated at Tikokino, about 50 kilometres west of Hastings. The historic Gwavas Station was owned by the Te Awahohonu Forest Trust. Te Awahohonu Forest Trust – Gwavas Station hosted their field day on 11 April 2019.

The third finalist was Kiriroa Station situated at 253 Motu Road, Matawai, about 70 kilometres north-west of Gisborne, owned by Eugene and Pania King. Kiriroa Station hosted its field day on 18 April 2019.

The Winner is Announced

Of the three exceptional finalists in 2019, Eugene and Pania King of Kiriroa Station were announced as the winners at an Ahuwhenua awards function held in Gisborne on 24 May 2019. The event was attended by more than 600 people including the Minister for Māori Development, Hon Nanaia Mahuta, and other dignitaries, agribusines leaders and whānau supporting the three finalists.

This win was particularly momentous for one family. By winning the award, Mr King became the second member of his family to do so, following his sister Nukuhia, who had also won the award with her husband Bart in 2015, whilst another brother, Ron, and his wife, Justine, were also finalists in 2017.

The chairman of the Ahuwhenua Trophy Management Committee, Mr Kingi Smiler, congratulated Eugene and Pania King, describing them as a great example of a couple who set challenging goals and then achieved them. He said that the King whānau worked so well together, helping each other to achieve farm ownership. They had now also earned a unique place in the legacy of the Ahuwhenua Trophy, comprising three finalist selections and two outright winners, an exceptional whānau achievement. Mr Smiler described Eugene and Pania as outstanding role models for Māori farming. 'All New Zealanders should take note of their achievements and that of their whānau.'[408]

Finalist Profiles

Kiriroa Station

Eugene and Pania King were the owners of Kiriroa Station in the Motu Valley, Matawai, about 72 kilometres north-west of Gisborne. Kiriroa was a 483 hectare (effective 357) sheep and beef property of which 60 hectares was flat, 200 hectares medium hill country and the balance consisting of steep hill country. The Kings traded cattle and finished all stock on their farm, and were at present withering 3800 stock units, being a mix of 40 percent cattle and 60 percent sheep.

Soil types were mainly pumice with some sedimentary, and the property had an annual rainfall of between 2.1 and 2.5 metres. Motu Valley was regarded as summer safe but did have long cold winters. Regular snow falls were not uncommon. Some parts of the farm were 732 metres above sea level.

Kiriroa was a special place for Eugene and Pania. They regarded the Motu River as taonga and saw themselves as kaitiaki of the 2.2 kilometres that flowed through their farm. The farm was also home to innumerable native birds, particularly the weka. Because of the declining weka population, the Kings had retired 2 hectares of land for them to live on, also establishing a wetland.[409]

When visiting Kiriroa Station, judges were impressed with the Kings' clear vision for the farm to attain whānau and environmental values as well as business profitability. Robust decision making processes were in place with trusted advisors actively consulted. The Kings were also clearly committed to the local community, as shown by their environmental leadership and support for local schools. The farm finances were well managed with appropriate budget setting and variance analysis to support decision making. Cash flow management systems were also robust.

The farm's focus on building soil fertility and subdivision for improved productivity were commended, with good understandings also shown as to what was needed to grow more feed and maintain pasture quality. Very good pasture utilisation and animal performance was also noted.[410]

Whangara Farms

Whangara Farms were located just north of Gisborne on State Highway 35 and comprised a partnership of three Māori incorporations. This allowed the respective parties to join as commercial partners but to retain their own identities.

Whangara Farms was a 8300 hectare (6900 effective) property running 75,000 stock units. Of these, 45,000 were sheep and the balance cattle. The farm itself was a mix of steep and flat land. Seventeen full-time staff were employed to run this extensive operation. Whangara Farms was also a significant contributor to agricultural research and development in the Gisborne region. The farm also contained a beef and lamb progeny test site.

In 2018, Whangara Farms became the first farm outside Europe to be awarded flagship status by the McDonald's Restaurant chain. The Chairman of Whangara Farms, Ingrid Collins, said their vision was to be an outstanding business delivering ongoing sustainable returns.

'We moved away from using the word "agribusiness" to "business" as we looked to develop

Whangara Farms Field Day, Whangara, 2019
(John Cowpland – Alphapix)

our commercial activities to be wider than just agriculture'. The farm operation was strongly focused on using science, technology and data collection to improve decision making.[411]

Many aspects of the Whangara Farms operation impressed the judges. The success of the partnership was clearly based on trustee relationships, quality information, regular contact and clear expectations. The background to establishing this commendable partnership was well documented – a partnership that clearly benefitted whānau, hapū and iwi, with all contributing to the setting of strategic and business objectives.

The budget setting, variance reporting and reforecasting process was robust with comprehensive monthly reporting by the well-performing business unit. Stock transfers between units were accounted for at market value. The farm benchmarks were seen as commendable and set against appropriate external data, with emphasis on gathering of data to drive proactive decision making in such key areas as climate, soil, pasture, stock and finances. Use of experts to facilitate enhanced feed production was noted, along with the use of cash cropping as a tool to develop land, including the use of deep ripping to improve aeration.[412]

Te Awahohonu Forest Trust – Gwavas Station

Historic Gwavas Station, located in Central Hawke's Bay, was owned by Te Awahohonu Forest Trust, which won the Ahuwhenua Trophy in 2013 with its other farm, Tarawera Station, sited on the Napier–Taupō Road. Gwavas Station comprised 1000 hectares with the owners leasing an additional 178 hectares, making the property 989 hectares effective. The property was of an irregular shape with State Highway 50 bisecting the farm at the north eastern end of the property.

About 70 percent of the area was flat to easy rolling country with the balance consisting of moderately steeper hills and steep faces connecting with lower terraces and riverbeds. Gwavas Station was located in a region that has experienced summer dry conditions, underscored by a complex range of mainly free draining soils. These soils comprised mainly Takapau silts and Tukituki gravelly sands on the flatter areas, with Poporangi and Mangatahi soils to be found on some intermediate terraces and rolling hills. Gwavas sandy loams featured on the remaining easy and steeper hills.

Gwavas Station wintered nearly 12,000 stock units, which comprised approximately 50 percent sheep. Gwavas was farmed as an intensive dry-land finishing

Bob Cottrell, Chairman of Te Awahohonu Forest Trust, speaking at Te Awahohonu Forest Trust – Gwavas Station Field Day, Central Hawke's Bay, 2019
(John Cowpland – Alphapix)

property that complemented the Tarawera Station's breeding operation. Between 14,000 and 16,000 lambs and approximately 800 cattle were finished annually, depending on the season. The farm was effectively run, with three full-time staff made up of the farm manager and two shepherds. According to Trust Chairman Bob Cottrell the aim of the farm was to finish off all stock bred off Tarawera Station and at the same time to maximise returns through optimising processing weights and specifications, alongside a profile of a broader out of season stock supply.[413]

When the judges visited Gwavas Station, they were impressed by the well-managed changes to the Trust Order, which enabled a rotation of trustees, postal voting, no proxies and votes cast according to shareholding, with the commendable aim of ensuring that the best people were able to make the best decisions. Connections with beneficial shareholders were strong, with regular newsletters, field days on the station and the use of social media. Grants for tangihanga, education and personal development were also commended.

The trustees also possessed a clearly defined granular and fluid process of planning, reviewing and financial reforecasting on a short-term basis. Financial budgets were based on clear key performance indicators with variance reporting on a monthly basis. The benefits of an intensive feed production programme were now being realised with improved pasture production and quality. The trust also showed a good understanding of the environmental implications of crop rotation with planning well in hand to address issues arising.[414]

2020 Horticulture

Ahuwhenua Trophy for Horticulture

In a new and exciting development for the Ahuwhenua Trophy, in 2018, horticulture was signed up as a new competition category, competing for a new special award, the Ahuwhenua Trophy Te Puni Kōkiri Excellence in Māori Farming Award in Horticulture.

Much earlier, Governor-General Lord Bledisloe had first presented a Bledisloe Cup for Horticulture in 1931 in order to encourage competition among orchardists at large for the best exhibit of New Zealand apples at the Imperial Fruit Show.[415] The cup was thereafter awarded annually to horticulturalists who attained outstanding merit in their horticultural endeavours.

Lyn Harrison and Kingi Smiler, Chairman of the Ahuwhenua Trophy Management Committee, with the whānau of four trophies on display at the Horticulture Conference, Mystery Creek, Hamilton in 2019. The trophies from left: Ahuwhenua Trophy for Dairy; Ahuwhenua Trophy for Horticulture, Ahuwhenua Trophy for Sheep and Beef, and, front, Bledisloe Cup for Horticulture
(John Cowpland – Alphapix)

With Māori growers in mind, the Chief Executive of Horticulture New Zealand, Mr Mike Chapman and Mr Kingi Smiler, Chairman of the Ahuwhenua Trophy Management Committee, signed an agreement on 20 August 2018 that would see an Ahuwhenua Trophy awarded for Horticulture in 2020. Thereafter, horticulture would feature on a three-year rotation, along with the exisiting categories of dairy, and sheep and beef.

In signing the agreement, Mike Chapman said that, throughout New Zealand, 'there were significant Māori holdings in horticulture, so it was great for the horticulture industry to join the legacy that the Ahuwhenua Trophy holds'. Mr Chapman was excited that success stories in horticulure would now be given a wider audience through this competition. Together with the Ahuwhenua Trophy Management Committee, Horticulture New Zealand could now encourage Māori horticulturalists everywhere to tell their stories.

Horticulture was a rapidly growing industry and Horticulture New Zealand was keen to promote the industry, especially to young people, as seen through their Young Growers' competitions, which had such a strong synergy with the Young Māori Farmer of the Year Award. 'We are very much about trusted people providing healthy, safe and sustainable food, so we feel a strong alignment to the Ahuwhenua Trophy. We are proud to be a part of this partnership', said Mr Chapman.

Mr Kingi Smiler said that he was delighted that horticulture was about to join the Ahuwhenua whānau by becoming involved in the staging of an award dedicated to recognising Māori excellence in horticulture. He said this added a new dimension to the Ahuwhenua competition and showed that the event was moving with the times whilst still embracing the vision and values of the trophy's two founders, Sir Apirana Ngata and Lord Bledisloe.

Māori had alway been great horticulturalists. It was therefore fitting that Māori growers now had the opportunity to be publicly recognised for their efforts, in the same way as Māori in sheep and beef, and dairy, he said.[416]

Ahuwhenua Trophy Competition Launched

The new Ahuwhenua Trophy competition for horticulture was launched at the HortNZ Gala Dinner on 2 August 2019 by Minister for Māori Development, Hon Nanaia Mahuta. Speaking at the launch, Mr Kingi Smiler reiterated that it was great to have the support of Horticulture New Zealand, along with the various sector groups now sponsoring and supporting the new Ahuwhenua Trophy competition for horticulture.

Mr Smiler said that it was logical and timely to have a separate event in order to recognise the Māori contribution to horticulture. For more than two centuries, Māori had exported horticultural products, often in challenging circumstances, a fact not widely recognised in New Zealand. In the last ten years in particular, Māori had become major investors in the hort culture sector, and this had resulted in good financial returns and jobs for whānau, hapū and iwi. 'But there is still much more to be done, and I am sure that, with the positive publicity that the Ahuwhenua Trophy generates, this will encourage more Māori to become involved in horticulture', he said.

Mr Smiler added that he was delighted that the Ministry of Primary Industries would become a platinum sponsor, along with Te Puni Kōkiri. This was a generous sponsorship that would ensure the financial viability of the new competition and enable it to maintain the high professional standards that have long characterised the Ahuwhenua competition. In addition, the Ministry of Primary Industries were offering a $30,000 prize in order to provide native seedlings to the winner and finalists. These seedlings were being donated by Te Uru Rākau – Forestry New Zealand and produced by Minginui Nursery.[417]

An Earlier Bledisloe Cup for Horticulture

When the Ahuwhenua Trophy for Excellence in Horticulture was introduced in 2020, the Chairman of the Ahuwhenua Trophy Management Committee, Mr Kingi Smiler, observed, interestingly, that in 1931 Ahuwhenua Trophy-founder Lord Bledisloe had also introduced an award for horticulture. The Bledisloe Cup for Horticulture had been introduced to encourage competition between orchardists 'for the best exhibit at the Imperial Fruit Show'.

Thereafter, the trophy had been awarded on an annual basis to orchardists and horticulturalists attaining excellence in their endeavours. For example, in 2013, the Bledisloe trophy was awarded to Fay and Joe Gok, a Chinese couple in their eighties from Auckland in recognition of their outstanding contribution to the development of horticulture for more than forty years.

The Goks had been the first growers to plant Brussels sprouts, alongside their peas, cauliflowers, carrots, parsnips and potatoes. They had also introduced commercial washing of vegetables and had developed a unique, seedless watermelon. They are also credited with first producing a commercial variant of 'backyard' rhubarb, successfully exporting large quantities to England and Japan.

What is interesting about the Goks is that they are also credited with reinvigorating the growing of kūmara, which had become somewhat defunct, not least because of a black rot epidemic. The Goks had developed a disease-resistant strain and had gifted their stock for the use of other farmers. They also experimented in storage methods and invested in a curing shed that dramatically reduced wastage. In 2016, Fay and Joe appeared in a television documentary entitled 'How Mr and Mrs Gok saved the kumara'.

The Goks also experimented with the planting of low-lying areas with taro, which they then successfully exported to the Pacific Islands for use in traditional umu cooking. By 2013, Fay and Joe were undoubtedly worthy receipients of the Bledisloe Cup for Excellence in Horticulture.[418]

A New Ahuwhenua Cup for Horticulture

With horticulture now added to the Ahuwhenua Trophy as a new competition category, the Ahuwhenua Trophy Management Committee commissioned noted jewellers The Village Goldsmith of Wellington to craft a new Ahuwhenua Trophy. Thereafter, the creation of the trophy was an extensive collaboration. As described by Mr Ian Douglas of The Village Goldsmith, developing the new Ahuwhenua cup was 'really complex but a fantastically creative process'.

Many daunting issues were faced when looking to create a copy of the original cup, given that the creators of the original, Walker and Hall in 1937, no longer existed. Nor were there any companies in New Zealand that had 'the capacity to spin the bowl sizes, in sterling silver, to the shape and size required', says Mr Douglas.

The Village Goldsmith then searched overseas for a company that could work with such a bowl size in silver. Eventually, Ottewill Silversmiths, based in Ashford, England, were discovered. The business had 'a reputation second to none' and, after extensive discussions, Ottewill confirmed their willingness to take on this precision project. Exacting measurements were determined by The Village Goldsmith, as the existing Ahuwhenua Trophy could not be sent overseas.

'Accurate templates and impressions [were] taken from the originals,' says Mr Douglas. These needed to be 'perfect' so that Ottewill could create an exact copy of the original. The Village Goldsmith's master jeweller Dan Palmer 'created all the necessary templates and shapes that were sent to the UK'. These templates, which included photographs, aluminium plates and pink wax mouldings, all contributed to the construction of the new Ahuwhenua Trophy for Horticulture. The engraving, which was the work of The Village Goldsmith's engraver Darren Orr, necessitated 'careful tracings of the original engraving', with tracings and photographs also conveyed to the UK.

From a jeweller's perspective, says Mr Douglas, the final product was not just a trophy, 'it was a work of art'. The whole piece had come together beautifully, which was a great testament to the skill of so many creative people at The Village Goldsmith, Wellington, and Ottewill, Kent, UK. 'I know this trophy will go on to be a part of an amazing legacy', said Mr Douglas. The Village Goldsmith was delighted and proud to have been a part of the creation of the new Ahuwhenua Trophy.[419]

The team of craftsmen and craftswomen at Ottewill Silversmiths, Ashford, England, who so skillfully created the new Ahuwhenua Trophy for Excellence in Māori Horticulture (foreground), first awarded in 2020

(Ottewill Family, Ashford, England)

Inaugural Māori Horticulture finalists 2020

From left: Ratahi Cross, Chairman, Ngai Tukairangi Trust; Hon Nanaia Mahuta, Minister for Māori Development; Norman Carter, Chairman, Te Kaha 15B Hineora Orchard; and Homman Tapsell, Advisory Trustee/Orchard Manager, Otama Marere

(John Cowpland – Alphapix)

The Finalists and COVID-19 Affected Field Days

Three finalists were selected to contest the inaugural Ahuwhenua Trophy Te Puni Kōkiri Excellence in Māori Farming Award in Horticulture. The first finalist was Te Kaha 15B Hineora Orchard, which was a Māori freehold block located at Te Kaha in the eastern Bay of Plenty. The second finalist was Otama Marere located at Paengaroa, near the Bay of Plenty town of Te Puke. The third finalist was the Ngai Tukairangi Trust, which was a very large kiwifruit operation with one of its orchards based at Matapihi, just a few kilometres from the centre of Tauranga.

Field days were also organised for each competitor. However, on 19 March 2020, the Chairman of the Ahuwhenua Trophy Management Committee, Mr Kingi Smiler, announced that the field days and awards dinner planned for the horticulure competition would need to be cancelled due to the unfortunate impacts of the COVID-19 pandemic.

Each of the finalists' field days would normally have attracted between 200 and 300 people, and the awards dinner normally hosted between 700 and 900 guests.

However, the health and safety of finalists, sponsors, whānau and others was paramount, said Mr Smiler. Therefore, it was important that the Ahuwhenua Trophy competition follow the advice of New Zealand's health authorities, especially the restraints being placed on large gatherings because of the pandemic. 'We are pleased that we can complete the judging process.' However, because of the very small number of people involved, a winner for this inaugural horticulture competition would be announced by the Minister for Māori Development, Hon Nanaia Mahuta, and the Minister of Agriculture, Hon Damien O'Connor. A special event for the presentation of trophies would be arranged once the COVID-19 restrictions were eased.[420]

From left: Dan Palmer and Ian Douglas of The Village Goldsmith crafted the new Horticulture Trophy in partnership with Ottewill Silversmiths, Ashford, England; Norman Carter, Chairman of Te Kaha 15B Hineora Orchard, inaugural winners of Haumiatiketike – the Ahuwhenua Trophy for Horticulture.

(John Cowpland – Alphapix)

Ahuwhenua Awards Dinner Back On!

With the national pandemic situation much improved, the Ahuwhenua Trophy Management Commitee announced in September 2020 that the Horticulture Awards dinner was now back on. 'Put Friday 20th November in your diary', advised the committee: 'This new date has been set to stage the awards dinner for the inaugural Ahuwhenua Trophy competition for Horticulture and the Young Māori Grower Award.' The dinner would be held at the Rotorua Energy Events Centre. Whilst acknowledging the continuing uncertainty of the months ahead, where there was still a risk that things might change at the last moment, the management committee was sure that it would be possible to continue to plan for the upcoming event.

With the impact of COVID-19, the final judging and field days for finalists scheduled for March and April 2020 were postponed. Judging for the competition, however, resumed in October. Of the finalists, Ngai Tukairangi Trust in Tauranga were still able to hold a field day on 30 October 2020.[421]

The Winner is Announced

The winner of the inaugural Ahuwhenua Trophy for Excellence in Horticulture was Te Kaha 15B Hineora Orchard. The announcement was made by Governor-General The Rt Hon Dame Patsy Reddy at the awards function held in Rotorua on 20 November 2020, attended by 750 people including the Minister for Māori Development, Hon Nanaia Mahuta, and the Minister of Agriculure, Hon Damien O'Connor, with other politicians, dignitaries, agribusiness leaders and whānau.

Governor-General The Rt Hon Dame Patsy Reddy presented the Ahuwhenua Trophy to Norm Carter, the Chairman of Hineora Orchard, while Hon Willie Jackson, Hon Damien O'Connor and Dave Samuels, Chief Executive of Te Puni Kōkiri, presented the replica trophy as well as medals, certificates and cash prizes to the winning trust.

Mr Kingi Smiler said that the Te Kaha 15B Hineora Orchard was a worthy winner of the new trophy, which, having been commissioned by the Ahuwhenua Trophy Management Committee, had been named

Haumiatiketike by Hon Nanaia Mahuta. The Te Kaha 15B Hineora Orchard trustees had shown great vision, persistence and resilience to establish their operation and to achieve some impressive results, which must have made their whānau feel proud of their efforts. This was the first time in the eighty-seven-year history of the Ahuwhenua competition that the trophy had been opened to Māori horticulturalists who were deserving of recognition for their substantial contribution to the Māori and New Zealand economy.

Mr Smiler also praised the other finalists, saying their operations were undoubtedly amongst the finest in the horticultural sector. All three Māori finalists had helped to set a benchmark for future events in this competition that would be challenging to eclipse. 'What makes it more impressive is the fact that they have done this in one of the most difficult times in the history of this country where uncertainty is now a way of life.' All Māori and the whole country could be proud of their efforts, he said.[422]

Finalist Profiles

Te Kaha 15B Hineora Orchard

The whenua of this orchard fell within the tribal rohe of Te Whānau a Apanui and more specifically was associated with Te Whānau a Te Ehutu hapū. Te Kaha 15B Hineora Orchard comprised 11.5 hectares on which the trust managed a kiwifruit joint venture operation, along with a commercial pack house facility which housed the local kiwifruit spray company, in which the trust held shares. A four bedroom home for accommodation was also managed on site.

Prior to the trust's creation in 1970, the land had been largely occupied by different whānau who farmed the block, maintaining subsistence living as well as growing a range of fruit and vegetables for a local community market. Later a citrus orchard was established but, given the small area, this unfortunately failed to provide an economic return for its owners. In 1998 the trustees recognised that, though asset rich, they lacked capital to develop their land. They then entered into an agreement with a number of eastern Bay of Plenty orchardists who proposed a 50/50 joint venture to develop the new 'gold' varieties of kiwifruit. This provided sufficient capital for developments to proceed.

The Hineora Orchard began in 1999 and was the last of six blocks to join the innovative joint venture, with decision making and profits shared over a 20-year period. The joint venture had remained and managed the orchard through the highs of good returns to the

Te Kaha 15B Hineora Orchard, Te Kaha, 2020
(John Cowpland – Alphapix)

lows of vine disease. The land was due to be returned to 100 percent trust ownership in 2023.

As a result of this journey, the trustees now managed 8.13 hectares of G3 SunGold kiwifruit, producing over 133,000 trays a year, working closely with a contracted orchard manager and a local cool-storage company. The trust had also contributed to establishing the local spray company, along with five other joint venture blocks. They had also assisted in setting up Te Whānau a Apanui Fruitgrowers Incorporated. The trust was now also able to offer annual kaumātua grants to shareholders, as well as asisting in tangi, health, education, sport, culture and travel.[423]

Visiting judges to Te Kaha 15B Hineora Orchard were impressed by the strong self-belief and excellent governance skills on show. A clear focus on engaging shareholders was evident, as was an array of innovative strategies to encourage whānau back to what was in reality still quite an isolated environment. However, clear success was evident in this regard. The trust had broadened its investments and had developed initiatives of great benefit to the community. The trust was also providing training and educational opportunities to whānau and shareholders and actively working to improve the employment prospects of younger whānau in particular.

A strong focus was observed on the health of the soil with organic fertiliser and compost used to encourage microorganisms and retain soil moisture. Capacity had also been developed for future water supply from a bore of 9000 litres per hour in order to meet future needs.[424]

Otama Marere

Otama Marere Orchard was located on Paengaroa North A5 Block in Paengaroa, near the Bay of Plenty town of Te Puke. The trust managed a total land area of 45,144 hectares with the orchard comprising 11.87 hectares of Hayward Green kiwifruit, 2.21 hectares of SunGold G3 and 7.08 hectares of SunGold G3. In addition to kiwifruit, the trust had planted 950 Gem avocados spanning four separate blocks, a total of 2.1 hectares.

Approximately 70 percent of the property was flat to easy rolling country spanning the Kaituna River. The balance consisted of moderately steeper hills and steep faces connecting lower terraces down to the river. There was a gully running through the orchard that had historically been planted in forestry. Lower wetter parts of the property had been planted out in wetlands. The property had good road access, with soil described as typically allophanic strongly influenced by clay minerals and areas of volcanic soil.

The Ahuwhenua judges who visited Otama Marere Orchard were impressed by the strong governance team that clearly understood the horticulture business, with clear succession planning in place and admirable strategies aimed at staying relevant with the environmental expectations of shareholders and consumers. Innovative brand reforms had also helped with widespread promotion of Otama Marere Orchard. The trust offered a range of grants in support of social, educational and cultural activities.

Strategies were also in place to improve the orchard's financial performance, enabling higher returns to be distributed to shareholders. A strong environmental focus was also apparent, moving into organics with demonstrated success with respect to approaches and yields. Diversification away from reliance on kiwifruit was also evident. Scenic wetlands were being restored in conjunction with local regional authorities. Restoration efforts were commendable and timely, with wetlands to be returned to their natural state and tied into the history of the whenua with its historic Pā site and early occupation accounts of whānau tūpuna.[425]

Ngai Tukairangi Trust

Ngai Tukairangi Trust was affiliated to Ngāi Tukairangi hapū, Ngāi Te Rangi iwi. The trust owned a kiwifruit orchard operation in two different locations. The Matapihi-based operation was mainly sited on the Matapihi peninsula, a small Māori community between Tauranga and Mt Maunganui. This included a small Te Puke orchard.

The Matapihi-based operation had a total planted area of 57 hectares of which 36.1 hectares were utilised for producing G3 kiwifruit, 12.7 hectares for Green Hayward, 6.9 hectares for non-producing G3 and a small trial variety kiwifruit. At 20 metres above sea level, the orchard was on flat land with very good volcanic soils. The orchard also benefitted from excellent road access, with local kura, marae, papa kāinga and Tauranga city all close by.

In 2017, the trust purchased a large kiwifruit orchard operation in Hawke's Bay, which saw the operations double as a result. The Heretaunga orchard operation had 60 hectares of G3 Gold kiwifruit, all under cloth. The orchards were about 70 metres

Otama Marere Orchard, Paengaroa, 2020
(John Cowpland – Alphapix)

Ngāi Tukairangi Trust Orchard, Matapihi Peninsula, 2020
(John Cowpland – Alphapix)

above sea level on flat land and were founded upon more difficult alluvial soils. The Heretaunga orchard was also dedicated to growing and producing the G3 Gold variety kiwifruit.

Many aspects of the Ngai Tukairangi Trust operations impressed the Ahuwhenua judges. A passionate and engaged board with a strong focus on the future was evident. Trustees and management clearly cooperated in determining all financial matters and overall business performance. The trust had established a very good financial performance and had approached investments cautiously, weighing the risk of significantly expanding their business into Hawke's Bay. The trust was the largest Māori grower of kiwifruit, with 95 hectares of Gold and 18 hectares of Green across the business, leading to an impressive $150 million evaluation.

The trust also offered scholarships, internships and training opportunities for its supporting whānau and community, with a commitment to enhance cultural outcomes alongside those of the commercial business. The trust also demonstrated a commendable awareness of the environment, taking steps to enhance the whenua, which included the development of organic farming and careful irrigation management.[426]

2021 Dairy

Prosperity for Māori

In welcoming Ahuwhenua competition participants and supporters in 2021, the Minister for Māori Development, Hon Willie Jackson, acknowledged the Ahuwhenua Trophy as 'the most prestigious of its kind'. The trophy showcased 'excellence in the pastoral and horticultural sectors', significantly contributing to the prosperity of the Māori economy. Māori agribusiness was now 'a powerful driver in the New Zealand economy', reaching $297 million in 2018, almost twice the figure in 2017. One sixth of all Māori authorities had now developed interests in agriculture; many were also running farms that were on average four times larger than the median New Zealand farm.

Minister Jackson also acknowledged the introduction of horticulture as an Ahuwhenua competition category in 2020, with its origins and cultural significance dating back many thousands of years, symbolising Māori adaptation, proficiency and, latterly, economic innovation.

This year's category – dairying – accounted for 100 million Māori shares in Fonterra, owned by some of the largest and most profitable corporations. Seen globally, the environmental challenges facing dairying, and agribusiness at large, were prodigious, not least climate change and the continuing impacts of the COVID-19 pandemic. However, kaitiakitanga and associated customary approaches to present needs were as relevant as ever, he said, as demonstrated each year by the Ahuwhenua competition.[427]

The Minister of Agriculture, Hon Damien O'Connor, spoke of the 'Accelerating our Economic Recovery' roadmap launched by the government in 2020 to ensure sustainability, reduction of methane emmissions and the restoration of freshwater resources. An essential component of the roadmap, he said, was the Te Taiao Framework, which, by combining mātauranga with genuine partnership between Māori and Pākehā in the primary sector, provided a pathway by which Māori agribusiness could improve the wellbeing of all, distinguishing New Zealand's primary sector on the world stage.[428]

The ever-present uncertainties of the COVID-19 pandemic were also alluded to by the Chairman of the Ahuwhenua Trophy Management Committee, Mr Kingi Smiler. Despite these uncertainties, the inaugural horticulture competition had been successfully introduced in 2020. The postive attitude evident in the primary sector 'and in particular the area of Māori agriculture' continued to be reflected in the number of Māori dairy farmers who entered this year's competition. 'We salute their ongoing commitment to uphold the vision and values of Sir Apirana Ngata and Lord Bledisloe.'

The Māori way of farming, said Mr Smiler, was rapidly becoming the New Zealand way of farming. Farmers at large were recognising how profitable Māori enterprises were and how well they fulfilled their responsibilities as kaitiaki of the environment. Māori agribusiness now enjoyed a high profile, not least because of the Ahuwhenua competition, significantly contributing to the 'powerhouse of the Māori economy'.[429]

Launching the Ahuwhenua Trophy Competition 2021

Three finalists for the Ahuwhenua competition in dairying for 2021 were announced in Parliament.

One of the finalists, Pouarua Farms Ltd, was situated on the Hauraki Plains near the small township of Ngātea. The Pouarua Farms platform – the largest in the Hauraki region – comprised ten dairy units and one drystock unit. A total of 4600 cows were milked

2021 finalists

From left: Hon Damien O'Connor, Minister of Agriculture; Jack Mihaere, Chairman, Tunapahore B2A Incorporation; Paki Nikora, Trustee Chair, Tataiwhetu Trust; and Wati Ngakoma, Kaumātua, Pouarua Farms

(John Cowpland – Alphapix)

across 1775 hectares, producing approximately 1.65 million kg of milk solids.

The second finalist was the Tataiwhetu Trust, which was located in the Ruatoki Valley south of Whakatane. The Tataiwhetu Trust managed 432 cross cows and carried 188 replacement stock on two supporting blocks. Tataiwhetu Trust was an organic farm milking once a day, their herd producing 129,140 kg of milk solids.

The third finalist in 2021 was Tunapahore B2A Incorporation, which managed a farm situated in the Bay of Plenty near Opotiki. The farm comprised 376 hectares with 385 cows producing 125,940 kg of milk solids.[430]

The Minister of Agriculture, Hon Damien O'Connor, applauded the finalists for their longstanding commitment to sustainably developing their lands and environments for future generations. 'You are leaders in the entire primary sector and people will look to you in the future,' he said. The Māori business sector was largely under-appreciated, but the official statistics underlined the enormous potential of Māori participation in the agribusiness sector.

The rise of Māori agribusiness was impressive, added the Minister of Māori Development, Hon Willie Jackson, 'especially given its unstinting commitment to all aspects of sustainability'. The introduction of horticulture as an Ahuwhenua category demonstrated how Māori were diversifying into all areas of the primary sector, 'playing an increasingly significant role in producing higher value food and fibre products'. Despite the challenges of the COVID-19 pandemic, the Māori farming sector

The Tataiwhetu Trust of Ruatoki Valley celebrates their winning of the 2021 Ahuwhenua Trophy for Dairy
(John Cowpland – Alphapix)

remained strong with landowners able to fulfil their aspirations for their whenua and their people.

The Associate Minister for Agriculture, Hon Meka Whaitiri, also congratulated the finalists, adding that the government was committed to working with Māori farmers to ensure a valuable future for the country.[431]

The Field Days

Three field days were scheduled during which each of the finalists was able to present all aspects of their operations to whānau, friends, supporters, interested members of the public and, not least, officials and judges representing the Ahuwhenua Trophy Management Committee.

The field day for Pouarua Farms was held on 25 March 2021 on their farm property at 180 Central Road South, Ngātea.[432] The field day for the Tataiwhetu Trust was held on 1 April 2021 on the trust's property at 359 Rūātoki Valley Road, south of Whakatane.[433] The field day for the Tunapahore B2A Incorporation was held on 8 April 2021 on the incorporation's property at 2558 State Highway 35, Hāwai, near Opotiki.[434]

The Winner is Announced

The winner of the Ahuwhenua Trophy for Dairy 2021 was the Tataiwhetu Trust, located in the Ruatoki Valley south of Whakatane. The winner was announced by the Minister of Agriculture, Hon Damien O'Connor, at an awards dinner attended by 800 people at the TSB Centre, New Plymouth. Guests also present included Hon Willie Jackson, Hon Meka Whaitiri, Kingi Tuheitia, Sir Tumu Te Heuheu, Hon Winston Peters, Hon John Luxton and other dignitaries, politicians, agribusiness leaders and whānau from all of the finalists.

The award was received by Paki Nikora, Chairman of the Tataiwhetu Trust who said he was elated at the news. As the minister presented Paki Nikora with the trophy, there were scenes of great jubilation from whānau and supporters of the Tataiwhetu Trust. When receiving the trophy, Paki Nikora said that the 'biggest driver now for the Trust [was] diversification'. As far as Tūhoe were concerned, he said, 'we are trying to lift the whole image of our iwi into another space, and winning this award is a launching pad for this initiative'.

All finalists were congratulated by Mr Kingi Smiler, Chairman of the Ahuwhenua Trophy Management Committee. The standard for the 2021 competition had been especially high, presenting the judges with a difficult decision. The three field days run by the finalists were extremely well managed, attesting to the 'quality and depth of Māori farming enterprises'. Finalists excelled and remained exceptional role models for Māori farmers. In the end, he said, the Tataiwhetu Trust Farm was 'very special' and was yet another example of Māori farmers working innovatively, focusing on key strategic objectives.[435]

Finalist Profiles

Pouarua Farms

The Pouarua Farms were located on the Hauraki Plains, 35 kilometres south-east of the Bombay Hills. The 2200-hectare platform was made up of ten farms, nine of which were dairy farms with one property managed as a drystock unit. Pouarua was the largest single dairy platform in the Hauraki region, with 4600 cows milked across 1775 heactares, producing approximately 1.65 million kg of milk solids.

Pouarua Farms were jointly owned by Ngāti Maru, Ngāti Tamaterā, Ngāti Tara Tokanui and Te Patukirikiri. The farm had been purchased by the owners in 2013 for $53 million from the Land Corporation of New Zealand, and was thus returned to its original owners, having been taken by the Crown during the nineteenth century.

Of the nine farms, Farm A had been established in 2017 after a reconfiguration of four of the farms in order to increase efficiencies. Currently, the farm was milking 600 cows on 217 hectares, supported by a fifty-four section rotary shed. According to the Pouarua Trust, 'the farm is the vision of its owners, with practical technologies and careful consideration for the environment'.

Throughout the 2019/2020 season, Farm A recorded an 18 percent increase per cow milk production to 390 kg of milks solids, whilst also recording a 20 percent increase in per hectare production to 1034 kg of milk solids despite a significant drought. 'This was achieved with careful utilisation of on-farm grown feeds and adjusting the stocking rate down.' In addition, 3-in-2 milking was utilised to conserve energy during the hotter months.

Nitrogen use was also capped to 150 units per hectare, across all farms. Farm environmental plans had been adopted as soon as full operation care was undertaken. A forever planting plan also

Pouarua Farms Field Day, Hauraki, 2021
(John Cowpland - Alphapix)

saw approximately 7500 native species planted annually across all farms, with riparian planting of the drains the main priority. Cultural gardens had also been developed, comprising specific species that might be utilised in weaving, medicines, food, bird habitat, water quality, soil conservation and landscape improvement.

Finally, the staff at Pouarua Farms were supported in multiple on and off farming opportunities, with the farm community active in local schools and the local community. A spend local policy also contributed to the continuing prosperity of the communities of which Pouarua Farms were an essential part.[436]

Many aspects of the Pouarua Farms operation impressed the judges. Senior managers and staff had a strong sense of where they had come from, forming 'an agile governance team with an excellent knowledge' of all aspects of relevant agribusiness. Following the land confiscations of the 1860s, local united Hauraki iwi had turned this severe disconnection into a positive strategy to regain and develop the ancestral land on which Pouarua Farms stood.

A strong focus on tikanga Māori had restored wider community and whakapapa connections, driving the utility of local customary knowledge in planting and planning decisions. Robust processes for budgeting, cashflow management and real-time reporting ensured excellent operating profits at over $3600 per hectare with low farm working expenses ($4.25 per kg of milk solids) and strong production levels (1044 kg of milk solids per hectare). Important off-farm investments included carbon sinks, mānuka honey and maize. Feed production for the winter months was exceptional, including turnips, sorghum and ryegrass. Animal performance per cow (386 kg of milk solids per cow) and hectare (1044 kg of milk solids per hectare) reflected excellent reproductive performance and animal health. Finally, 21 percent of the staff were Māori; with good training schemes in place to ensure that levels of highly skilled Māori staff might increase, going forward.[437]

Tataiwhetu Trust

The origins of the Tataiwhetu Trust can be traced back to a Land Consolidation meeting held at Tauarau Marae in 1921. At that time, Sir Apirana Ngata was encouraging Māori to subdivide their lands into productive units in order to sustain Tūhoe whānau. But by the 1950s, the units were seen to be too small, leaving the lands under-utilised. 'Finally, in 1986, six Ngatirongo families agreed to combine their lands to form the Ngatirongo Trust Farm.' Nine blocks were aggregated – A40B, 41, 44, 45, 46, 47, 48C, 50 and 74 – giving a start-up area of 97,689 hectares with a useable dairy platform of 80 hectares. Over time, adjoining blocks were leased, giving a total dairy platform of 184 hectares.

The initial development of the Ngatirongo Trust Farm was led between 1986 and 2009 by kaumātua Frank Vercoe. By 2009, sharemilking was achieved at 600 cows twice a day, supplying Fonterra all year round. When Paki Nikora was appointed Chairman in 2009, he initiated the purchase of 400 in-calf cross heifers, milking once a day, transitioning back to seasonal milking. The farm's sharemilkers then excelled through Primary ITO Levels 1, 2, 3 and 4 over the next ten years.

The Tataiwhetu Trust now stocked 432 Kiwi-cross cows, at a stocking rate of 2.5 per hectare. Production per cow was at 273 kg of milk solids, with production per hectare at 694 kg of milk solids. Imported supplementary feed per cow was 0.8 t per cow, with purchased nitrogen surplus set at 12 kgN per hectare. Finally, greenhouse gas emissions were 7736 kg CO_2 per hectare.

In 2010, the trust received the Ballance Farm Environment Award, recognising the trust's creation of wetlands, landscape features and historical sites. A year later, the trust was awarded the Green Ribbon Award for protecting biodiversity, in collbaoration with Tūhoe Pūtaiao.

In 2014, the name of the Ngatirongo Trust Farm was changed to Tataiwhetu Trust, recognising the input of the six original families. In that year, also, the trust was awarded with the Fonterra Grade Free Certificate, which recognised the continuing high quality of their milk supplies. In 2015, the Tataiwhetu Trust transitioned from conventional milking to organic and was fully certified by AsureQuality, receiving Fonterra Organic certification in 2019.[438]

The judges commented positively on many aspects of the Tataiwhetu Trust's operations. The trust was operated as a whānau-based business 'with a very strong connection to the whenua, with a central focus on utilising their ancestral lands to support whānau and community outcomes'.

A diversification investment portfolio and the purchase of other farms also aimed to offset the risks associated with uncertainties regarding leases. The trustees came from a wide variety of important backgrounds, with active learning a continuing focus. Local community connections

Desma Feakins and Whareauahi Teepa, Managers of Tataiwhetu Dairy Farm, Tataiwhetu Trust Field Day, Taneatua, 2021
(John Cowpland – Alphapix)

were excellent, especially as emphasised through grants and scholarships with local schools. A strong commitment to employing local people was also evident, with tangata whenua playing key roles in the trust's management. The move to organics had been driven by a strong desire to lessen the farm's impacts on the wider enviroment, thus improving animal and human wellbeing.

The conversion of the farm to organics, and the reducing of stock rates, had undoubtedly impacted upon the trust's operating profit. However, the current net return of $1946 per hectare for 2019/2020 was closer to the Bay of Plenty (Owner Operator) benchmark of $2215 per hectare. Pasture production was also an important continuing function, with new pastures tested regularly.[439]

Tunapahore B2A Incorporation

Tunapahore was originally one whole land mass located between two coastal features of Hāwai. The land mass was divided by the Māori Land Court in 1953 into Tunapahore B2A and Tunapahore B2B. Thereafter, Tunapahore B2A was managed for a time as a dairy and dry stock unit, before reverting to a leasehold to adjoining Pākehā farmers. Tunapahore B2A was incorporated in 1959.

Tunapahore B2A comprises 400.83 hectares located at Hāwai and Tōrere 'on the famous State Highway 35 on the East Coast of the North Island'. A runoff area comprised 10.93 effective hectares, with two other leased areas forming part of a milking platform (16.5 hectares) and a maize silage facility (10.1 hectares). Tunapahore B2A's overall milking platform was 132 hectares with 385 cows producing 125,940 kg of milk solids.

Following many years as a leasehold property, the farm was returned to the management of a

Tunapahore B2A Incorporation Field Day, Hāwai, East Coast, 2021
(John Cowpland – Alphapix)

committee in the late 1980s. The farm was then operated as a dairy unit with 50:50 sharemilking, acquiring the adjoining Tawaputa Block under lease in order to increase the milking platform. A twenty-aside herringbone shed was also built at this time. In 2018, the management committee purchased the dairy herd and plant and began hiring staff. In 1985, part of the farm was leased for pine forest, with an additional 6 hectares added to the dairy platform. The farm also contained native forests, which included important archaeological sites.

A 5.54 kiwifruit orchard was purchased in 2006. 'The orchard consistently produces great fruit and is within the top producers for both EastPack and Zespri.'[440]

Many aspects of the Tunapahore B2A operation impressed the judges. The Committee of Management clearly had 'a strong understanding of the business, driven in part by the unexpected shift from 50:50 sharemilking to full herd ownership and the recent purchase of a kiwifruit orchard'.

The committee had also demonstrated a strong focus on benchmarking using information gathered by farm consultants. Tunapahore B2A was also part of an MPI Māori agribusiness cluster. Strong community support and involvement was also evident, including grants to local schools, marae, churches and iwi. Staff also showed a deep knowledge of cultural and wāhi tapū sites, having put in place an active management plan in order to safeguard these sites.

Tunapahore B2A also had a strong focus on reporting and budget management, taking account of innovative accounting measures to reflect the farm's diversification aspirations. Pastures were managed well, with good knowledge shown of pasture covers and feed intakes. Regular re-grassing programmes had also been introduced to improve the overall feed grown on the milking platforms.[441]

Pākarae Whāngārā B5 Field Day, 14 May 2009
(Ahuwhenua Archive, Field Day-51)

Tunapahore B2A Incorporation
(John Cowpland – Alphapix)

Chapter Twelve

Looking to the Future – Young Māori Farmers and Growers

E tipu e rea, mō ngā rā o tō ao
Ko to ringaringa ki ngā rākau a te Pākehā hei oranga mō tō tīnana
Ko tō ngākau ki ngā taonga a ō tīpuna hei tikitiki mō tō māhuna
Ko tō wairua ki tō Atua, nāna nei mea katoa

Grow tender shoots for the days of your world
Turn your hands to the tools of the Pākehā for the well-being of your body
Turn your heart to the treasures of your ancestors as a crown for your head
Give your soul unto God the author of all things
                                            Sir Apirana Ngata

Weaving our collective efforts for the benefit of our whānau and communities is what it means to be a young Māori farmer today – we celebrate your success as part of a long line of leaders, role models and mentors for those who follow. This is about whakapapa, the land and the people.
Ahuwhenua Trophy Management Committee

Ahuwhenua Young Māori Farmer of the Year
The Young Māori Farmer of the Year award was introduced by the Ahuwhenua Trophy Management Committee in 2012 in order to acknowledge and celebrate young Māori working in the agricultural sector or undergoing agricultural training. According to the Ahuwhenua Trophy Management Committee, the concept of encouraging young Māori to attain excellence in practical farming aligned perfectly with the vision of Sir Apirana Ngata and Lord Bledisloe in 1933.

2012 Inaugural Winner
The inaugural winner of the Young Māori Farmer award in dairying in 2012 was Tangaroa Walker (Ranginui/Pukenga). Tangaroa began milking cows at age thirteen and strongly believes the dairy industry needs more Māori leaders. As a twenty-two-year-old lower order sharemilker for an equity partnership in Southland, Tangaroa aspired one day to own his own farm, hoping that his success might motivate other young Māori 'down the path I have followed'.

Left to right: finalist Mark Coughland; winner of the inaugural Ahuwhenua Young Māori Farmer of the Year Award 2012, Tangaroa Walker; and finalist Tyson Kelly
(John Cowpland – Alphapix)

'Winning the competition fast-tracked my career by three years at least. It's opened up a lot of opportunities for me, and I'm looking forward to helping someone else through the process so they can achieve their own goals', says Tangaroa. Tangaroa says he would encourage young Māori to take the opportunities that are out there and be in a position where they can eventually run their own farms and become good role models in business.

'I think the key to being successful in the dairy industry is to focus on the goals ahead.' It could be tough, moving from a farm hand into farm management. 'But if you have the goal of becoming a contract milker or sharemilker, and keep focused on that, you will get there', he says. The other two outstanding finalists in 2012 were Mark Coughlan (Tuhoe/Ngai Tai) and Tyson Kelly (Tuhoe Whakatohea).[442]

2013 Young Māori Sheep and Beef Farmer
In 2013, the winner of the Young Māori Sheep and Beef Farmer of the Year award was Jordan Smith. 'It was a really positive experience for me', said Jordan, when accepting his award at the Ahuwhenua awards function held in Taradale on 7 June 2013. 'I would recommend it to anybody because it will help them go further in the industry.'[443]

At the time, says Jordan, he had never heard of the Ahuwhenua Award. However, he received a call from one of his agricultural training tutors, asking if he was interested in entering the competition.

'Initially, I said no as I had never really done anything like this before and I wasn't very confident that I would be good enough to even be considered for even entering the competition.' However, following further discussions with tutors and whānau, he was persuaded to enter 'and it snowballed from there'.

Fast-forward to the awards evening, and Jordan was still convinced that his entering the competition must have been a mistake. 'When they read my name as the winner, it was the first time in my life that I was being recognised for something that I was extremely passionate about.'

It was a great moment that Jordan would remember with pride as it set the foundation for his farming career. 'Through the process, I realised the importance Māori have in the agriculture industry, as a lot of the practices are that of manaki whenua.' Winning the award 'awakened a sense of pride in myself, my heritage and my culture which has lasted the years since'.[444]

2014 Young Māori Farmer Positive about His Win

At the 2014 Ahuwhenua Trophy awards evening in Tauranga, the Young Māori Dairy Farmer of the Year for 2014 was awarded to Wiremu Reid, a fourth generation dairy farmer originally from Whangarei now working with his wife Bettina sharemilking 1150 cows in Ranfurly. Aged twenty-four, Wiremu viewed his success as a positive step towards his life's goal of owning his own farm before age thirty. 'I'm confident we'll get there', said Wiremu. 'Ultimately, we want a run off, or a lease block, with conversion possibilities; but, for the short term, we would like somewhere to graze our own stock.'

Wiremu was 'humbled to have won' the competition and described the other finalists as 'very, very impressive'. The other finalists in 2014 for Young Farmer were 50/50 sharemilker Joshua MacDonald from Waikato, and James Matheson, a Herd Manager working in Gore.[445]

Winner of the Young Māori Farmer of the Year 2013, Jordan Smith (centre) with finalist Te Moana Sidney (right) with sponsors
(John Cowpland – Alphapix)

Winner of the Young Māori Farmer of the Year Award 2014, Wiremu Reid

(John Cowpland – Alphapix)

'From the start,' says Wiremu, 'it was a bit daunting to put myself out there to be judged but with some gentle persuasion from my Primary ITO trainer Michelle Phelan and my whānau it was a really enjoyable experience'. At the time, Wiremu and his wife, Bettina, had a young family and were 'flat out trying to survive our first year being self-employed'. Taking part in the Ahuwhenua Young Farmer competition therefore provided a great opportunity to reflect on where they were, and how far they had come.

'We were only twenty-three and had staff, debt and stock we were raising, outside of our day-to-day business to try and build our equity faster.' Wiremu and Bettina could tell the judges were impressed at what they had taken on and were achieving, which was 'a really positive end to a difficult year'.

Spending time with the other Young Māori Farmer competitors was a highlight of the experience, says Wiremu. 'To be able to spend time with peers your own age who have the same drive to achieve but were facing similar challenges was awesome.' Being introduced to a network of successful Māori 'movers and shakers in the Māori economy' was also 'incredibly humbling'.

Left to right: finalist Taane-nui-a-rangi Hubbard; winner of the Ahuwhenua Māori Young Farmer of the Year Award 2015, Hannah Wallace; finalist Hemoata Kopa

(John Cowpland – Alphapix)

Wiremu says that he feels 'emboldened by the relationships that I made during the competition and look forward to using my skill set to contribute to Māori excellence. Whanaungatanga is a popular word used by corporate New Zealand today, but it best describes the rewards I received when I participated in the Ahuwhenua Young Farmer Competition.'[446]

2015 Hannah from Wairoa, First Female Young Māori Farmer

Hannah Wallace from Wairoa was crowned the 2015 Young Māori Sheep and Beef Farmer of the Year during the 2015 Ahuwhenua Awards dinner at Whanganui. Aged twenty-two, Hannah was the first female to win the Young Māori Farmer Trophy. Her award was presented to her by Te Tumu Paeroa Chief Executive Jamie Tuuta, who acknowledged the strong female presence in the 2015 competition, signalling a new direction in leadership.

Hannah worked as a Shepherd General, dividing her time between the family farm, which ran 5600 stock units across 1500 hectares, and Rotanui Station at Te Whakaari. Hannah hoped that one day she would own her own farm, but for the present, she was wishing to enrol in tertiary study focusing on the financial side of farming.

'Winning the Young Māori Farmer competition was indescribable,' says Hannah. 'I felt hugely honoured to have even made it through to the final three, let alone winning the Young Māori Farmer trophy.' At the time of her success, women were not hugely recognised as farmers 'so it was just an amazing feeling to be an "understudy", so to speak, and to win a competition where the industry is male dominated!' she says.

The competition was an amazing experience to be a part of, says Hannah. 'We met some really amazing people who I never thought I would ever meet. At the same time, it was a very nerve-racking experience for me.'

Now Hannah Wallace-Bright, Hannah says the Young Māori Farmers have started an alumni group 'and there are some really cool things happening there'. For the present, Hannah was being kept busy raising two young children and running a 600 hectare

Winner of the Ahuwhenua Young Māori Farmer of the Year Award 2016, Jack Raharuhi (centre), with (from left) John Rutherford, North Island Manager of Allflex; Mark Jefferies, CEO Primary, ITO; Harepaora Ngaheu, finalist; Ash-Leigh Campbell, finalist; Michelle Hippolite, CEO Te Puni Kōkiri; and Tiaki Hunia, Deputy Māori Trustee, Te Tumu Paeroa

(John Cowpland – Alphapix)

lease farm with her husband Jeremy. Prior to entering the Ahuwhenua Young Māori Farmer competition, she had never put herself 'out there' and public speaking was not her forte. 'But I soon learned,' says Hannah. 'Winning the competition gave me a real confidence boost'.[447]

The two other outstanding finalists for Young Māori Farmer in 2015 were Taane-nui-a-rangi Hubbard, a so from Wairoa, and Hemoata Kopa from Matawaia in the Bay of Islands.[448]

2016 Young Māori Farmer in the Right Crowd

The Young Māori Dairy Farmer for 2016 was announced at the 2016 Awards dinner in Hamilton. A young Māori farmer, Jack Raharuhi, who once 'got into the wrong crowd' was crowned as the sixth Young Māori Farmer since the award's inception in 2012. Of Ngāti Kahu origins, Jack grew up on the West Coast, leaving school at fifteen to work on a farm. At present, he managed a 450-hectare dairy unit owned by Landcorp in Westport. He managed five full-time staff and looked after 1100 cows, which were milked twice daily through most of the season. Supported by his fiancee and two children, Jack said he was passionate about the need for education, and one day he hoped to supervise a much larger property, perhaps with as many as 5000 cows.

The two other finalists for Young Māori Dairy Farmer in 2016 were Ash-Leigh Campbell from Lincoln, currently a student at Lincoln University, and Harepaora Ngaheu from Te Teko, currently a dairy farm manager working in Ruatoki, just south of Whakatane.[449]

2017 Young Māori Shepherd Farmer from Wairoa

The winner of the 2017 Ahuwhenua Young Māori Farming Award was Jordan Biddle aged twenty-one of Ngāti Pāhauwera, near Wairoa. Jordan was a shepherd on the 2000 hectare Pihanui Station, south of Wairoa, owned by Ngāti Pāhauwera. The station managed 500 Angus cows and 2200 Romdale breeding ewes.

Jordon's winning of the Young Farmer Trophy was announced by Jamie Tuuta, Chief Executive of Te Tumu Paeroa, at the Ahuwhenua awards dinner at Whangarei in May 2017. Jordan was born in Raupunga where his father worked in forestry. Jordan says he had always enjoyed the farming life, along with hunting, diving and playing rugby.

The other two finalists were Dylan Ruki-Fowlie of Te Atihaunui a Paparangi, working as a General Shepherd on Tawanui Station, south of Raetihi; and Hemoata Kopa of Ngā Puhi/Matawaia who worked as a General Shepherd on Pukemiro Station just south of Dannevirke.[450]

2018 Young Māori Farmer from Te Teko

The winner of the 2018 Ahuwhenua Young Māori Farmer Award was twenty-six-year-old Harepaora Ngaheu from Te Teko in the eastern Bay of Plenty. The announcement was made by the Māori Trustee and Chief Executive of Te Tumu Paeroa, Jamie Tuuta, at the 2018 Ahuwhenua awards function in Christchurch.

Harepaora was from Ngāti Awa and Te Whānau a Āpanui. He was a manager on a dairy farm near Te Teko, though he was planning to go into contract milking for his same employer. Harepaora had also been a finalist in the 2016 competition. He was born in Porirua but returned to the Bay of Plenty with his family. His first real taste of farming came at the age of twenty-one, during a training course. Now with partner Aiesha and two daughters, Harepaora says he enjoyed the farming life – which he saw as a life saver – and he hoped to help others into similar

Winner of the Young Māori Farmer of the Year Award 2017, Jordan Biddle
(John Cowpland – Alphapix)

Winner of the Ahuwhenua Young Māori Farmer of the Year Award 2018, Harepaora Ngaheu (centre) with finalists Cheyenne Wilson (left) and Mathew Pooley (right)
(John Cowpland – Alphapix)

Winner of the Young Māori Farmer of the Year Award 2019, Kristy Marai Roa (centre) with finalists Tumoanakotore-i-whakairioratia Harrison-Boyd (left) and Taane-nui-a-rangi Hubbard (right)
(John Cowpland – Alphapix)

career opportunities.[451] The two finalists in 2018 were Cheyenne Wilson and Mathew Pooley.

2019 Kristy from Tolaga Bay Young Māori Farmer

The winner of the 2019 Ahuwhenua Young Māori Farmer Award was Kristy Marai Roa, aged twenty years, from Ngāti Maniapoto and Ngāti Apakura. Kristy worked as a shepherd on Iwinui Station near Tolaga Bay on the East Coast, a 2100-hectare property, which ran 5300 ewes, 3700 ewe lambs, 450 cows and 1000 trading bulls. The announcement of Māori Young Farmer was made by the Māori Trustee and Chief Executive of Te Tumu Paeroa, Dr Charlotte Severene, at the Ahuwhenua awards evening held in Gisborne in May 2019.

Kristy described herself as a city girl who grew up in Hamilton with no family connections to farming, though the family owned an engineering business that built milk tankers for Fonterra. Kristy later attended a farm cadet course run by Waipaoa Cadet Training Trust on the scenic East Coast, which inspired her to choose agriculture as a career. Thereafter, she worked on farms in the Waikato before moving to Tolaga Bay. Kristy loved the farming life, she said, and was particularly attracted to the business and commercial aspects of farming. Her long-term aim was to manage a large scale sheep and beef farm on the East Coast, and one day own her own farm.[452]

2020 First Young Māori Grower of the Year

The winner of the inaugural Ahuwhenua Young Māori grower of the Year was twenty-six-year-old Maatutaera Akonga, of Ngāti Porou, Ngāi Tahu and Ngāti Kahungunu descent. Maatutaera was a Senior Leading Hand at Llewellyn Horticulture based in Hastings.

The announcement on Maatutaera's success was made by Dr Charlottee Severne, the Māori Trustee and Chief Executive of Te Tumu Paeroa, at the Ahuwhenua Trophy awards dinner in Rotorua on 20 November 2020.

Award judge Aaron Hunt from Te Tumu Paeroa said the standard of entrants for this inaugural award was very high, reflecting the number of exceptional young Māori making successful careers in horticulture. 'Māori had always been involved in the horticulture sector, and in recent years they had been behind a number of significant new enterprises.'

Winner of the inaugural Ahuwhenua Young Māori Grower of the Year Award 2020, Maatutaera Akonga (centre) with finalists Brandon Cross (left) and Finnisha Tuhiwai (right)
(John Cowpland – Alphapix)

Horticulture, as a consequence, was seen as an industry making an enormous contribution to the national economy, with Māori horticulture being integral to that contribution. By their outstanding efforts, successful young role models like Maatutaera helped encourage young people into the industry, enhancing the positive future propsects for Māori horticulture.[453]

2021 Young Māori Farmer off to a Good Start

The 2021 Ahuwhenua Young Māori Dairy Farmer Award was taken out by twenty-six-year-old Quinn Morgan – a remarkable achievement, given that Quinn had only been in the dairy industry for just over a year.

Quinn was in his first year of farming, working as a Farm Assistant on a 155 hectare dairy farm in Otakiri, near Whakatane, running 570 cross breed cows. The two other finalists were Anahera Hale, also from Whakatane, and Ben Purua from Tokoroa.

Quinn said he felt humbled at winning the award. Not everyone got such a good start in their careers as he had, he said, with such great employers, Sam and Kate Moore. The announcement was made by Māori Trustee and Chief Executive of Te Tumu Paeroa, Dr Charlotte Severne. Quinn accepted the award with his wife Samantha at the Ahuwhenua Trophy awards dinner in New Plymouth.

Quinn was born in Taumarunui before moving to Tongariro and Whanganui, where he went to school. He then moved with his family to Australia, working for a time in the fitness industry. His farming career came about following a relation offering him some dairy work experience. Since then, Quinn has completed day courses such as Calving Smart with Dairy NZ, First Aid for Cattle with Bayvets and a Lameness Course with the Dairy Hoofcare Institute. He is currently studying level 3 with Primary ITO and is a member of the Farm4Life Education Hub.

Quinn's wife Samantha is from Porirua but grew up in Queensland. She is currently studying for her Diploma in Agribusiness.

'The dairy industry suits me especially being home for breakfast and seeing the kids off to school,' says Quinn. 'I like the early morning starts and the hard work.' Working in the dairy industry was a real treat, says Quinn, and as a young father, it gave him stability, helping him and Samantha 'to develop our aspirations as a collective'.[454]

When reflecting on the opportunity to take part in the Ahuwhenua Young Māori Farmer Award, Quinn said that the overall experience was 'mind-blowing'.

Winner of the Ahuwhenua Young Māori Farmer of the Year Award 2021, Quinn Morgan (centre) with finalists Anahera Hale (left) and Ben Purua (right)

(John Cowpland – Alphapix)

The effort that Ahuwhenua organisers had put into the finalists was 'second to none'. 'Over the course of the three-day weekend, we were educated in a lot of areas that wouldn't just help us on the farm, but off it as well.' A strong bond was built up with the other finalists, Anahaera and Ben. 'We were able to draw so much insight from our judges Trevor, Matiu and Aaron, who made the process very educational for us.'

Winning the award was an incredible feeling, says Quinn. 'I honestly didn't think my name would be called due to the calibre of Anahaera and Ben. To this day, I'm humbled for what the Ahuwhenua whānau have done for me.'

The awards catapulted Quinn's career and presented him with so many new opportunities. 'I still reap the rewards with help from the sponsors such as Te Tumu Paeroa, who have invested a lot in me. Winning the award made me realise that it's all about reaching back to pull the next young farmer forward, and that's what I intend to do', says Quinn.[455]

Four Young Māori Farmers Talk about Participating in the Ahuwhenua Competition[456]

Wiremu Reid
Te Arawa

2014 Dairy winner

Cheyenne Wilson
*Ngāi Tūhoe, Ngāti Awa,
Ngāti Tūwharetoa,
Te Arawa*

2018 Dairy finalist

Maatutaera Akonga
*Ngāi Tahu, Ngāti Porou,
Ngāti Kahungunu*

2020 Horticulture winner

Kristy Roa
*Ngāti Maniapoto,
Ngāti Apakura*

2019 Sheep and Beef winner

Wiremu
What makes it special to me is the history. As Māori, we're naturally humble; we don't like to talk ourselves up. But, whenua-based mahi is very important to our people.

Cheyenne
It's about our people being recognised as a finalist; it wasn't about me being recognised, it was everyone in my whānau, you know, my ancestors, everyone who has helped me on my journey.

Maatutaera
What kinda made it special for me was celebrating Māori in a whole new way and just showed the high level of achievements that are out there within the Māori industry, and Māori working within all of the industries.

Kristy
This award not only opened me up to networks that will be hugely important going forward but set me up on a path and set me up for the future. It's just an amazing opportunity being a part of and to be standing up for your whānau.

Wiremu
The biggest thing I got out was the networking, just to get out and to be pushed out of my comfort zone and to be introduced to all these amazing people who are just keen to make you grow and make you successful and help you with your journey is awesome.

Maatutaera
If you're a young Māori out there and shy or not, put your name down, you never know where you could end up. Just don't limit yourself to your current position, so keep yourself open and don't be afraid to give it a go.

Kristy
I would say if you're seeing this, this is your sign to apply for the Ahuwhenua Young Māori Farmer of the Year 2022.

Wiremu
And go for it. I challenge you to step outside your comfort zone. It'll be amazing for you, not only just for your farming career but personal development.

Acknowledgements

I would like once again to thank Kingi Smiler for his assistance and support during the compilation of the additional pieces, and revisions, that comprise this book's continuing recounting of the astonishing Ahuwhenua history, for which of course Kingi can take no small credit.

I am also very grateful to Lyn Harrison, whose prodigious gathering of photographs and past participants' stories has added so much more to this project, contributing depth and substance that would otherwise have been missing.

The support of the Ahuwhenua staff was also much appreciated, especially from Jean Rangiwai, Allan Frazer and Marama Steele, who compiled the lists of Ahuwhenua judges from 2003 to 2022. I am also grateful to John Cowpland whose amazing photographs illustrate and elucidate this publication.

Thanks also to Eboni Waitere and her wonderful team at Huia Publishers, especially Te Kani Price and Michaela Tapp, who were all always supportive and helpful in guiding this publication to fruition.

Finally, I'm very grateful to my family who continue to accept with patience and good humour the stresses that come with having an inveterate writer in their midst, especially my wife Gaylene for her endless forbearance and aroha.

Danny Keenan
WHANGANUI

Appendix 1

Ahuwhenua Trophy Judging Criteria 2013

JUDGING PROCESS

First Round Judging:

Initial judging of this competition will occur during February 2014 when three finalists will be selected.

Prior to judging, the entrant will be contacted to make arrangements for the farm visit. This visit will take up to three hours and will include a tour of key farm features. The judging will relate solely to the dairy components of the farm business.

The Chairman (or other elected representative), Supervisor and Manager are required to meet with the First Round Judges at the beginning of the visit. It is important to allocate your time wisely and to demonstrate clearly to the judges how your dairy business meets the criteria established for the competition on page 5.

Each finalist is required to hold a field day. Guidance and considerable financial assistance is provided to finalists to assist in staging the field day.

Final Round Judging:

The judging to select the national winner from the three finalists will take place in late March/early April and the process involves:

- At approximately 11.00am on the morning prior to the field day the judges will meet with the Chairman, other members of the governance team, the Farm Supervisor/Consultant and Manager may participate at owners discretion. This meeting will focus on governance, management and financial aspects and will include a farm visit. It will continue into the early evening with an opportunity for informal discussion over a meal.
- Judges will continue their judging assessment through attendance at the field day and the content and standard of presentation during the field day will be taken into consideration.

Judging will be based on:

A. THE EFFICIENCY WITH WHICH THE PROPERTY IS FARMED RELATIVE TO ITS POTENTIAL.

This will not be based solely on financial measures such as profit per hectare or return on business capital. These measures will be taken as a guide but consideration will also be given to other factors such as:
- The physical resources available to the farmer (e.g. local climate, soil types, water, region of location, contour etc).
- Stage of development, financial structure.

B. FINANCIAL RESULTS WHERE:
- Profit will be determined by the calculation of the operating profit per hectare; that is the gross income, net of stock purchases (adjusted for changes in livestock numbers) less working expenses. Interest, development, capital expenditure, drawings, dividends and taxation are not included in the calculation of operating profit.
- Financial performance will be determined from annual financial statements for the three years ending at the farm balance date in 2013.

C. THE EFFECTIVENESS OF THE FARMS GOVERNANCE PROCEDURES AND INITIATIVES, IN AREAS THAT INCLUDE:
- The adoption of, innovative farming systems and reinvestment in the business.
- The pursuit of sustainable management strategies including the up-skilling of all farm personnel.
- Keeping up to date with new farming methods and ways to monitor performance.
- The level of recognition given to kaitiakitanga and nga tikanga Māori in the operation of the farming enterprise.

CONSIDERATIONS

The organisers note that in recent times a number of new measures have been introduced to assess the performance of farming and other businesses. These include:

- Triple Bottom Line Reporting which focuses a business on its economic value, added or lost, as well as its environmental and social value. Entrants are encouraged to outline their efforts in these areas to the judges during their visits.
- Cost of Production Analysis – calculating the cost of production per unit of output. This encourages the setting of goals for improved performance and allows comparisons to be made between different types of farming businesses. We encourage all farmers to discuss the benefits of adopting such an approach with their advisors.

The judges will also look for best practice in relation to people management and career development; a consideration in this regard is the extent to which the governance team and management encourage staff participation in the Ahuwhenua Young Māori Farmer of the Year competition where staff meet the competition entry criteria.

CRITERIA

CRITERIA	MAX. POINTS AWARDED	FACTORS TAKEN INTO ACCOUNT INCLUDE THE FOLLOWING (WHERE POSSIBLE JUDGES WILL COMPARE WITH INDUSTRY BENCHMARKS AND BEST PRACTICE)
Governance and Strategy	17	<ul><li>Strong leadership</li><li>Good strategy</li><li>Monitoring of strategy</li><li>Implementation of strategy</li></ul>
Social/Community/ Ngā Tikanga Māori	10	<ul><li>Contribution to, and participation in, communities of interest to the organisation; support for local hapū, Marae, and wider local community</li><li>Governance or management team's ability to include tikanga Māori in aspects of the business</li><li>Identification and protection of cultural sites</li></ul>
MANGEMENT AND PERFORMANCE		
Financial and Benchmarking	20	<ul><li>Economic Farm Surplus (EFS)</li><li>GFR/HA</li><li>FWE as a % of GFR</li><li>Consistency over time</li><li>Wealth creation – Leveraging Asset Base, Internal Capital Investment/Development ROC</li><li>Understanding the Financials – Budgeting, Variance Reports, KPIs</li></ul>
Feed Production	10	<ul><li>Development and sustainability of soil fertility</li><li>Quality of permanent pastures [composition and nutritive value]</li><li>Forage crop yields and integrated use</li><li>Use of least cost supplements and tactical use of nitrogen</li><li>Feed budgeting and grazing plans</li></ul>
Animal Performance	10	<ul><li>Stock health and welfare</li><li>Genetic improvement</li><li>Reproductive and growth performance</li><li>Supply of products to market specifications</li><li>Purchasing and marketing skills</li></ul>
Farm Staff	11	<ul><li>Employment agreements and job specifications</li><li>Performance review approach</li><li>Training support and career development</li><li>Team culture and attitude</li><li>Health and safety plans and implementation</li></ul>
Environment/ Sustainability Goals and Strategies	12	<ul><li>Environmental plans in place</li><li>Environmental plans are being implemented</li><li>Environmental performance is being monitored and promoted</li><li>Biodiversity is being enhanced</li></ul>
Entrepreneurship and Innovation	10	<ul><li>Demonstrate innovative thinking and application of new technology and management approaches</li></ul>
TOTAL	100	

Appendix 2

Winners of the Ahuwhenua Trophy 1933–2021

1933	**1st**	William Swinton	*Raukokore*
	2nd	Jack Black	*Rūātoki*
	3rd	Tawera Kopae	*Rūātoki*
	Highly Commended	James Swinton	*Raukokore*

1934–1935 No competition

1936	**1st**	Henry J Dewes	*Tikitiki*
	2nd	Robert Clarke	*Rotorua*
	3rd equal	Jack Black	*Rūātoki*
		Pat Raharuhi	*Horohoro*

1937 No competition

1938	**1st equal**	Whareparoa Rewharewha	*Tōrere*
		Jack Black	*Rūātoki*
	3rd	Johnny Edwards	*Rotorua*
	4th	Huinga Nepia	*Tikitiki*
1939	**1st**	Johnny Edwards	*Rotorua*
	2nd	Tatai Hall	*Te Teko*
	3rd	Ngarangi Kohere	*Rangitukia*
	4th	Charles Sergeant	*Te Awamutu*
1940	**1st**	Tatai Hall	*Te Teko*
	2nd	Fred Amoamo	*Ōpōtiki*
	3rd	Ngarangi Kohere	*Rangitukia*
	4th	Tame Pukunui	*Te Kūiti*
1941	**1st**	Fred Amoamo	*Ōpōtiki*
	2nd	J D Jones	*Te Kūiti*
	3rd	Eruera Hoera	*Takahue*
	4th	Tihema Kingi	*Rotorua*
1942	**1st**	Tame Pukunui	*Te Kūiti*
	2nd	H and T Paraone	*Clevedon*
	3rd	Tihema Kingi	*Rotorua*
	4th	Charles Wells	*Whakatane*
1943	**1st**	Tihema Kingi	*Rotorua*
	2nd	Whare Moke	*Kāwhia*
	3rd	Bunny Otene	*Hastings*
	4th	John Savage	*Ōpoutere*
1944	**1st**	Bunny Otene	*Hastings*
	2nd	Henry Parone	*Clevedon*
	3rd	Joe Wharekura	*Rotorua*
	4th	Wiremu Karaka	*Pāmapuria*

AHUWHENUA

1945
- **1st** — Joe Wharekura — *Rotorua*
- **2nd** — Whare Mill — *Tikitiki*
- **3rd** — Eruera Hoera — *Takahue*
- **4th** — Robert Tapa — *Rānana*

1946
- **1st** — Henry Paraone — *Clevedon*
- **2nd** — Robert Tapa — *Rānana*
- **3rd** — Wiremu Matene Naera — *Waiotemarama*
- **4th** — Heemi Lawson — *Tōrere*

1947
- **1st** — Wiremu Matene Naera — *Waiotemarama*
- **2nd** — John Savage — *Ōpoutere*
- **3rd** — Mihi Stevens — *Rangiāhua*
- **4th** — Tiaka Tamaki — *Pirongia*

1948
- **1st** — Tikirau Callaghan — *Raukokore*
- **2nd** — Patuwahine Albert — *Ōmaio*
- **3rd** — Bunny Otene — *Hastings*
- **4th** — E F Clements — *Tauranga*

1949
- **1st** — Eruera Hoera — *Takahue*
- **2nd** — Waerata Harris — *Inglewood*
- **3rd** — Irihapeti Tuhakaraina — *Omokoroa*
- **4th** — Paul Toroa — *Tikitiki*

1950
- **1st** — G Thompson — *Ōtorohanga*
- **2nd** — Paul Toroa — *Tikitiki*
- **3rd** — P Hei — *Te Kaha*
- **4th** — K I Waihi — *Ruatōria*

1951
- **1st** — Kopua Waihi — *Ruatōria*
- **2nd** — C H de Thierry — *Pirongia*
- **3rd** — K A McGregor — *Waihī*
- **4th** — Manu Stainton — *Hicks Bay*

1952
- **1st** — C R Beasley — *Pōkeno*
- **2nd** — Tapuae Rogers — *Tōrere*
- **3rd** — M Reid — *Mangonui*
- **4th** — C H de Thierry — *Pirongia*

1953
- **1st** — Rohe Takiari — *Kāwhia*
- **2nd** — Aumihi Davis — *Okoroire*
- **3rd** — Sam Shelford — *Whakaparai*
- **4th** — Noti Tiopira — *Te Kaha*

1954 — **Sheep and cattle**
- **1st** — P Raharuhi — *Horohoro*
- **2nd** — J W Thompson — *Glen Murray*
- **3rd** — M Morehu — *Rotoiti*
- **4th** — R Vercoe — *Horohoro*

Dairy

1st	Mihi Stevens	*Rangiāhua*
2nd	Tapuae Rogers	*Tōrere*
3rd	Aumihi Davis	*Ōkoroire*
4th	M and D Wikaira	*Te Awamutu*

1955

Sheep and cattle

1st	John Chadwick	*Taumarunui*
2nd	Edward C Pohia	*Ōtaramarae*
3rd	Robert T Kingi	*Ōtaramarae*

Dairy

1st	J W Hedley	*Morrinsville*
2nd	Ilet Hemi	*Kaeo*
3rd	Foley Eru	*Horohoro*

1956

Sheep and cattle

1st	R T Kingi	*Ōtaramarae*
2nd	Edward C Pohia	*Ōtaramarae*

Dairy

1st	Foley Eru	*Horohoro*
2nd	W J Swinton	*Whangamatā*
3rd	James Nelson	*Kōpua*

1957

Sheep and cattle

1st	Henry Davis	*Rotorua*
2nd	John Tahuri	*Ruatāhuna*
3rd	Desmond Royal	*Rotorua*

Dairy

1st	Rehua Cairns	*Tauranga*
2nd	Tom Haeata	*Mangakino*
3rd	James Nelson	*Te Awamutu*

1958

Dairy

1st	Tom Haeata	*Mangakino*
2nd	W J Swinton	*Whangamatā*
3rd	John Peterson	*Mangonui*

1959

Sheep and cattle

1st	Jack Steedman	*Tauranga*
2nd equal	Desmond Royal	*Rotorua*
	Desmond and Pikihuia Manning	*Manunui*

Dairy

1st	W J Swinton	*Whangamatā*
2nd	J W Hedley	*Hoe o Tainui*
3rd	Wiremu Mauriohooho	*Te Awamutu*

AHUWHENUA

1960 — **Sheep and cattle**
1st   W Waaka .... *Kaikohe*
2nd   Parekura Raroa .... *Tokaanu*
3rd   Aperehama Whata .... *Rotoiti*

Dairy
1st   Mihi Stevens .... *Rangiāhua*
2nd equal   J W Hedley .... *Morrinsville*
   Wallace Mangu .... *Ōtorohanga*

1961 — **Sheep and cattle**
1st   Parekura Raroa .... *Tokaanu*
2nd   Aperehama Whata .... *Rotoiti*
3rd   George Kuru .... *Tokaanu*

Dairy
1st   Wallace Mangu .... *Hangatiki*
2nd equal   J W Hedley .... *Hoe o Tainui*
   Tapua Pita Heperi .... *Rangiāhua*

1962 — **Sheep and cattle**
1st   Kingi Grace .... *Tokaanu*
2nd   John J Reid .... *Kaikohe*
3rd   Aperehama Whata .... *Rotorua*

Dairy
1st   W Maki .... *Takahiwai*
2nd   G and R Rutledge .... *Te Kōpuru*
3rd   J W Hedley .... *Hoe o Tainui*

1963 — **Sheep and cattle**
1st   Alec McAllister .... *Kāwhia*

Dairy
1st   J W Hedley .... *Hoe o Tainui*
2nd   Alfred Parker .... *Mangakino*
3rd   Bernie Parker .... *Ōpōtiki*

1964 — **Sheep and cattle**
1st   John Steedman .... *Tauranga*
2nd   Joseph W Thompson .... *Tuakau*

Dairy
1st   Rawson Wright .... *Tapora*
2nd   Edward R Tamati .... *New Plymouth*
3rd   Tongawhiti Manu .... *Ōpunake*

1965 — **Sheep and cattle**
1st   J W Thompson .... *Glen Murray*
2nd   John Tahuri .... *Rotorua*
3rd   Aperehama Whata .... *Rotorua*

	Dairy		
	1st	Edward R Tamati	*New Plymouth*
	2nd	Wiremu Mauriohooho	*Te Awamutu*
	3rd	T Manu	*Ōeo*
1966	**Dairy**		
	1st	Wiremu Mauriohooho	*Ōwairaka*
	2nd	Joseph Niwa	*Pūniho*
	3rd	George Taurua	*Poutū*
1967	**Sheep and cattle**		
	2nd	A A Alexander	*Ōkaihau*
	Dairy		
	1st	G C Hopa	*Tauhei*
	2nd	Tongawhiti Manu	*Ōpunake*
	3rd	George Taurua	*Poutū*
1968	**Sheep and cattle**		
	1st	Jack Steedman	*Tauranga*
	Dairy		
	1st	Jack Karatau	*Whangaehu*
	2nd	Tongawhiti Manu	*Ōpunake*
	3rd	Wharehuia Eri	*Te Puke*
1969	**Sheep and cattle**		
	1st	Reihana F Apatu	*Hastings*
	2nd	Waka Konui	*Manunui*
	Dairy		
	1st	T L Jones	*Dargaville*
	2nd	J Edwards	*Horohoro*
	3rd	G Simeon	*Ōwairaka*
1970	**Sheep and cattle**		
	1st	Waka Konui	*Manunui*
	2nd	John Foley	*Rotorua*
	Dairy		
	1st	Charles Bailey	*Waitara*
	2nd	Charles Berry Wells	*Ōturu*
	3rd	John Klaricich	*Ōmapere*
1971	**Sheep and cattle**		
	1st	A A Alexander	*Ōkaihau*
	Dairy		
	1st	Edward R Tamati	*New Plymouth*
	2nd	Tongawhiti Manu	*Ōeo*
	3rd	J Edwards	*Horohoro*

1972 — **Sheep and cattle**
- 1st A A Alexander .... *Ōkaihau*
- 2nd C H and C D Boulter .... *Pāhia*
- 3rd A R Austin .... *Thornbury*

Dairy
- 1st J Edwards .... *Horohoro*
- 2nd Tongawhiti Manu .... *Ōeo*
- 3rd E and R Walden .... *Rahotū*

1973 — **Sheep and cattle**
- 1st Reihana F Apatu .... *Hastings*
- 2nd C H and D C Boulter .... *Pāhia*

Dairy
- 1st N J Ormsby .... *Pirongia*
- 2nd E and R Walden .... *Rahotū*
- 3rd John Klaricich .... *Ōmapere*

1974 — **Sheep and cattle**
- 1st C H and D C Boulter .... *Riverton*
- 2nd Robert T Kingi .... *Rotorua*
- 3rd Norman Hoete .... *Tuakau*

Dairy
- 1st Monte Retemeyer .... *Te Awamutu*
- 2nd E and R Walden .... *Rahotū*
- 3rd Charles Bailey .... *Waitara*

1975 — **Dairy**
- 1st Claude Edwards .... *Ōpōtiki*
- 2nd Tutere Hohepa .... *Te Teko*
- 3rd Charles Bailey .... *Waitara*

1976 — **Sheep and cattle**
- 1st Thomas Hawira .... *Raetahi*

Dairy
- 1st Charles Bailey .... *Waitara*
- 2nd Tutere Hohepa .... *Te Teko*
- 3rd Tony Edwardson .... *Ōpōtiki*

1977 — **Sheep and cattle**
- 1st J R and S C Stewart .... *Wyndham*
- 2nd J H Morris .... *Masterton*
- 3rd T P and N Te Aika .... *Mason Bay*

Dairy
- 1st Jack Karatau .... *Whangaehu*
- 2nd A T Edwardson .... *Ōpōtiki*
- 3rd P and A Hemi .... *Kaeo*

1978 ——— **Sheep and cattle**
    1st   R D, P B and H J Paewai .... *Dannevirke*
    2nd   M R Mohi .... *Waipukurau*

    Dairy
    1st   Maurice Anderson .... *Whakatāne*
    2nd   W P Peachey .... *Ōtorohanga*
    3rd   C B Wells .... *Kaitāia*

1979 ——— **Sheep and cattle**
    1st   Reihana F Apatu .... *Hastings*
    2nd   J R and S C Stewart .... *Wyndham*

    Dairy
    1st   Raumoa Amoamo .... *Reporoa*

1980–1981   **No competition**

1982 ——— **Sheep and cattle**
    1st   Martin Kingi .... *Whāngārā*
    2nd   Paewai Bros Partnership .... *Dannevirke*
    3rd   D W W Hawkins .... *Te Haukē*

    Dairy
    1st   W C and C M Edwards .... *Te Kūiti*
    2nd   D Edwards .... *Ōpunake*
    3rd   G Brons .... *Reporoa*

1983–1989   **No competition**

1990 ——— **Sheep and cattle**
    1st   Paewai Bros Partnership .... *Dannevirke*

1991 ——— **Sheep and cattle**
    1st   Parekarangi Trust .... *Rotorua*

    Dairy
    1st   N and S Armitage .... *Atiamuri*

1992–2002   **No competition**

2003 ——— **Sheep and beef**
     Kapenga M Trust .... *Rotorua*

2004 ——— **Sheep and beef – two regional finalists in lieu of a national winner**
     Waituhi Kuratau Trust .... *Tūrangi*
     Tarawera Station/Te Awahohonu Forest Trust .... *Napier*

2005	Sheep and beef		
		Te Pou a Kani Farms/ Wairarapa Moana Inc	*Mangakino*
2006	Dairy		
	Winner	Parininihi ki Waitōtara Incorporation	*New Plymouth*
	Finalists	Kōkako Trust	*Ngātira*
		Aotearoa Trust	*Wharepūhunga*
2007	Sheep and beef		
	Winner	Pah Hill Station/Ātihau-Whanganui Incorporation	*Ōhakune*
	Finalists	Tūaropaki Trust	*Taupō*
		Matariki Partnership	*Ruatōria*
2008	Dairy		
	Winner	Dean and Kristen Nikora/Cesped Lands Ltd	*Takapau*
	Finalists	Parekarangi Trust	*Rotorua*
		Hauhungaroa Partnership	*Taupō*
2009	Sheep and beef		
	Winner	Pakarae Whangara B5 Partnership	*Gisborne*
	Finalists	Morikau Station	*Rānana*
		Hereheretau Station	*Wairoa*
2010	Dairy		
	Winner	Waipapa 9 Trust	*Taupō*
	Finalists	Hanerau Farms Ltd	*Paparoa*
		Rangatira 8A 17 Trust	*Reporoa*
2011	Sheep and beef		
	Winner	Waipapa 9 Trust	*Taupō*
	Finalists	Ōtakanini Tōpū Incorporation	*Helensville*
		Pākihiroa Station	*Tapuaeroa Valley*
2012	Dairy		
	Winner	Kapenga M Trust	*Rotorua*
	Finalists	Tauhara Moana Trust	*Taupō*
		Waewaetutuki 10 – Wharepi Whanau Trust	*Te Puke*
2013	Sheep and beef		
	Winner	Te Awahohonu Forest Trust – Tarawera Station	*Te Hāroto*
	Finalists	Te Uranga B2 Incorporation – Upoko B2	*Taumarunui*
		Te Hape B Trust – Te Hape Station	*Te Kūiti*
2014	Dairy		
	Winner	Te Rua o Te Moko Ltd	*South Taranaki*
	Finalists	Putauaki Trust – Himiona Farm	*Te Teko*
		Ngāti Awa Farms Ltd – Ngakauroa Farm	*Whakatane*
2015	Sheep and beef		
	Winner	Mangaroa Station	*Ruakituri Valley*
	Finalists	Maranga Station	*Tiniroto*
		Paua Station	*Te Kao*

2016 — **Dairy**
    Winner   The Proprietors of Rakaia Incorporation .... *Rakaia*
    Finalists   Ngāi Tahu Farming Limited .... *Oxford*
                Tewi Trust .... *Tirau*

2017 — **Sheep and beef**
    Winner   Omapere Taraire E & Rangihamama
                X3 Ahu Whenua Trust .... *Kaikohe*
    Finalists   R A & J G King Partnership, Puketawa Station ...... *Paihiatua*
                Pukepoto Farm Trust .... *Taumarunui*

2018 — **Dairy**
    Winner   Onuku Māori Lands Trust .... *Rotomahana*
    Finalist   The Proprietors of Mawhera Incorporation .... *Hokitika*

2019 — **Sheep and beef**
    Winner   Eugene and Pania King, Kiriroa Station .... *Matawai*
    Finalists   Whangara Farms .... *Whangara*
                Te Awahononu Forest Trust – Gwavas Station ..... *Napier*

2020 — **Horticulture**
    Winner: Te Kaha 15B Hineora Orchard .... *Te Kaha*
    Finalists: Otama Marere Orchard .... *Paengaroa*
                Ngai Tukairangi Trust Orchard .... *Tauranga*

2021 — **Dairy**
    Winner: Tataiwhetu Trust .... *Taneatua*
    Finalists: Pouarua Farms .... *Hauraki*
                Tunapahore B2A Incorporation .... *Te Kaha*

Appendix 3

Winners of the Young Māori Farmer/Grower of the Year 2012–2021

2012	**Dairy**	
	Young Māori Farmer	Tangaroa Walker
	Finalists	Mark Coughlan
		Tyson Kelly
2013	**Sheep and beef**	
	Young Māori Farmer	Jordan Smith
	Finalist	Te Moana Sidney
2014	**Dairy**	
	Young Māori Farmer	Wiremu Reid
	Finalists	Joshua MacDonald
		James Matheson
2015	**Sheep and beef**	
	Young Māori Farmer	Hannah Wallace
	Finalists	Taane Hubbard
		Hemoata Kopa
2016	**Dairy**	
	Young Māori Farmer	Jack Raharuhi
	Finalists	Ash-Leigh Campbell
		Harepaora Ngaheu
2017	**Sheep and beef**	
	Young Māori Farmer	Jordan Biddle
	Finalists	Dylan Ruki-Fowlie
		Hemoata Kopa
2018	**Dairy**	
	Young Māori Farmer	Harepaora Ngaheu
	Finalists	Cheyenne Wilson
		Mathew Pooley
2019	**Sheep and beef**	
	Young Māori Farmer	Kristy Marai Roa
	Finalists	Tumoanakotore-i-whakairioratia Harrison-Boyd
		Taane-nui-a-rangi Hubbard
2020	**Young Māori Grower**	Maatutaera Akonga
	Finalists	Brandon Cross
		Finnisha Tuhiwai
2021	**Dairy**	
	Young Māori Farmer	Quinn Morgan
	Finalists	Anahera Hale
		Ben Purua

Appendix 4

Judges of the Ahuwhenua Trophy Competition 1933–2022

From 1933 to 1953, the Ahuwhenua Trophy competition was judged by a sole judge who evaluated all aspects of the nominated Māori farms.

From 1954, judging was divided into two categories, sheep and cattle, and dairy, with a sole judge continuing to assess both categories.

From 1977 to 1991, however, two judges were appointed, one to evaluate sheep and cattle and another to evaluate dairy.

The era of the 'sole judge' ended in 1991.

Following a recess from 1992 to 2002, the Ahuwhenua competition resumed in 2003, with significant changes to judging. In 2003, four regional finals were introduced, with regional judging and finalists, followed by the adjudication of a national winner.

In 2005, it was decided that annual judging would alternate between sheep and beef; and dairy. There were regional finals, field days and the adjudication of a winner, with detailed and multifaceted judging regimes introduced and thereafter developed. In 2020, horticulture was added to the Ahuwhenua Trophy competition as a third judging category.

The Sole Judges from 1933–1991

From 1933, a sole judge was appointed to judge all aspects of Māori farming, even then (in 1933) a significant and far-reaching undertaking, as the years that followed would confirm.

Year	Judge	Location
1933	W Dempster	*Hamilton*
1934–1935 Competition on hold		
1936	C Walker	*Tauranga*
1937 Competition on hold		
1938	C Walker	*Tauranga*
1939	C Walker	*Tauranga*
1940	C Walker	*Tauranga*
1941	C R Taylor	*Hamilton*
1942	E B Glanville	*Auckland*
1943	K M Montgomery	*Auckland*
1944	J F Shepherd	*Hamilton*
1945	A D Mercer	*Auckland*
1946	T E Rodda	*Hamilton*
1947	E H Arnold	*Whangarei*
1948	G A Blake	*Matamata*
1949	A V Allo	*Tauranga*
1950	K M Montgomery	*Auckland*
1951	C R Taylor	*Hamilton*
1952	C R Taylor	*Hamilton*
1953	C R Taylor	*Hamilton*

From 1954, following Judge Taylor's recommendation, the Ahuwhenua Trophy competition would now be judged in two categories – sheep and cattle, and dairy – with sole judges continuing their formidable and impressive duties.

Year	Judge	Location
1954	C R Taylor	*Hamilton*
1955	E H Arnold	*Whangarei*
1956	E H Arnold	*Whangarei*
1957	E H Arnold	*Whangarei*
1958	A V Allo	*Tauranga*
1959	J R Murray	*Hamilton*
1960	J R Murray	*Hamilton*
1961	J R Murray	*Hamilton*

1962	H Macmillan Bull	*Auckland*
1963	H Macmillan Bull	*Auckland*
1964	H Macmillan Bull	*Auckland*
1965	J R Murray	*Hamilton*
1966	J R Murray	*Hamilton*
1967	J R Murray	*Hamilton*
1968	J R Murray	*Hamilton*
1969	J R Murray	*Hamilton*
1970	J R Murray	*Hamilton*
1971	J R Murray	*Hamilton*
1972	J R Murray	*Hamilton*
1973	J R Murray	*Hamilton*
1974	A N Hall	*Wellington*
1975	A N Hall	*Wellington*
1976	A N Hall	*Wellington*

From 1977, it was decided that two judges would now be appointed, one to judge the category of dairy, and another to adjudge the category of sheep and cattle.

1977	***Sheep and cattle***	J D McNaught
	Dairy	A McKenzie
1978	***Sheep and cattle***	J D McNaught
	Dairy	J G Simmonds
1979	***Sheep and cattle***	*judges not recorded*
	Dairy	*judges not recorded*

1980–1981 Competition on hold

| **1982** | ***Sheep and cattle*** | *judges not recorded* |
| | ***Dairy*** | *judges not recorded* |

1983–1989 Competition on hold

| **1990** | ***Sheep and cattle only*** | *judges not recorded* |

| **1991** | ***Sheep and cattle*** | *judges not recorded* |
| | ***Dairy*** | *judges not recorded* |

1992–2002 Competition on hold

2003–2022

Ahuwhenua Trophy competition continues and greatly expands, commensurate with significant changes affecting Māori agriculture and, after 2020, horticulture, with detailed judging regimes also being developed.

List of 2003–2022 Judges

Compiled by Marama Steele

2003–2009 John Acland *Chief Judge Sheep and Beef Competition*
2006–2012 Doug Leeder *Chief Judge Dairy Competition*

2012 **Dairy Finalist Judges**
Doug Leeder *Chief Judge*	Independent
Stephen Veitch	Bank of New Zealand
Paul Klee	Fonterra
Tafadzwa Manjala	Dairy New Zealand
Dean Nikora	Mangatewai

2013 **Sheep and Beef Finalist Judges**
Dana Blackburn *Chief Judge*	Independent
Gavin Sheath	Independent
Rob Davison	Beef & Lamb
Sam Johnson	Bank of New Zealand

2014 **Dairy Finalist Judges**
Dean Nikora *Chief Judge*	Independent
David Handley	Bank of New Zealand
Tafadzwa Manjala	Dairy New Zealand
Paul Klee	Fonterra

2015 **Sheep and Beef Finalist Judges**
Dana Blackburn *Chief Judge*	Independent
Gavin Sheath	Independent
Rob Davison	Beef & Lamb
Sam Johnson	Bank of New Zealand
Abe Seymour	Primary ITO

2016 **Dairy Finalist Judges**
Dean Nikora *Chief Judge*	Independent
Abe Seymour	Competition kaumātua
David Handley	Bank of New Zealand
Katrina Knowles	Dairy New Zealand
Paul Radich	Fonterra

2017 **Sheep and Beef Finalist Judges**
Dana Blackburn *Chief Judge*	Independent
Abe Seymour	Competition kaumātua
Chris Garland	Beef & Lamb/Independent
Sam Johnson	Bank of New Zealand

2018 **Dairy Finalist Judges**
Dean Nikora *Chief Judge*	Independent
Abe Seymour	Competition kaumātua
David Handley	Bank of New Zealand
Paul Bird	Dairy New Zealand
Paul Radich	Fonterra

2019	**Sheep and Beef Finalist Judges**	
	Dana Blackburn *Chief Judge*	Independent
	Abe Seymour	Competition kaumātua
	Chris Garland	Beef & Lamb/Independent
	Sam Johnson	Bank of New Zealand
	Gavin Sheath	Independent
2020	**Horticulture Finalist Judges**	
	Julian Raine *Chief Judge*	Independent
	Rito Tapuke	Taha Māori/Independent
	Abe Seymour	Taha Māori/Independent
	Dr Ian Scott	Plant & Food Research
	David Proudfoot	Bank of New Zealand
	Rachel Jones	Te Puni Kōkiri
2021	**Dairy Finalist Judges**	
	Dean Nikora *Chief Judge*	Independent
	Rito Tapuke	Taha Māori/Independent
	Duncan Mathews	Bank of New Zealand
	Wilma Foster	Dairy New Zealand
	Paria King	Te Puni Kōkiri
2022	**Sheep and Beef Finalist Judges**	
	Dana Blackburn *Chief Judge*	Independent
	Rito Tapuke	Taha Māori
	Pat O'Brien	Beef & Lamb/Independent
	Rob Gemmell	Bank of New Zealand
	Paula Rawiri	Te Puni Kōkiri

Endnotes

INTRODUCTION
The Ahuwhenua Trophy

1. Ahuwhenua Trophy Management Committee, field day handbook 2013, p. 33.
2. Letter, Lord Bledisloe to Prime Minister M J Savage, 10 July 1939, AAMK W3074 868 Box 1325 (MA 60/8), Archives New Zealand, Wellington.
3. Memorandum, Private Secretary H R N Balneavis, 1 August 1939, AAMK W3074 868 Box 1325 (MA 60/8), Archives New Zealand, Wellington.
4. Report on Ahuwhenua Trophy Competition, 1969, ABRP W4598 6844 Box 244, Archives New Zealand, Wellington.
5. Kingi Smiler addressing attendees at the Ahuwhenua Trophy awards dinner at the Pettigrew Green Arena in Taradale, 2013.
6. K Smiler, *Mana Magazine*, Issue 112, June–July 2013, pp. 48–49.
7. Ahuwhenua Trophy Management Committee, 'The Ahuwhenua Trophy RoundUp' newsletter, November 2010, p. 2.
8. Ibid, p. 2.
9. 'The Ahuwhenua Trophy RoundUp', March 2010, p. 1.
10. Field day handbook 2013, p. 2.

CHAPTER ONE
Retaining the Land – to 1933

11. Ahuwhenua Trophy Management Committee, field day handbook 2007, p. 9.
12. *Wanganui Chronicle*, 21 June 2007, p. 10.
13. S Lambert, 'Te Ahuwhenua and the "Sons" of the Soil: A History of the Māori-Farmer-of-the-Year Award', unpublished paper, Te Whanake, Lincoln University, Christchurch, 2010, p. 2.
14. Field day handbook 2007, p. 12.
15. R Boast, *Buying the Land, Selling the Land: Governments and Maori Land in the North Island 1865–1921*, Victoria University Press, Wellington, 2008, pp. 284–285.
16. Ibid, pp. 284–287.
17. H Petrie, *Chiefs of Industry: Māori Tribal Enterprise in Early Colonial New Zealand*, Auckland University Press, Auckland, 2006, pp. 240–241.
18. G V Butterworth, *Nga Take i Neke ai te Maori: Maori Mobility*, Manatu Maori, Wellington, 1991, pp. 20–21, as discussed in Boast, *Buying the Land*, p. 286.
19. E Best, *Maori Agriculture*, Government Printer, Wellington, 1976 (first published 1925), pp. 21–23.
20. Ibid, pp. 26–29.
21. Ibid, pp. 32–34.
22. Petrie, *Chiefs of Industry*, pp. 33–34.
23. A R Parsonson, 'The Pursuit of Mana' in W H Oliver and B R Williams (eds), *The Oxford History of New Zealand*, Oxford University Press, Wellington, 1981, pp. 140–167.
24. Petrie, *Chiefs of Industry*, p. 82.
25. Ibid, pp. 100–101.
26. Ibid, pp. 100–101.
27. Ibid, pp. 100–101.
28. J Cowan, *The New Zealand Wars: A History of the Maori Campaigns and the Pioneering Period*, Government Printer, Wellington, 1922 (reprinted 1955), pp. 351–364.
29. Appendix I (Memorandum by Mr Richmond), AJHR 1861, E-3, pp. 35–36.
30. E Stokes, 'Contesting Resources: Maori, Pakeha and a tenurial revolution' in E Pawson and T Brooking (eds), *Environmental Histories of New Zealand*, Oxford University Press, Melbourne, 2002, p. 35.
31. D V Williams, 'Te Kooti Tango Whenua': The Native Land Court 1864–1909, Huia, Wellington, 1999, p. 1.
32. Stokes, 'Contesting Resources', p. 35.
33. Boast, *Buying the Land*, pp. 258–259.
34. Ibid, pp. 260–261.
35. J A Williams, *Politics of the New Zealand Maori: Protest and Cooperation, 1891–1909*, Auckland/Oxford University Press, Auckland, 1969, p. 123.
36. Boast, *Buying the Land*, pp. 270–273.
37. Williams, *Politics of the New Zealand Maori*, p. 147.
38. Boast, *Buying the Land*, pp. 287–291.
39. R S Hill, *State Authority, Indigenous Autonomy: Crown–Maori Relations in New Zealand/Aotearoa 1900–1950*, Victoria University Press, Wellington, 2004, pp. 108–109.
40. Hill, *State Authority*, pp. 108–109.
41. Williams, *Politics of the New Zealand Maori*, pp. 147–143.
42. Hill, *State Authority*, pp. 108–109.
43. Ibid, pp. 108–109.
44. Ibid, pp. 108–109.

CHAPTER TWO
The Early Years of the Ahuwhenua Trophy – 1933–1939

45. H M Bull, 'Ahuwhenua Trophy for Maori Farming', paper located in Te Ahuwhenua Archives, Office of the Māori Trustee, Wellington.
46. R Walker, *He Tipua: The Life and Times of Sir Apirana Ngata*, Penguin, Auckland, 2001, pp. 240–241.
47. *Auckland Star*, 29 November 1932, p. 9.

48. Letter, Minister of Native Affairs to Minister of Agriculture, 13 December 1933, AAMK W3074 869 Box 1325 (MA 60/8), Archives New Zealand, Wellington.
49. List of Competitors in the Waiariki District 1933, Te Ahuwhenua Trophy, AAMK W3074 869 Box 1326, Archives New Zealand, Wellington.
50. Letter, Dempster to Dairy Division, 1933.
51. *Evening Post*, 22 May 1933, p. 9.
52. *Evening Post*, 18 July 1933, p. 13.
53. Letter, Under Secretary to All Registrars, Ahuwhenua Trophy, 28 August 1939, AAMK W3074 869 Box 1326, Archives New Zealand, Wellington.
54. Memorandum, M J Savage, Native Minister, 19 May 1936, AAMK W3074 869 Box 1326, Archives New Zealand, Wellington.
55. Report on Ahuwhenua Trophy Competition, 1936, ABRP W4598 6844 Box 244, Archives New Zealand, Wellington.
56. Ibid.
57. Ibid.
58. Letter, Under Secretary, Native Affairs to Registrars of Native Land Court Districts, 24 September 1936, ABRP W4598 6844 Box 244, Archives New Zealand, Wellington.
59. H Patete and Others, Report to Dr Pomare, AJHR, 1906, H-31, pp. 77–79; E Best, Report to Dr Pomare, AJHR, 1906, H-31, pp. 75–76.
60. *Auckland Star*, 20 March 1937, p. 8.
61. Letter, Native Land Court Registrar (Tairawhiti) to Under Secretary, Native Affairs, 15 April 1937, ABRP W4598 6844 Box 244, Archives New Zealand, Wellington.
62. Letter, Secretary to the Treasury to Under Secretary of Native Affairs, 20 July 1937, ABRP W4598 6844 Box 244, Archives New Zealand, Wellington.
63. Letter, Walker & Hall to Under Secretary, Native Department, 6 August 1937, ABRP W4598 6844 Box 244, Archives New Zealand, Wellington.
64. Report on Ahuwhenua Trophy Competition, 1938, ABRP W4598 6844 Box 244, Archives New Zealand, Wellington.
65. Report on Ahuwhenua Trophy Competition, 1939, ABRP W4598 6844 Box 244, Archives New Zealand, Wellington.
66. Ibid.

CHAPTER THREE
The War Years – 1940–1945

67. M Soutar, 'Ngā pakanga ki tāwāhi – Māori and overseas wars – Second World War: the Māori war effort' in *Te Ara – the Encyclopedia of New Zealand*: www.teara.govt.nz/en/nga-pakanga-ki-tawahi-maori-and-overseas-wars/page-4 (accessed 31 July 2013).
68. Report on Ahuwhenua Trophy Competition, 1940, ABRP W4598 6844 Box 244, Archives New Zealand, Wellington.
69. Letter, J M Smith, Fields Superintendent to Under Secretary, Native Affairs, 18 March 1941, ABRP W4598 6844 Box 244, Archives New Zealand, Wellington.
70. Report on Ahuwhenua Trophy Competition, 1941, ABRP W4598 6844 Box 244, Archives New Zealand, Wellington.
71. Report on Ahuwhenua Trophy Competition, 1940.
72. Ibid.
73. Letter, Under Secretary, Native Affairs to Registrars of Native Land Court Districts, 30 October 1941, ABRP W4598 6844 Box 244, Archives New Zealand, Wellington.
74. Ibid.
75. Report on Ahuwhenua Trophy Competition, 1943, ABRP W4598 6844 Box 244, Archives New Zealand, Wellington.
76. Ibid.
77. Report on Ahuwhenua Trophy Competition, 1944, ABRP W4598 6844 Box 244, Archives New Zealand, Wellington.

CHAPTER FOUR
Māori 'Economic Advancement' and Farming – 1945–1961

78. M Soutar, 'Ngā pakanga ki tāwāhi – Māori and overseas wars – Second World War: the Māori war effort' in *Te Ara – the Encyclopedia of New Zealand*: www.teara.govt.nz/en/nga-pakanga-ki-tawahi-maori-and-overseas-wars (accessed 31 July 2013); 'Maori Battalion Returns' newsreel: www.nzonscreen.com/title/maori-battalion-returns-1946 (accessed 31 July 2013).
79. Report on Ahuwhenua Trophy Competition, 1944–1945, ABRP W4598 6844 Box 244, Archives New Zealand, Wellington.
80. Report on Ahuwhenua Trophy Competition, 1945–1946, ABRP W4598 6844 Box 244, Archives New Zealand, Wellington.
81. Ibid.
82. *New Zealand Herald*, 11 September 1946.
83. Report on Ahuwhenua Trophy Competition, 1947, ABRP W4598 6844 Box 244, Archives New Zealand, Wellington.
84. Ibid.
85. G V Butterworth and H R Young, *Maori Affairs*, Iwi Transition Agency/GP Books, Wellington, 1990, p. 123.
86. Letter, C H Tidswell, Supervisor to Under Secreatry of Native Affairs, 29 December 1947, ABRP W4598 6844 Box 244, Archives New Zealand, Wellington.
87. Report on Ahuwhenua Trophy Competition, 1947–1948, ABRP W4598 6844 Box 244, Archives New Zealand, Wellington.
88. Letter, Under Secretary, Department of Māori Affairs, 12 October 1948, ABRP W4598 6844 Box 244, Archives New Zealand, Wellington.

89. R Boast, *Buying the Land, Selling the Land: Governments and Maori Land in the North Island 1865–1921*, Victoria University Press, Wellington, 2008, pp. 287–291.
90. Report on Ahuwhenua Trophy Competition, 1949, ABRP W4598 6844 Box 244, Archives New Zealand, Wellington.
91. Ibid.
92. Ibid.
93. Report on Ahuwhenua Trophy Competition, 1950, ABRP W4598 6844 Box 244, Archives New Zealand, Wellington.
94. Ibid.
95. Report on Ahuwhenua Trophy Competition, 1951, ABRP W4598 6844 Box 244, Archives New Zealand, Wellington.
96. Ibid.
97. S Lambert, 'Te Ahuwhenua and the "Sons" of the Soil: A History of the Māori-Farmer-of-the-Year Award', unpublished paper, Te Whanake, Lincoln University, Christchurch, 2010, p. 4.
98. Report on Ahuwhenua Trophy Competition, 1953, ABRP W4598 6844 Box 244, Archives New Zealand, Wellington.
99. Letter, Lord Bledisloe to Minister of Native Affairs, 2 December 1953, AAMK W3074 869 Box 1325 (MA 60/8), Archives New Zealand, Wellington.
100. Report on Ahuwhenua Trophy Competition, 1955, ABRP W4598 6844 Box 244, Archives New Zealand, Wellington.
101. Report on Ahuwhenua Trophy Competition, 1954, ABRP W4598 6844 Box 244, Archives New Zealand, Wellington.
102. Ibid.
103. Ibid.
104. Ibid.
105. Report on Ahuwhenua Trophy Competition, 1955.
106. Ibid.
107. Ibid.
108. Report on Ahuwhenua Trophy Competition, 1956, ABRP W4598 6844 Box 244, Archives New Zealand, Wellington.
109. Ibid.
110. Report on Ahuwhenua Trophy Competition, 1957, ABRP W4598 6844 Box 244, Archives New Zealand, Wellington.
111. Ibid.
112. Report on Ahuwhenua Trophy Competition, 1958, ABRP W4598 6844 Box 244, Archives New Zealand, Wellington.
113. Ibid.
114. Report on Ahuwhenua Trophy Competition, 1959, ABRP W4598 6844 Box 244, Archives New Zealand, Wellington.
115. Ibid.
116. Report on Ahuwhenua Trophy Competition, 1960, ABRP W4598 6844 Box 244, Archives New Zealand, Wellington.
117. Ibid.
118. Report on Ahuwhenua Trophy Competition, 1961, ABRP W4598 6844 Box 244, Archives New Zealand, Wellington.
119. Report on Ahuwhenua Trophy Competition, 1960.
120. Ibid.

CHAPTER FIVE
The 'Spirit of Friendly Rivalry'– 1962–1972

121. Report on Ahuwhenua Trophy Competition, 1962, ABRP W4598 6844 Box 244, Archives New Zealand, Wellington.
122. Ibid.
123. Report on Ahuwhenua Trophy Competition, 1963, ABRP W4598 6844 Box 244, Archives New Zealand, Wellington.
124. Ibid.
125. Report on Ahuwhenua Trophy Competition, 1964, ABRP W4598 6844 Box 244, Archives New Zealand, Wellington.
126. Ibid.
127. Report on Ahuwhenua Trophy Competition, 1965, ABRP W4598 6844 Box 244, Archives New Zealand, Wellington.
128. Ibid.
129. Ibid.
130. Report on Ahuwhenua Trophy Competition, 1966, ABRP W4598 6844 Box 244, Archives New Zealand, Wellington.
131. Ibid.
132. R Boast, 'Te tango whenua – Māori land alienation – 20th-century developments' in *Te Ara – the Encyclopedia of New Zealand*: www.teara.govt.nz/en/te-tango-whenua-maori-land-alienation/page-7 (accessed 31 July 2013).
133. Report on Ahuwhenua Trophy Competition, 1967, ABRP W4598 6844 Box 244, Archives New Zealand, Wellington.
134. Ibid.
135. Report on Ahuwhenua Trophy Competition, 1968, ABRP W4598 6844 Box 244, Archives New Zealand, Wellington.
136. Ibid.
137. Report on Ahuwhenua Trophy Competition, 1969, ABRP W4598 6844 Box 244, Archives New Zealand, Wellington.
138. Ibid.
139. Report on Ahuwhenua Trophy Competition, 1970, ABRP W4598 6844 Box 244, Archives New Zealand, Wellington.

140. Memorandum Secretary to Hamilton, 18 June 1970, MA/DO file22/17, AAMK W3074 869 Box 1326, 60/8/69, Archives New Zealand, Wellington.
141. Memorandum, District Field Officer (Wanganui) to Head Office, 14 August 1970, MA/DO file2/175, AAMK W3074 869 Box 1326, 60/8/69, Archives New Zealand, Wellington.
142. Memorandum, District Officer (Rotorua) to Head Office, 12 June 1970, AAMK W3074 869 Box 1326, 60/8/69, Archives New Zealand, Wellington.
143. Memorandum, District Officer (Whāngarei) to Head Office, 24 June 1970, AAMK W3074 869 Box 1326, 60/8/69, Archives New Zealand, Wellington.
144. Report on Ahuwhenua Trophy Competition, 1970.
145. Ibid.
146. Report on Ahuwhenua Trophy Competition, 1972, ABRP W4598 6844 Box 244, Archives New Zealand, Wellington.
147. Ibid.
148. Ibid.

CHAPTER SIX
Challenges – 1973–2002

149. Report on Ahuwhenua Trophy Competition, 1973, ABRP W4598 6844 Box 244, Archives New Zealand, Wellington.
150. Ibid.
151. Ibid.
152. Ibid.
153. Report on Ahuwhenua Trophy Competition, 1974, ABRP W4598 6844 Box 244, Archives New Zealand, Wellington.
154. Ibid.
155. Ibid.
156. Ibid.
157. Ibid.
158. Report on Ahuwhenua Trophy Competition, 1975, ABRP W4598 6844 Box 244, Archives New Zealand, Wellington.
159. Report on Ahuwhenua Trophy Competition, 1964, ABRP W4598 6844 Box 244, Archives New Zealand, Wellington.
160. Letter, A M Hall, New Zealand Dairy Board to Secretary of Māori Affairs, 5 March 1976, ABRP W4598 6844 Box 244, Archives New Zealand, Wellington.
161. Report on Ahuwhenua Trophy Competition, 1975.
162. Ibid.
163. Report on Ahuwhenua Trophy Competition, 1976, ABRP W4598 6844 Box 244, Archives New Zealand, Wellington.
164. Ibid.
165. Ibid.
166. Ibid.
167. Letter, Assistant Secretary, Māori Affairs to 'New Zealand Dairy Exporter', 14 February 1977, ABRP W4598 6844 Box 244, Archives New Zealand, Wellington.
168. Report on Ahuwhenua Trophy Competition, 1977, ABRP W4598 6844 Box 244, Archives New Zealand, Wellington.
169. Report on Ahuwhenua Trophy Competition, 1978, ABRP W4598 6844 Box 244, Archives New Zealand, Wellington.
170. Ibid.
171. Memorandum, Māori Trustee Head Office, Wellington, 7 April 1978, File MA 4/1/1/437, Department of Māori Affairs, Archives New Zealand, Wellington.
172. Memorandum, Minister of Māori Affairs, 22 February 1979, File MA 4/1/1/460, Department of Māori Affairs, Archives New Zealand, Wellington.
173. Report on Ahuwhenua Trophy Competition, 1979, ABRP W4598 6844 Box 244, Archives New Zealand, Wellington.
174. Ibid.
175. Memorandum, Minister of Maori Affairs, 7 May 1980, File MA 4/1/1/497, Department of Māori Affairs, Archives New Zealand, Wellington.
176. Memorandum, Secretary for Māori Affairs, 23 November 1981, File MA 4/1/1/500, Department of Māori Affairs, Archives New Zealand, Wellington.
177. Memorandum, Secretary of Māori Affairs, 13 April 1982, MA 60/8/82, Archives New Zealand, Christchurch.
178. Report on Ahuwhenua Trophy Competition, 1982, ABRP W4598 6844 Box 244, Archives New Zealand, Wellington.
179. Ibid.
180. Memorandum, 18 November 1985, MA 4/1/1 (Vol 12), Te Puni Kōkiri, Wellington.
181. Memorandum, Minister of Māori Affairs, 16 February 1990, File 4/1/1 (Vol 12), Archives New Zealand, Wellington.

CHAPTER SEVEN
Ahuwhenua Continuing – 2003–2013

182. Meat New Zealand, 'Meat Matters' newsletter, 6 December 2002, p. 2.
183. *Wanganui Chronicle*, 5 May 2003.
184. Ahuwhenua Trophy Management Committee, 'The Ahuwhenua Trophy RoundUp', newsletter, 2011, p. 1.
185. *Gisborne Herald*, 11 June 2003.
186. Meat New Zealand, 'First Regional Winner Announced for Maori Farmer of the Year', media release, 29 March 2004.
187. Meat New Zealand, 'East Coast Māori Farmer of the Year Winner Announced', media release, 15 April 2004.
188. Office of the Māori Trustee, 'Discussion Papers – Review of 2004 Maori Farmer of the Year', Te Ahuwhenua Archives, Office of the Māori Trustee, Wellington.
189. Ahuwhenua Trophy Management Committee, field day handbook 2011, p. 5.

190. 'The 2005 Māori Farmer of the Year Award', booklet, 2005, p. 2.
191. Ibid.
192. Meat & Wool New Zealand, field day handbook, 2005, pp. 4–7.
193. Ibid, pp. 6–7.
194. H Petrie, *Chiefs of Industry: Māori Tribal Enterprise in Early Colonial New Zealand*, Auckland University Press, Auckland, 2006, pp. 100–101.
195. Kingi Smiler, quoted in Meat & Wool New Zealand, field day handbook, 2005, p. 29.
196. Ibid, pp. 35–36.
197. Ibid, pp. 51–56.
198. Ahuwhenua Trophy Management Committee, field day handbook 2006, p. 5.
199. AgResearch, information sheet and publicity release on Ahuwhenua Trophy, Māori Dairy Excellence Awards, 2006; field day handbook 2006, pp. 11–23.
200. AgResearch, information sheet and publicity release on Ahuwhenua Trophy, Māori Dairy Excellence Awards, 2006; field day handbook 2006, pp. 39–51.
201. AgResearch, information sheet and publicity release on Ahuwhenua Trophy, Māori Dairy Excellence Awards, 2006, pp. 39–51.
202. Ahuwhenua Trophy Management Committee, field day handbook 2007, p. 1.
203. Letter, Minister of Native Affairs to Minister of Agriculture, 13 December 1933, AAMK W3074 869 Box 1325 (MA 60/8), Archives New Zealand, Wellington.
204. Field day handbook 2007, p. 9.
205. Ahuwhenua Trophy Management Committee, field day handbook 2008, p. 9.
206. Field day handbook 2007, p. 1.
207. Field day handbook 2008, pp. 8–10.
208. Field day handbook 2007, p. 52.
209. Ibid, pp. 30–31.
210. Ahuwhenua Trophy Management Committee, field day handbook 2009, pp. 6–7.
211. Field day handbook 2008, p. 24.
212. Field day handbook 2009, pp. 6–7.
213. Ibid, p. 8.
214. Field day handbook 2008, p. 40.
215. Field day handbook 2009, p. 4.
216. Ibid, p. 47.
217. Ibid, p. 48.
218. Ibid, p. 15.
219. Ibid, p. 31.
220. Ibid, p. 32.
221. 'The Ahuwhenua Trophy RoundUp', March 2010, p. 15.
222. Ibid, p. 15.
223. Ibid, p. 31.
224. Ibid, p. 33.
225. Ibid, p. 48.
226. Ahuwhenua Trophy Management Committee, field day handbook 2011, p. 4.
227. Ibid, p. 5.
228. 'The Ahuwhenua Trophy RoundUp', March 2011, pp. 11–13.
229. 'The Ahuwhenua Trophy RoundUp', March 2010, p. 12.
230. Field day handbook 2011, p. 43.
231. Ibid, p. 44.
232. Letter, Lord Bledisloe to Prime Minister M J Savage, 10 July 1939, AAMK W3074 869 Box 1325, Archives New Zealand, Wellington.
233. Ahuwhenua Trophy Management Committee, field day handbook 2012, p. 12.
234. Field day handbook 2011, p. 28.
235. www.ahuwhenuatrophy.maori.nz/young.php (accessed 31 July 2013).
236. Ahuwhenua Trophy Management Committee, field day handbook 2013, p. 15.
237. Ibid, p. 33.
238. Ibid, p. 50.
239. Ibid, p. 51.

CONCLUSION TO PART ONE
'Administering the Policy Effectively'

240. Letter, A T Ngata to Lord Bledisloe, 28 February 1934, AAMK W3074 868 Box 1325 (MA 60/8), Archives New Zealand, Wellington.
241. DVD compilation of interviews with finalists of the BNZ Māori Excellence in Farming Award 2013 (interviews conducted by Ms Lyn Harrison, May 2013).
242. Ibid.
243. W Pere, 'Meeting Between the Premier and Chiefs and Others of the Ngāti Kahungunu Tribe at Waipatu 30 March 1898', AJHR, 1899, G-2, pp. 4–5.
244. DVD compilation of interviews with finalists of the BNZ Māori Excellence in Farming Award 2013.
245. DVD compilation of interviews with past winners (2003–2012) of the Ahuwhenua Trophy (interviews conducted by Ms Lyn Harrison, May 2013).
246. Ibid.
247. Ibid.

PART TWO

INTRODUCTION
Growing the Trophy 2014–2021

248. Native Land Amendment and Native Claims Adjustment Act 1929 [7 November 1929], 20 GEO V 1929, No. 19.
249. Ahuwhenua Trophy Management Committee, 'Horticulture signs up to prestigious Ahuwhenua Trophy', media release, 20 August 2018, p. 1.

250. Ahuwhenua Trophy Management Committee, 'The Ahuwhenua Trophy RoundUp' newsletter, November 2019, pp. 1–3.

CHAPTER EIGHT
Origins of Modern Māori Horticulture 1840–1860

251. H Petrie, *Chiefs of Industry: Maori Tribal Enterprise in Early Colonial New Zealand*, Auckland University Press, Auckland, 2006, p. 46.

252. Mr Carroll, 'Note' appended to Report of the Commission Appointed to Inquire into the Subject of the Native Land Laws 1891, *Appendices to the House of Representatives*, Session II, G-1, pp. xxvi-xxx.

253. R P Hargreaves, 'The Maori Agriculture of the Auckland Province in the Mid-Nineteenth Century', *Journal of the Polynesian Society*, Volume 68, No. 2, p. 61 [*JPS 68,2*] http://www.jps.auckland.ac.nz/document//Volume_68_1959/Volume_68%2C_No._2/The_Maori_agricuture_of_the_Auckland_Province_in_the_mid-nineteenth_century%2C_by_R._P._Hargreaves%2C_p_61-79/p1 (accessed 3 December 2020).

254. Petrie, *Chiefs of Industry*, pp. 240–241.

255. I Pool, *Te Iwi Māori: A New Zealand Population Past, Present & Projected*, Auckland University Press, Auckland, 1991, p. 51.

256. J B Bennett, Registrar-General, Report to Colonial Secretary, 'Table No. 1 Showing The Estimated European Population of New Zealand in December 1860, Statistics New Zealand Ta Tauranga Aotearoa. https://www3.stats.govt.nz/historic_publications/1860-statistics-nz/1860-statistics-nz.html#idsect2_1_84 (accessed 5 December 2020).

257. Hargreaves, 'The Maori Agriculture of the Auckland Province', *JPS 68,2*, p. 64.

258. Petrie, *Chiefs of Industry*, p. 6.

259. Hargreaves, 'The Maori Agriculture of the Auckland Province', *JPS 68,2*, p. 64.

260. A Ward, *A Show of Justice: Racial 'amalgamation' in nineteenth century New Zealand*, Australian National University Press, Melbourne, 1973, reprinted with corrections by Auckland University Press, Auckland, 1995, p. 86.

261. Wesleyan Native Industrial School, Kai Iwi; and Church of England Native Industrial School, Whanganui 'Reports', [10 October 1857], *Appendices to the Journals of the House of Representatives*, 1858, E-1, pp. 49–50.

262. M Roberts, 'Revisiting 'The Natural World of the Māori', in Danny Keenan (ed), *Huia Histories of Māori Ngā Tāhuhu Kōrero*, Huia Publishers, Wellington, 2012, p. 34.

263. J Polack, *Manners & Customs of the New Zealanders*, James Madden and Leadenhall Street, London, 1830, reprinted by Capper Press, Christchurch, 1976, p. 212.

264. Ibid, pp. 288–289.

265. Hargreaves, 'The Maori Agriculture of the Auckland Province', *JPS 68,2*, p. 64.

266. E Best, *Maori Agriculture*, Government Printer, Wellington, 1976 (first published 1925), pp. 45–46.

267. Hargreaves, 'The Maori Agriculture of the Auckland Province', *JPS 68, 2*, p. 65.

268. Best, *Maori Agriculture*, p. 193.

269. Hargreaves, 'The Maori Agriculture of the Auckland Province', *JPS 68, 2*, p. 66.

270. Civil Commissioner of the Bay of Islands 'No. 2 Report', [3 April 1862], *Appendices to the Journals of the House of Representatives*, 1862, Session I, E-05A : https://paperspast.natlib.govt.nz/parliamentary/AJHR1862-I.2.1.6.6 (accessed 5 December 2020).

271. Ibid.

272. Ibid.

273. W H Skinner, *Reminiscences of a Taranaki Surveyor*, Thomas Avery and Son, New Plymouth, 1946; reprinted by Capper Press, Christchurch, 1977, pp. 50–51.

274. Hargreaves, 'The Maori Agriculture of the Auckland Province', *JPS 68,2*, p. 69.

275. Ibid., pp. 69–70.

276. Ibid., p. 73.

277. Ibid.

CHAPTER NINE
Dispossession of Māori Lands 1858–1912

278. R Boast, *Buying the Land, Selling the Land: Governments and Maori Land in the North Island 1865–1921*, Victoria University Press, Wellington, 2008, pp. 32–33.

279. Letter, Judge W Dempster to Director of Dairy Division, Wellington, 7 March 1933, AAMK W3074 869, Box 1326, Archives New Zealand, Wellington.

280. A T Ngata, 'Maori Land Settlement' in I.L.G. Sutherland (ed), *The Maori People Today: A General Survey*, The New Zealand Institute of National Affairs & The New Zealand Council for Educational Research, Wellington, 1940, pp. 144–154.

281. A Ward, National Overview, Volume II, Waitangi Tribunal Rangahaua Series, 1997, pp. 10–14, Prelims.fm (waitangitribunal.govt.nz) (accessed 16 February 2022).

282. H C Evison, *Te Wai Pounamu, The Greenstone Island: A History of the Southern Maori during the European Colonisation of New Zealand*, Christchurch: Aoraki, 1993, p. 444, as cited in https://teara.govt.nz/en/ngai-tahu/page-8 (accessed 16 February 2022).

283. T Sole, *Ngāti Ruanui: A History*, Huia Publishers, Wellington, 2005, p. 218.

284. Preamble, The New Zealand Settlements Act 1863 [3 December 1863].

285. H Riseborough, *Days of Darkness. The Government and Parihaka*, Penguin Books, Auckland, 2002 [first published 1989], p. 28.
286. Native Lands Act 1862 [15 September 1862], 26 Victoriae 1862 No. 42.
287. Mr Fox, Native Lands Bill (No. 1) [22 July 1862], *New Zealand Parliamentary Debates*, Volume D, p. 422.
288. Section IV, Native Rights Act 1865 [26 September 1865].
289. Native Lands Act 1865 [30 October 1865], 29 Victoriae 1865 No. 71.
290. Maori Representation Act 1867 [10 October 1867], 31 Victoriae 1867 No. 47.
291. T Te Moananui, Policy of the Government [4 August 1868], *New Zealand Parliamentary Debates*, Volume 2, pp. 270–271.
292. M K Paetahi, Policy of the Government [4 August 1868], *New Zealand Parliamentary Debates*, Volume 2, p. 272.
293. C W Richmond, Hawkes Bay Native Lands Alienation Commission 1872 [31 July 1873], *Appendices to the Journals of the House of Representatives*, 1873, G-7, p. 19.
294. H Riseborough, 'Politics, Law and the Land: The West Coast Settlement Reserves Act 1881 and its Amendments', unpublished research report, 2002, pp. 1–15; Return of West Coast Settlement Reserves District Leases, 1890, *Appendices to the Journals of the House of Representatives*, 1890, G-7, pp. 2–8.
295. Riseborough, 'Politics, Law and the Land', pp. 15–23; the West Coast Settlement Reserves Act, passed 24 September 1892.
296. D Tuuta, 'Chapter II Perpetual Leasing in Taranaki, 1880–2008', in Richard Boast and Richard S. Hill (eds), *Raupatu: The Confiscation of Maori Land*, Victoria University Press, Wellington, 2009, pp. 239–240.
297. AgResearch, information sheet and publicity release on Ahuwhenua Trophy, Māori Dairy Excellence Awards, 2006, pp. 39–51.
298. J A T K Pere et al., *Wiremu Pere: The Life and Times of a Maori Leader, 1837–1915*, Libro International Publishers, Gisborne, 2010.
299. Ward, *Show of Justice*, p. 290.
300. A Ward, National Overview, Waitangi Tribunal, pp. 10–14.
301. T Brooking, ''Busting Up' The Greatest Estate of All: Liberal Maori Land Policy, 1891–1911', *New Zealand Journal of History*, Vol. 26, No. 1, April 1992, p. 78.
302. Mr Seddon, Ministerial Portfolios [6 September 1893], *New Zealand Parliamentary Debates*, Vol. 82, p. 32.
303. Native Land Court Act 1894 [23 October 1894], 58 VICT, 1894, No. 43.
304. Mr Carroll, Native Land Court Bill [28 September 1894], *New Zealand Parliamentary Debates*, pp. 380–383.
305. T Brooking, *Richard Seddon King of God's Own: The Life and Times of New Zealand's Longest Serving Prime Minister*, Penguin Books, Auckland, 2014, pp. 144–145.
306. Mr Carroll, Urewera District Native Reserve Bill [24 September 1896] *New Zealand Parliamentary Debates*, Volume 96, p. 157.
307. Native Affairs Committee, Native Lands Settlement and Administration Bill, *Appendices to the Journals of the House of Representatives*, AJHR, 1898, I - 3a, see summary of petitioners on p. 1.
308. Meeting Between the Premier and the Chiefs and others of the Ngati Kahungunu Tribe, Waipatu, 30 March 1898, http://nzetc.victoria.ac.nz/tm/scholarly/tei-BIM1562Engl-t1-body1-d2.html (accessed 9 July 2019).
309. Mr Seddon, Maori Lands Administration Bill [19 October 1899], *New Zealand Parliamentary Debates*, Volume 110, p. 740.
310. Brooking, ''Busting Up' The Greatest Estate of all', p. 78.
311. G V Butterworth and H R Young, *Maori Affairs*, Iwi Transition Agency/GP Books, Wellington, 1990, pp. 58–59.
312. Ibid, p. 62.
313. J A Williams, *Politics of the New Zealand Maori: Protest and Cooperation, 1891–1909*, Auckland/Oxford University Press, Auckland, 1969, p. 18.
314. Ward, National Overview, Waitangi Tribunal, pp. 10–14.
315. Ibid.
316. Boast, *Buying the Land*, p. 32.
317. Ibid.
318. M P K Sorrenson, 'Ngata, Apirana Turupa', *Dictionary of New Zealand Biography (1996), Te Ara – the Encyclopedia of New Zealand*; https://teara.govt.nz/en/biographies/3n5/ngata-apirana-turupa (accessed 11 April 2022).
319. R S Hill, *State Authority, Indigenous Autonomy: Crown–Maori Relations in New Zealand/Aotearoa 1900–1950*, Victoria University Press, Wellington, 2004, p. 128.
320. Mr Carroll, 'Note' appended to Report of the Commission Appointed to Inquire into the Subject of the Native Land Laws 1891, pp. xxvi–xxx.
321. T Kingi, Maori Land Ownership and Land management in New Zealand, Institute of Natural Resources Massey University, p. 132: https://www.dfat.gov.au/sites/default/files/MLW_VolumeTwo_CaseStudy_7.pdf (accessed 11 April 2022).

CHAPTER TEN
Revisiting the Early Ahuwhenua Farmers 1933–1963

322. Report on the Ahuwhenua Trophy Competition 1948–1949, ABRP W4598 6844 Box 244, Archives New Zealand, Wellington.
323. Report on the Ahuwhenua Trophy Competition, 1936, ABRP W4598 6844 Box 244, Archives New Zealand, Wellington.
324. Report on the Ahuwhenua Trophy Competition, 1941, ABRP W4598 6844 Box 244, Archives New Zealand, Wellington.

325. Report on the Ahuwhenua Trophy Competition, 1938, ABRP W4598 6844 Box 244, Archives New Zealand, Wellington.
326. Report on the Ahuwhenua Trophy Competition, 1940, ABRP W4598 6844 Box 244, Archives New Zealand, Wellington.
327. Report on the Ahuwhenua Trophy Competition, 1948.
328. Report on the Ahuwhenua Trophy Competition, 1952, ABRP W4598 6844 Box 244, Archives New Zealand, Wellington.
329. Report on the Ahuwhenua Trophy Competition, 1954, ABRP W4598 6344 Box 244, Archives New Zealand, Wellington.
330. Report on the Ahuwhenua Trophy Competition, 1946, ABRP W4598 6344 Box 244, Archives New Zealand, Wellington.
331. Report on the Ahuwhenua Trophy Competition, 1950, ABRP W4598 5844 Box 244, Archives New Zealand, Wellington.
332. Report on the Ahuwhenua Trophy Competition, 1951, ABRP W4598 5844 Box 244, Archives New Zealand, Wellington.
333. Report on the Ahuwhenua Trophy Competition, 1954.
334. Report on the Ahuwhenua Trophy Competition, 1947, ABRP W4598 6844 Box 244, Archives New Zealand, Wellington.
335. Report on the Ahuwhenua Trophy Competition, 1949.
336. Report on the Ahuwhenua Trophy Competition, 1950.
337. Report on the Ahuwhenua Trophy Competition, 1951.
338. Report on the Ahuwhenua Trophy Competition, 1965, ABRP W4598 6844 Box 244, Archives New Zealand, Wellington.
339. Report on the Ahuwhenua Trophy Competition, 1966, ABRP W4598 6844 Box 244, Archives New Zealand, Wellington.
340. Report on the Ahuwhenua Trophy Competition, 1949.
341. Report on the Ahuwhenua Trophy Competition, 1950.
342. Report on the Ahuwhenua Trophy Competition, 1951.
343. Report on the Ahuwhenua Trophy Competition, 1947.
344. Report on the Ahuwhenua Trophy Competition, 1944, ABRP W4598 6844 Box 244, Archives New Zealand, Wellington.
345. Report on the Ahuwhenua Trophy Competition, 1946.
346. Report on the Ahuwhenua Trophy Competition, 1948.
347. Report on the Ahuwhenua Trophy Competition, 1951.
348. Report on the Ahuwhenua Trophy Competition, 1966.
349. Report on the Ahuwhenua Trophy Competition, 1946.
350. Report on the Ahuwhenua Trophy Competition, 1946.
351. Report on the Ahuwhenua Trophy Competition, 1949.
352. Report on the Ahuwhenua Trophy Competition, 1948.
353. Report on the Ahuwhenua Trophy Competition, 1950.
354. Report on the Ahuwhenua Trophy Competition, 1951.
355. Report on the Ahuwhenua Trophy Competition, 1954.
356. Report on the Ahuwhenua Trophy Competition, 1962, ABRP W4598 6844 Box 244, Archives New Zealand, Wellington.
357. Report on the Ahuwhenua Trophy Competition, 1945, ABRP W4598 6844 Box 244, Archives New Zealand, Wellington.
358. Report on the Ahuwhenua Trophy Competition, 1948.
359. Report on the Ahuwhenua Trophy Competition, 1951.
360. Report on the Ahuwhenua Trophy Competition, 1954.
361. Report on the Ahuwhenua Trophy Competition, 1949.
362. Report on the Ahuwhenua Trophy Competition, 1950.
363. Report on the Ahuwhenua Trophy Competition, 1953, ABRP W4598 6844 Box 244, Archives New Zealand, Wellington.

CHAPTER ELEVEN
Ahuwhenua Trophy Competition Continues 2014–2021

364. Ahuwhenua Trophy Management Committee, 'The Ahuwhenua Trophy RoundUp' newsletter, March 2014, pp. 1–3.
365. Ibid.
366. Ahuwhenua Trophy Management Committee, 'Taranaki Dairy Farm Wins Ahuwhenua Trophy', media release, 13 June 2014.
367. 'The Ahuwhenua Trophy RoundUp', March 2014, p. 3.
368. Ahuwhenua Trophy Management Committee, field day handbook, 2014, p. 33.
369. 'The Ahuwhenua Trophy RoundUp', March 2014, p. 3.
370. Field day handbook 2014, p. 13.
371. 'The Ahuwhenua Trophy RoundUp', March 2014, p. 3.
372. Field day handbook 2014, p. 13.
373. Ahuwhenua Trophy Management Committee, 'The Ahuwhenua Trophy RoundUp', April 2015, p. 1.
374. 'The Ahuwhenua Trophy RoundUp', April 2015, pp. 1–2.
375. Ahuwhenua Trophy Management Committee, 'Ahuwhenua Trophy winner announced', media release, 29 May 2015.
376. 'The Ahuwhenua Trophy RoundUp', April 2015, pp. 1–2.
377. Ahuwhenua Trophy Management Committee, Field day handbook 2015, p. 13.
378. 'The Ahuwhenua Trophy RoundUp', April 2015, pp. 1–2.
379. Field day handbook 2015, p. 51.
380. Ahuwhenua Trophy Management Committee, 'Ahuwhenua Trophy winner announced', media release, 29 May 2015.
381. Field day handbook 2015, p. 31.
382. 'The Ahuwhenua Trophy RoundUp', November 2015, p. 1.
383. Ibid., p.2.
384. Ahuwhenua Trophy Management Committee, 'The Ahuwhenua Trophy RoundUp', March 2016, pp. 1–3.
385. Ahuwhenua Trophy Management Committee, 'Ahuwhenua Trophy winner makes history', media release, 20 May 2016, p. 1.
386. 'The Ahuwhenua Trophy RoundUp', March 2016, pp. 1–3.

387. Ahuwhenua Trophy Management Committee, Field day handbook 2016, p. 51.
388. 'The Ahuwhenua Trophy RoundUp', April 2016, p. 2.
389. Field day handbook 2016, p. 13.
390. 'The Ahuwhenua Trophy RoundUp', March 2016, pp. 1–3.
391. Field day handbook 2016, p. 33.
392. Ahuwhenua Trophy Management Committee, 'The Ahuwhenua Trophy RoundUp', April 2017, pp. 1–3.
393. Ibid.
394. Ahuwhenua Trophy Management Committee, 'Far North Farm Wins Ahuwhenua Trophy', media release, 27 May 2017, p. 1.
395. Ahuwhenua Management Committee, 'Far North Farm Wins Ahuwhenua Trophy', media release, 27 May 2017, p. 2.
396. Ahuwhenua Trophy Management Committee, Field day handbook 2017, p. 13.
397. Ibid, pp. 1–3.
398. Ibid, p. 13.
399. 'The Ahuwhenua Trophy RoundUp', April 2017, pp. 1–3.
400. Field day handbook 2017, p. 49.
401. Ahuwhenua Trophy Management Committee, 'The Ahuwhenua Trophy RoundUp' newsletter, March 2018, pp. 1–3.
402. Ahuwhenua Trophy Management Committee, 'The Ahuwhenua Trophy RoundUp' newsletter, July 2018, pp. 1–3.
403. 'The Ahuwhenua Trophy RoundUp' newsletter, March 2018, pp.1–3.
404. Ahuwhenua Trophy Management Committee, Field day handbook 2018, p. 39.
405. 'The Ahuwhenua Trophy RoundUp', March 2018, pp. 1–3.
406. Field day handbook 2018, p. 39.
407. Ahuwhenua Trophy Management Committee, 'The Ahuwhenua Trophy RoundUp', March 2019, pp. 1–3.
408. Ahuwhenua Trophy Management Committee, 'Ahuwhenua Trophy 2019 Winner Announced', media release, 24 May 2019, pp. 1–2.
409. 'The Ahuwhenua Trophy RoundUp', March 2019, pp. 1–3.
410. Ahuwhenua Trophy Management Committee, Field day handbook 2019, p. 64.
411. 'The Ahuwhenua Trophy RoundUp', March 2019, pp. 1–3.
412. Field day handbook 2019, p. 14.
413. 'The Ahuwhenua Trophy RoundUp', March 2019, pp. 1–3.
414. Field day handbook 2019, p. 40.
415. Horticulture New Zealand, 'Mike Chapman Wins Bledisloe Cup for Horticulture', 5 August 2021: https://www.hortnz.co.nz/news-events-and-media/media-releases/mike-chapman-wins-bledisloe-cup-for-horticulture/ (accessed 12 April 2022).
416. Ahuwhenua Trophy Management Committee, 'Horticulture signs up to prestigious Ahuwhenua Trophy', media release, 20 August 2018, p. 1.
417. Ahuwhenua Trophy Management Committee, 'Ahuwhenua Trophy Competition for Horticulture Launched', media release, 2 August 2019, p. 1.
418. *Manukau Courier*, 8 August 2013: http://www.stuff.co.nz/auckland/local-news/manukau-courier/9012161/Bledisloe-Cup-for-service-to-horticulture (accessed 19 April 2022).
419. I Douglas, The Village Goldsmith, Wellington, Ahuwhenua Trophy Management Committee, 31 March 2022.
420. Ahuwhenua Trophy Management Committee, '2020 Field Days', media release, 12 August 2020, p. 1.
421. Ahuwhenua Trophy Management Committee, 'The Ahuwhenua Trophy RoundUp', September 2020, p. 1.
422. Ahuwhenua Trophy Management Committee, 'Ahuwhenua Trophy 2020 Winner Announced', media release, 20 November 2020, pp. 1–2.
423. Ibid, p. 2.
424. Ahuwhenua Trophy Management Committee, Field day handbook 2020, p. 12.
425. Ibid.
426. Ibid.
427. Ahuwhenua Trophy Management Committee, Te Kawerongo March 2021, p. 1.
428. Ibid.
429. Ibid.
430. Ibid.
431. Ibid.
432. Ahuwhenua Trophy Management Committee, Field day handbook, p. 11.
433. Ibid, p. 31.
434. Ibid, p. 53.
435. Ahuwhenua Trophy Management Committee, Te Kawerongo June 2021, p. 1.
436. Field day handbook 2021, p. 11.
437. Ibid, pp. 14–15.
438. Ibid, p. 33.
439. Ibid, p. 34.
440. Ibid, p. 55.
441. Ibid, p. 56.

CHAPTER TWELVE
Looking to the Future – Young Māori Farmers and Growers

442. Ahuwhenua Trophy Management Committee, 'Tangaroa Walker Wins First Ahuwhenua Young Māori Farmer Award', media release, 13 June 2014, p. 1.
443. Ahuwhenua Trophy Management Committee, 'Ahuwhenua Young Māori Farmer Award Entries', media release: 7 June 2013 www.ahuwhenuatrophy.maori.nz/young.php (accessed 11 December 2020).
444. J Smith, personal communication [email], 5 February 2022.
445. Ahuwhenua Trophy Management Committee, '2014 Ahuwhenua Young Māori Dairy Farmer Award', media release, 13 June 2014.

446. W Reid, personal communication (email), 15 May 2022.
447. H Wallace-Bright, personal communication (email), 17 April 2022.
448. Ahuwhenua Trophy Management Committee, 'Ahuwhenua Young Māori Farmer winner announcement', media release, 30 May 2015, pp. 1–2.
449. Ahuwhenua Trophy Management Committee, 'Young Māori dairy farmer Jack Raharuhi changes direction and wins award', media release, 20 May 2016, www.ahuwhenuatrophy.maori.nz/young.php (accessed 11 December 2020).
450. Ahuwhenua Trophy Management Committee, 'Ahuwhenua Young Māori Farmer Award Winner', media release, 27 May 2017, pp. 1-2.
451. Ahuwhenua Trophy Management Committee, 'Ahuwhenua Young Māori Farmer Winner 2018', media release, 25 May 2018, pp. 1–2.
452. Ahuwhenua Trophy Management Committee, 'Winner of the Ahuwhenua Young Māori Award, media release, 24 May 2019, pp. 1–2.
453. Field day handbook 2020, p. 12.
454. 'The Ahuwhenua Trophy RoundUp', June 2021, p. 4.
455. Q Morgan, personal communication (email), 18 April 2022.
456. Ahuwhenua Young Māori Farmer of the Year 2022, https://www.youtube.com/watch?v=d1OOOnOy48g

Bibliography

Primary Materials

Appendices to the House of Representatives & New Zealand Parliamentary Debates
Civil Commissioner Reports
Native Lands Legislation / Acts of Parliament
Māori Parliamentarian's Addresses in the House
Royal Commission Reports / Native Lands Alienation Commissions
Statistics New Zealand Ta Tauranga Aotearoa Reports
Wesleyan Native Industrial School Reports

Ahuwhenua Trophy Management Committee
Ahuwhenua Trophy Roundup Newsletters 2014–2021
Field Day Handbooks 2014–2021
Media Releases 2014–2021

Papers, Letters and Memoranda – Archives New Zealand
Competitior Lists / Reports on Ahuwhenua Trophy Competition, 1933–1979 (judges' reports)
Dempster, W (Judge)
Fields Superintendent, Department of Native Affairs
Hall, A N (Judge)
Lord Bledisloe
Minister of Maori Affairs
Minister of Native Affairs
Private Secretary, Native Minister
Registrar, Native Land Court, Gisborne
Savage, M J
Secretary to the Treasury
Supervisors/district field officers, Department of Native Affairs
Under Secretary, Department of Native Affairs
Walker, C A (Judge)
Walker & Hall

Papers – Appendices to the Journals of the House of Representatives (AJHR)
Best, E, Report to Dr Pomare
Ministerial memoranda
Patete, H, and Others, Reports to Dr Pomare

Newspapers and Media Releases
Auckland Star
Evening Post
Gisborne Herald
Mana Magazine
Manukau Courier
Meat New Zealand media releases
New Zealand Herald
Wanganui Chronicle

Film

'Maori Battalion returns' newsreel: www.nzonscreen.com/title/maori-battalion-returns-1946
Memorandum, 18 November 1985, MA 4/1/1 (Vol 12), Te Puni Kōkiri, Wellington.

Secondary Materials

Articles and Chapters in Books

Boast, R, 'Te tango whenua – Māori land alienation – 20th-century developments' in *Te Ara – the Encyclopedia of New Zealand*: www.teara.govt.nz/en/te-tango-whenua-maori-land-alienation/page-9

Brooking, T, '"Busting Up" The Greatest Estate of All: Liberal Maori Land Policy, 1891–1911', *New Zealand Journal of History*, Vol. 26, No. 1, April 1992.

Hargreaves, R P, 'The Maori Agriculture of the Auckland Province in the Mid-Nineteenth Century', *Journal of the Polynesian Society*, Volume 68, No. 2, 1959.

Ngata, Apirana T, 'Maori Land Settlement' in I L G. Sutherland (ed), *The Maori People Today: A General Survey*, The New Zealand Institute of National Affairs & The New Zealand Council for Educational Research, Wellington, 1940.

Parsonson, A R, 'The Pursuit of Mana' in W H Oliver and B R Williams (eds), *The Oxford History of New Zealand*, Oxford University Press, Wellington, 1981.

Roberts, M, 'Revisiting 'The Natural World of the Māori', in D Keenan (ed), *Huia Histories of Māori Ngā Tāhuhu Kōrero*, Huia Publishers, Wellington, 2012.

Soutar, M, 'Ngā pakanga ki tāwāhi – Maori and overseas wars – Second World War: the Māori war effort' in *Te Ara – the Encyclopedia of New Zealand*: www.teara.govt.nz/en/nga-pakanga-ki-tawahi-maori-and-overseas-wars/page-4

Stokes, E, 'Contesting Resources: Maori, Pakeha and a tenurial revolution' in E Pawson and T Brooking (eds), *Environmental Histories of New Zealand*, Oxford University Press, Melbourne, 2002.

Tuuta, D, 'Chapter II Perpetual Leasing in Taranaki, 1880–2008', in R Boast and R S Hill (eds), *Raupatu: The Confiscation of Maori Land*, Victoria University Press, Wellington, 2009.

Papers

Bull, H M, 'Ahuwhenua Trophy for Maori Farming', paper located in Te Ahuwhenua Archives, Office of the Māori Trustee, Wellington.

Lambert, S, 'Te Ahuwhenua and the "Sons" of the Soil: A History of the Māori-Farmer-of-the-Year Award', unpublished paper, Te Whanake, Lincoln University, Christchurch, 2010.

Office of the Māori Trustee, 'Discussion Papers – Review of 2004 Maori Farmer of the Year', Te Ahuwhenua Archives, Office of the Māori Trustee, Wellington.

Books

Best, E, *Maori Agriculture*, Government Printer, Wellington, 1976 (first published 1925).

Boast, R, *Buying the Land, Selling the Land: Governments and Maori Land in the North Island 1865–1921*, Victoria University Press, Wellington, 2008.

Brooking, T, *Richard Seddon King of God's Own: The Life and Times of New Zealand's Longest Serving Prime Minister*, Penguin Books, Auckland, 2014.

Butterworth, G V, *Nga Take i Neke ai te Maori: Maori Mobility*, Manatu Maori, Wellington, 1991.

Butterworth, G V and Young, H R, *Maori Affairs*, Iwi Transition Agency/GP Books, Wellington, 1990.

Cowan, J, *The New Zealand Wars: A History of the Maori Campaigns and the Pioneering Period*, Government Printer, Wellington, 1922 (reprinted 1955).

Evison, Harry C, *Te Wai Pounamu, The Greenstone Island: A History of the Southern Maori During the European Colonisation of New Zealand*, Christchurch: Aoraki, 1993.

Hill, R S, *State Authority, Indigenous Autonomy: Crown–Maori Relations in New Zealand/Aotearoa 1900–1950*, Victoria University Press, Wellington, 2004.

Pere, J A T K et al., *Wiremu Pere: The Life and Times of a Maori Leader, 1837–1915*, Libro International Publishers, Gisborne, 2010.

Petrie, H, *Chiefs of Industry: Māori Tribal Enterprise in Early Colonial New Zealand*, Auckland University Press, Auckland, 2006.

Polack, J, *Manners & Customs of the New Zealanders*, James Madden, London, 1830, reprinted by Capper Press, Christchurch, 1976.

Pool, I, *Te Iwi Māori. A New Zealand Population Past, Present & Projected*, Auckland University Press, Auckland, 1991.

Riseborough, H, *Days of Darkness. The Government and Parihaka*, Penguin Books, Auckland, 2002 [first published 1989].

Riseborough, H, 'Politics, Law and the Land: The West Coast Settlement Reserves Act 1881 and its Amendments', unpublished research report, 2002.

Skinner, W H, *Reminiscences of a Taranaki Surveyor*, Thomas Avery and Son, New Plymouth, 1946; reprinted by Capper Press, Christchurch, 1977.

Sole, T, *Ngāti Ruanui: A History*, Huia Publishers, Wellington, 2005.

Walker, R, *He Tipua: The Life and Times of Sir Apirana Ngata*, Penguin, Auckland, 2001.

Ward, A, *A Show of Justice: Racial 'amalgamation' in nineteenth century New Zealand*, Australian National University Press, Melbourne, 1973, reprinted with corrections by Auckland University Press, Auckland, 1995.

Williams, D V, '*Te Kooti Tango Whenua: The Native Land Court 1864–1909*, Huia, Wellington, 1999.

Williams, J A, *Politics of the New Zealand Maori: Protest and Cooperation, 1891–1909*, Auckland/Oxford University Press, Auckland, 1969.

Government/University Reports

Kingi, T, Maori Land Ownership and Land Management in New Zealand, Institute of Natural Resources, Massey University.

Ward, A, National Overview, Volume II, Waitangi Tribunal Rangahaua Series, 1997.

Other Publications

'The 2005 Maori Farmer of the Year Award', booklet, 2005.

AgResearch, information sheets and publicity releases on Ahuwhenua Trophy, Māori Dairy Excellence Awards, 2003–2006.

Ahuwhenua Trophy Management Committee, field day handbooks, 2006–2013.

Ahuwhenua Trophy Management Committee, 'The Ahuwhenua Trophy RoundUp' newsletter, 2010–2011.

Meat New Zealand, 'Meat Matters' newsletter, 2002.

Meat & Wool New Zealand, field day handbook, 2005.

Websites/Electronic Sources
Victoria University Electronic Centre
www.ahuwhenuatrophy.maori.nz
www.nzhistory.net.nz
www.teara.govt.nz

Personal Communications
Jordan Smith
Wiremu Reid
Hannah Wallace-Bright
Quinn Morgan

Index

Entries in *italics* refer to illustrations. Māori *te*, like its English equivalent 'the' is ignored in alphabetisation.

2-4-5-T, 109

28th (Māori) Battalion, 45, 59, 69

A

Acland, John, xix, *167*
AFFCO New Zealand, 143
AgITO, 143, 150
AgResearch, 143, 150
Ahikōuka A6B, 152
Ahuwhenua medals, *xvii*, *34*, *102*
Ahuwhenua Trophy, x–xi, *30*
    after 2003, 143–4
    design and safeguarding, 30
    destruction of original, *35*, 37–8
    going missing, 64
    for horticulture, *192*, 268
    replica of, *35*, *87*, 151, 240
Ahuwhenua Trophy competition
    in 1930s, 34, 36–8, 40
    1945–61, 59–93
    1962–1972, 97–117
    1973–1991, 121–38
    2003–2013, 143–78, 182–4
    2014–2021, 191
    appointment of judges, 228
    during Second World War, 45–9, 52, 54
    early entrants in, 227
    extension to whole country, 34
    fall in entries, 127, 129, 134–5
    first judge's report, *31–2*
    Flowers' report on, 73
    focus on small farmers, 22–3
    inaugural, 29, 34
    iwi and individual winners, xvi–xvii
    judges, xvii, 310–15
    judges' suggestions in, 112
    judging criteria, 296–7
    and land loss, 209
    launch of, xv–xvi, 19, 27
    nominations for, 227
    points systems, 49–52, *50–1*, 65, 72–3, 80, 123
    recesses, xv, 34, 38, 136, 138, 143
    split into sheep and dairy categories, 79, 197
    sponsors of, 143
    today's, xviii–xxi
    winners list, 298–307
    see also dairy farming; horticulture competition; sheep and cattle competition
Ahuwhenua Trophy Management Committee, xviii, 149, 175, 191, 239, 285
Ahuwhenua Trust Board, 27, 127, 151, 175
Ahuwhenua Young Māori Farmer Award, 175, 191, 193, 247, 267, 285–93, 308–9
Ahuwhenua Young Māori Grower Award, 193, 270, 291, 309
Te Aika, T P and N, 130
Ainsley, Parata, 99–100, 117, 123
Akonga, Maatutaera, *291*, 293
Albert, Patuwahine, 67, 231
Alexander, A A, 105, 108, 114, 116
Allan, K, 93
Allflex, 143
Allo, A V, 68–70, 86, 234
Amoamo, Fredrick, 45, 47–8
Amoamo, Raumoa, 136
Anaru, T, 61
Anderson, Maurice C, 134
Angus cattle, 158, 175, 246, 257, 289
animal husbandry, 5, 8, 227
te ao Māori, x
Aotea Land Board, 161
Aotea Māori Land Council, 3
Aotearoa Trust, 149–50

Apatu, Reihana F, 110, 121, 135
Api, H, 107
Arahura Māori Reserve, 260–1
Te Araroa, 18, *37*, 230
Te Arawa, 6, 33, 61, 149, 173, 210
Ardern, Jacinda, x
Te Ariki Pā, *7*, *11*
Armitage, N and S, 133
Arnold, E H, 65, 81–5
Ashby, J W, 117
AsureQuality, 278
Ata, Teku, 104
Te Ātiawa, 5
Ātihaunui-a-Paparangi iwi, 151–2, 161, 289
Ātihau-Whanganui Incorporation, xix, 3–4, 10, 127, 152, 182
Auckland, sale of Māori produce in, 203–4
Auckland province, Māori horticulture in, 197–8
Te Aupōuri scheme, 17
Austin, A R, 116
Te Aute College, 18, 136, 148
Te Aute Trust Board Farm Ltd, 146, 148–9
Te Awahohonu Forest Trust, xv, xvii, *xviii*, 144, 175, *225*, 263, 265
Te Awamutu, 17
Āwhina Group for the Volcanic Plateau, 170

B

Bailey, Charles, *112*, *113*, 125–6, *128*, 129
Bailey, Taitoko, 231
Ballance, John, 214–17
Ballance Agri-Nutrients, 143, 150
Ballance Farm Environment Awards, 176, 278
Balneavis, H R N, xv
Bank of New Zealand (BNZ)
    shares in, 151
    as sponsor, 143, 150, 239–40, 244, 249
barley, 41, 199
barley grass, 69, 105, 107, 135
Bay Milk Products, xx
BDO, 143
Beasley, C R, 74, 80, 82, 228–9
Beef and Lamb New Zealand, 143, 176

Bell, P, 111
benchmarking, 143, 178, 181, 191, 243, 248, 250–1, 271, 279–80
Benevides, Hari, 161
Bernard Matthews Sheep for Profit programme, 160
Best, Elsdon, 5, 7, 37
Biddle, Jordan, *289*
Bidois, M S, 124
Biggs, Reginald, 13
Bishop, N, 113, 115
Black, Jack, 29, 35, 38
Black, Tiwai, 127
Black, Tiwi, 103
blackberry, 41, 64, 67, 137, 231, 233
    McMillan Bull on, 98, 100
    Montgomery on, 71
    Murray on, 89, 109, 111, 115
    Taylor on, 76, 78
Blackburn, Dana, xix–xx, *xxi*, 182
Blake, G A, 65–6, 69, 228, 231, 233
Bledisloe, Lady, *28*, *30*, 62
Bledisloe, Lord, ix, *62*
    and Ahuwhenua Trophy, xv–xvi, 23, 27, 30, 63, 191, 209, 227
    on dairy farming, 171
    and Ngata, 181
    and second Trophy, xvii, 79
Bledisloe Cups, *see* Ahuwhenua Trophy
Boast, Richard, 5, 14, 69, 221
Boulter, C H and C D, 116, 121, 123, 136
Boundary Road Farm, 259–60
boxthorn hedges, 103, 105, 231
bracken, xvii, 93, 98, 103, 109, 130
breeding cows, xv, 149, 151, 176
breeding ewes, 83, 107, 114, 149, 151
breeding sows, 54
Bright, Jeremy, 289
Brons, G, 137–8
Brown, George, 125
Brown, Waitangi, 103
bullock teams, *14*, *110*
bulls, pedigree, 40
bush, native, 67–9, 149, 152, 161, 178, 232

business model, 240
business plans, 145, 159, 162
Butler, John, 7
butterfat, 17
    in Ahuwhenua nominations, 227
    Allo on, 88
    Arnold on, 82–3, 86
    Blake on, 67
    Hall on, 125–6
    McMillan Bull on, 98–103
    Mercer on, 61
    Montgomery on, 70
    Murray on, 89–91, 105–7, 109, 111, 115, 117, 123
    Taylor on, 48, 77–8
    in Taylor's points system, 72
    Walker on, 35, 38
Butterworth, G V, 5

C

cabbages, 80, 199
Cairns, Rehua, 85
Callaghan, James, 69
Callaghan, John, 100
Callaghan, Tikirau, xvii, 66, 233
Callaghan, Urukakengarangi, 66
Campbell, Ash-Leigh, *288*, 289
Cape Runaway, 29, 34, 111, 228
carbon sinks, 278
Carr, P M, 113
Carr, Spencer, 150
Carroll, James, 13, 15–16, 19, 197, 218–19, 222
Carroll, R T, 61
Carter, David, 162
Carter, Norman, *193*, *269*, *270*
Carter Holt Harvey, 175
Cartwright, Sylvia, 144
cash prizes, 129, 131, 151, 240, 270
cattle farming, 3–4, 8
cattle yards, 85, 121, 130
Cesped Lands Ltd, 154–5
Chadwick, John, 71, 76, 81
Chapman, Mike, 266–7

Charteris, Marty and Janice, *243*, 246
Christianity, 8, 10
Church of England, 199
Clark, Jim, 88, 90, 103, 106
Clarke, Robert, 35, 38
Clemens, Elizabeth F, 69, 74, 76, 228–9
climate change, ix–x, 253, 274
clovers, 124, 152, 158, 161, 169
    red, 147, 154
    subterranean, 63, 65, 74
    white, 53–4, 71, 83–4, 161
coastal vessels, 203
Coates, Gordon, 19, 108
cocksfoot, 83–4, 107, 109
Collins, Ingrid, 157, 183, 264
Compensation Court, 214
Cook, James, 6
Corbett, E B, 79
corn, 9, 39, 199, 218
Coromandel, 143, 145, 202
coronavirus pandemic, xi, 269
Cottrell, Bob, xx, 149, *265*, 266
Couch, Ben, 136
Coughland, Mark, *285*
Courtenay, J, 91
COVID-19 pandemic, 269–70, 274–5
cowsheds
    automated, 156
    Blake on, 69
    Dempster on, 30, 209
    Hall on, 124
    Montgomery on, 71
    Murray on, 89, 92
    Taylor on, 48, 74, 77–8
Cribb, A G, 128
crop rotation, 200, 266
cropping programme, 163, 167, 178
Cross, Brandon, *291*
Cross, Ratahi, 269
Crozet, Jules, 7
Cullen, E L, 54
customary land, 15, 209–10, 214, 221

D

dairy farming, 9, 121, 124, 126, 197, 274
    in 2006 competition, 149–50, 215
    in 2008 competition, 154–7
    in 2010 competition, 162–7
    in 2012 competition, 171–5, 285
    in 2014 competition, 239–43, 286, 288
    in 2016 competition, 247–52, 289
    in 2018 competition, 257–61, 289
    in 2021 competition, 274–80, 292
    in inaugural Ahuwhenua competition, 29
    and Ngata's land development schemes, 17, 69

DairyBase analysis, 240, 243
Dairying and Clean Streams Accord, 155
dairying section, 80
DairyNZ, 143
Dalton, Ben, 243
Dargaville, Charles, 86, 91, 93, 99, 101, 104, 106
Darwin, Charles, 9
Davis, Aumihi, 76, 80
Davis, Henry Mathieson, 84
Davison, Rob, *xxi*
DDT, 82, 103
de Thierry, C H, 73–4
deer, 144, 147, 153, 156
Dempster, W D, 29–30, 209, 228
    first report from, 31–2
Department of Māori Affairs/Native Affairs
    change of name, 65
    and land development, 27
Department of Survey and Land Information, 162, 173
development plans, 160, 163
Dewes, Campbell, 34
Dewes, Henry J, 35, 37
    headstone, 34
Dewes, Mereheni, 34
difficult properties, 127, 230–1
diversification, 113, 145, 147, 197, 272, 276
dogs, 6, 202
Douglas, Ian, 268
drainage
    McMillan Bull on, 98
    Murray on, 92, 107
    pre-colonial, 6
    Taylor on, 80
    Walker on, 36, 40
droughts, 149, 152, 155–6, 168, 178, 230
    Arnold on, 82
    McMillan Bull on, 98, 101–4
    Montgomery on, 53
    Murray on, 89, 106–7, 111, 113–14, 121–2
    Rodda on, 64
    Shepherd on, 54
    Taylor on, 79–80
du Fresne, Marc-Joseph Marion, 7

E

earthwork defences, 6
East Coast, *xvi*, *xix*
    and land tenure reform, 20
    Māori farmers from, 17
    Ngata's contributions to agriculture, 18–19
    sheep and cattle farming on, 197, 199
    stock rearing on, 5
East Coast Commissioner, 16, 221
economic self-sufficiency, 198
eczema, 114, 137
    facial, 82, 158
Edmonds, John, 150
Edwards, Claude, 122, 125–6
Edwards, D, 137
Edwards, J, 111, 115–16
Edwards, Johnny, 38, 40
Edwards, W C and C M, 137
Edwardson, Tony, 129–30
effluent handling, 163, 167, 250
electric fences, 74, 77, 90, 109
English, Bill, *xviii*, 253, 254
Enoka, Matua, 218
environmental issues, 148, 150, 152, 243, 257, 260
environmental strategy, 156, 241
Eri, Wharehuia, 103, 109, 111
erosion, 85, 89–90, 134, 153, 170, 257
Eru, Foley, 82–3

Erueti, Tupu, 64, 230
Essential Freshwater and Resource Management Reforms, x
Ewe Hogget Competition, 245
ex-servicemen, xvii, 19, 60, 81–3, 109, 123, 134, 233–4

F

family farming, 233
family home, pride in, 232–3
farm balloting, 19
Farm Environmental Plans, 277
farm records, 72, 80, 82–3, 86, 89, 107, 124–6, 137
Farmers Union, 17
farming exports, 69
Feakins, Desma, 279
Federation of Māori Authorities, 145, 247
fencing, 5, 23, 136–7, 147, 149, 156, 161, 163, 168, 170, 227
    Allo on, 88
    Arnold on, 81–6
    Blake on, 66–7
    early Māori attitudes to, 200–2
    Hall on, 128
    McMillan Bull on, 102–3
    McNaught on, 130, 132, 134
    Montgomery on, 71
    Murray on, 89, 91–2, 105–7, 111, 113, 116
    riparian, 247
    Rodda on, 64–5
    Taylor on, 48, 72, 74, 76, 78–80
    upkeep of, 35
    Walker on, 35–6, 38
fencing regulations, 12
fertiliser, 17, 27, 135, 147, 149, 158, 163, 167, 178, 227, 246
    Allo on, 70
    Arnold on, 65, 84
    early Māori use of, 200
    Hall on, 123
    McMillan Bull on, 98–9
    Murray on, 90–2, 105–7, 109
    organic, 272
    Rodda on, 63–4
    Taylor on, 73, 80
field days, xv, 69
    for Ahuwhenua finalists, 135, 144–5, 239, 243–4, 248, 253, 258, 262–3, 269–70, 276
finance, lack of Māori access to, 13–14, 16, 21, 61, 83
Finlayson, Chris, 223
First World War, 19, 60, 109
FitzRoy, Robert, 10
Flavell, Te Ururoa, 243, 244, 247–9, 248, 252, 253, 254
flooding, 61
    Blake on, 69
    disrupting competition, 144
    McMillan Bull on, 103
    Murray on, 89, 91–2, 109, 116
    Taylor on, 74
flour, 11–12
flour and sugar policy, 198
flour mills, 12, 202–3
flower gardens, 48, 53
Flowers, J H, 73
Foley, John, 113
Fonterra, xx
    Māori shares in, 274
    as sponsor, 143
Fonterra Grade Free Certificate, 278
forests, native, xv, 178, 280
forward planning, 129, 136–7, 146
Fox, Richard, 146
Fox, William, 9
Frazer, Allan, 143–4
Freyberg, Bernard, xvii, 69, 233
Friesian cows, 98, 111, 166, 172, 174, 241

G

Galway, Lord, 40
gardens, cultural, 278
genetics programme, 166
Glanville, E B, 49–52
goats, 197, 202, 246
Gok, Fay and Joe, 267

Golden Shears competitions, 257
Goldie, G F, 33
Gore Browne, Thomas, 210
gorse, 41, 63, 233, 245, 257
    Allo on, 88
    Arnold on, 83, 85
    McKenzie on, 130
    McMillan Bull on, 101, 103
    McNaught on, 130
    Montgomery on, 71
    Murray on, 91–3, 105–6, 109, 121
    Taylor on, 73–4
governance, 144–5, 147–8, 161, 170, 172–3
    in Ahuwhenua competition, xx
    and management, 149, 159
governance skills, 272
governance teams, 150, 152, 163, 174, 251, 272, 278
Government Advances to Settlers Act 1894, 14, 16
Grace, Kingi, 90, 97–8
grass
    native, 130, 152, 158, 161
    permanent, 49, 228
grass grub, 82–3, 98, 103, 111
grazing, rotational, 49, 70, 81
grazing area, effective, 147, 149, 169
Green Ribbon Award, 278
greenhouse gas emissions, 278
Grey, George, 10, 193, 210
Guy, Nathan, 239, 240, 243, 244, 248, 249, 252, 253
Gwavas Station, 261, 263, 265–6
    field day, 265

H

Haa (hapū), 153
Haa, Anthony Tohu, 162, 184
Hadfield, Bart and Nukuhia, 243, 244, 245, 257, 263
Haeata, Tom, 86, 87
Hale, Anahera, 292
Hall, A N, 123–9
Hall, I D, 86, 113
Hall, Tatai, xvi, 40, 45, 48, 228

Hanerau Farms Ltd, 162–3, 166
    field day, 164
Te Hape B Trust, 153, 175, 178–9
Te Hapuku, 148
Harris, J C, 106
Harris, Waerata, 69, 230
Harrison, Lyn, 266
Harrison-Boyd, Tumoanakotore-i-whakairioratia, 290
harrowing, 38, 69, 77
Hau, Kuratau, 76, 231
Hauhungaroa Partnership, 154, 156
Haumiatiketike, 192, 270–1
Hauraki, Peter, 69
Hauraki iwi, 278
Hautu, Raka, 6
Hauturu, 76
Hāwera, 17, 111, 113
Hawira, Kelly, 82
Hawira, Thomas, 127
Hawke's Bay
    acquisition of Māori land in, 10, 15, 147
    Māori farmers in, 4–5, 17
Hawkins, D W W, 137
hay, 103, 108
    Arnold on, 86
    Blake on, 69
    McMillan Bull on, 98
    Montgomery on, 71
    Murray on, 89–90, 92, 105, 107, 109, 111
    Shepherd on, 54, 63
    Taylor on, 48, 73–4, 77, 79
    Walker on, 40–1
hay barns, 115, 128, 130
He Waka eke Noa, x
Healy, Anthony, 240
Hedley, Joseph William, 82, 89, 91–2, 98–9, 233
Hei, P, 71, 234
Hemi, Ilet (R and I), 78, 82, 107
Hemi, P and A, 130
hengahenga, 200
Henry, R, 136

Heperi, Tapua Pita, 92
herd management, 124–5, 129
herd testing
    Arnold on, 65
    Glanville on, 52
    Montgomery on, 53, 70
    Taylor on, 49
    Walker on, 35, 38, 40, 46
Hereheretau Station, 157, 159–61
    field day, 95, *119*, *159*
Herewini, Hauraki, 100
herringbone sheds, 109, 127, 166, 173, 241, 243, 251, 280
Heta, A T, 98
Hetet, A T, 78, 231
Te Heu Heu, 12
Te Heu Heu, Hepi, 153
Te Heuheu, Tumu, 276
Hicks Bay, 71, 74
Hikurangi maunga, 170–1
hill country, 137, 144, 151–2, 158, 161, 245–6, 255, 257
    Arnold on, 65, 83
    Hall on, 127
    McMillan Bull on, 104
    Montgomery on, 71
    Murray on, 115
Himiona Farm, 239–41
    field day, *242*
Hippolite, Michelle, *243*, *288*
Te Hira, H P, 93
Hoata Station, 159
Hobson, William, 10
Hoe o Tainui, 92, 98–9
Hoera, Eruera, 49, 61, 64–5, 69–70, 232–3
Hoera, Tewi, 251
Hoete, Norman, 116, 124
Hohepa, G, 125
Hohepa, Tutere, 111, 126, 129
Holden, J T, 111
Holyoake, Keith, *131*, 133
homestead and garden, 65, 67, 233
Hona, S, 91
Hongi Hika, 9

Hopa, G C, 109
Hori, T, 107
Horohoro Block, 27, 29
Horomia, Parekura, 144–5, *146*
Horopapere, Tom, 83
horses, 5, *37*, 67, *75*, 85, 88, *104*, *122*, 200–2, *201*
horticulture, 153
    2020 competition, *189*, *193*, 266–74, 291
    Bledisloe Cup for, x, 266–7
    and land dispossession, 209
    Māori history of, 197, 199–204
hot springs, 153, 248, 251
Houpapa, Traci, 175, 181
Hubbard, Taane-nui-a-rangi, *287*, 289, *290*
Hughes, W H, 115, 122
human elements, 181
Hunia, Erina, 70
Hunia, Tiaki, *288*
Hunt, Aaron, 291
Huriwai, Katene, 64, 231

I

ICI Farmer of the Year, 130
IFL (Integrated Foods Ltd), 16
Te Ihingarangi, 257
illness, 92, 105, 107, 123, 231–2
Industrial Schools, 199
Inspector of Native Mills, 203
interest return, 105, 112–13, 116–17, 121–2
isolation, 37, 76, 171, 191, 227, 231
    Allo on, 88
    Arnold on, 65, 82, 84
    and Māori lands administration, 220
    McMillan Bull on, 104
    McNaught on, 130
    Mercer on, 59, 61
    Murray on, 91, 111
    Taylor on, 74, 78

J

Jackson, K R, 123
Jackson, Willie, 270, 274–6

James Carroll Walkway, *218*
Janssen, John, *243*, 244, 249
Jefferies, Mark, *288*
Jersey cows, 29, 64–5, 86, 88–9, 98, 171–2, 233
Johnson, Sam, *xxi*
Jones, J D, 48
Jones, T L, 111

K

Te Kaha 15B Hineora Orchard, *193*, 269–72
Te Kaha Dairy Company, 17, 29, 69
kaheru, 200
Kahu Est. Ltd, 144
Kaio, H, 88
Kaipara Harbour, 163
kaitiakitanga, x, 154, 162, 274
Kapenga M Trust, 144, 171–2, 184
Karaka, Wiremu, 61
Karatau, Jack, 109, 130, *131*
Karioi, 3
Te Katene, 29
Kaui, G, 88
kaumātua, xv, 166, 176, 212, 241, 278
kaumātua grants, 149, 170, 272
kāuru, 6
Kawepo, Renata, 5
Kāwhia, 52, 76
Kawiti, Walter, 86
Kepa, Henare, 5
Kepa, Rauri, 218
ketu, 200
key performance indicators, 4, 150, 152, 159, 257, 266
kikuyu grass, 106–7, 122, 168
King, Pania and Eugene, *261*, *262*, 263–4
King, Ron, 78
King, Ronald and Justine, *253*, *255*, 263
King, T, 115
King, W, 113
King Country, xix, 12, 71, 217, 220
Kingi, Ani, 5
Kingi, E, 233
Kingi, Robert Tu, 80, 82, 123–4
Kingi, Tihema, 52
Kīngitanga, 210, 213, 217
Kiriroa Station, 261–4
    field day, *263*
Kirkwood, D R, 111, 113, 115
Kiwi-cross cows, 240–1, 249–50, 278
kiwifruit, 269, 271–2, 274, 280
Klaricich, John, 113, 122
Knox, Abel, 18
ko, 200
Te Kōhera, 153
Kohere, Mokena, 5
Kohere, Ngarangi, 40, 45, 48, 228
Kōkako Trust, 149
koko, 200
Kokohinau Marae, 241, *242*
Konui, Waaka, 110, 112, 144
Konui, William, 144
Koopu, Tohi, 69
Kopa, Hemoata, *287*, 289
Kopae, Tawera, 29
Kotua, K, *103*
Te Kūiti, 17, 52, 137, 151, 175
kūmara, 4, 6–7, *41*, 155
kūmara houses, 199
Kuratau Trust, xv, 144
Kuru, George, 92

L

labour, division of, 200
Lake Taupō, 17, 147, 156, 163, 170
lamb, frozen, 5
Land Based Training, 240
land clearances, by burning, 200
land consolidations, 21, 182
Land Corporation of New Zealand, 277
land development blocks, 23, 147, 150, 230
Land Development Encouragement Fund, 136
land development schemes
    and Ahuwhenua Trophy, 27, 29, 47, 73, 209, 227
    Ngata and, 19, 22, 182, 222
land reform, 18–20

Land Wars, 5, 12, 204
    memorials, *212*
Lands Improvement and Native Lands Acquisition Act 1894, 217
LandVision Ltd, 159
Latimer, Graham, 138
Lawson, Heemi, 64, 69
Leeder, Doug, xx, *167*
Lepperton, 17
Liberal Party, 209, 216–18
LIC, 143
lime, 65, 74, 76
Lingman, C K, 86
livestock holdings, Māori, 202
Locke, John, 8
lucerne, 38, 71, 98, 124
Luxton, John, 276

M

Maaka, Dion, *239*, *240*
Macdonald, J B, 125
MacDonald, Joshua, 286
Macmillan Bull, H, 97
Mahupuku, Tamahau, 218
Mahuta, Nanaia, 257–9, *258*, 262, 267, 269, 270
Maketū-Ōpape area, 29
Maki, Wiremu (Ngawati), 98, 233
Te Mana te Wai, x
mana whenua, 172, 240, 246, 250
management structures, 182
Manawaroa, T, 117
Mang, Weo, *253*
Mangaroa Station, 243–5, 255, 257
    field day, *245*
Mangatewai, 155
Mangatū Incorporation, 16
Mangu, Wallace, 89, 91–2
Maniapoto, Rewi, 12
Manning, Desmond and Pikihuia, 89
Manu, Tongawhiti, 103, 105, 109–10, 113, 115, 117, 121, 125, 231
mānuka, 53, 74, 130, 152, 229, 231, 257, 260

mānuka honey, 278
manuring, 35, 38, 124
Māori Affairs Act 1953, 108, 144, 254
Māori Affairs Amendment Act 1967, 107, 260
Māori Affairs Supervisors, 65, 228
Māori agribusiness, x, 4, 244, 249, 274–5, 280
Māori Agricultural College, *36*, *129*
Māori agriculture, 123
    decline of, 11, 197
    financing, 14
    Pākehā promotion of, 8–10
    pre-colonisation, 6–7
Māori Appellate Court, 220
Māori children, *75*, *77*, *81*
Māori District Land Councils, 219–20
Māori economy, xx
    1930–1950, 39
    asset base of, 252
    and land losses, 10
    post-war, 59
    powerhouse of, 274
    and urbanisation, 45
Māori Education Fund, 130
Māori Education Trust, 150
Māori Excellence in Export Award, 244
Māori farmers
    consistent supply for, 35
    economic aspirations of, 4
    judges supporting, 234
Māori farms, x, 145
    1930–1950, 39
    Glanville's observations on, 52
Māori Incorporations Constitution Regulations 1994, 161
Māori land
    colonists' interest in, 10, 12
    commodification of, 13
    confiscation of, ix, 17, 45, 69, 209–11, 214, 278
    Crown acquisition of, 19, 108, 147, 209–10, 215–17
    developing, *100*
    dispossession of, 182, 193, 204, 209–22
    leasing of, 16, 213–15, 219, 221

quality of, 47
reforming title to, 15
relative costs of purchases 221
today, 222
Māori Land Administration Act 1900, 218–19
Māori Land Boards, 19, 21–2, 152, 161, 219–21
Māori Land Claims Adjustment and Laws Amendment Act 1904, 219
Māori Land Court, 65
and Hereheretau Station, 161
and Tūaropaki Trust, 153
see also Native Land Court
Māori Land Settlement Act 1905, 219
Māori Land Settlement Amendment Act 1906, 161
Māori land titles, 12–13, 15–16, 18–19, 158, 209–12, 214–15
Māori Lands Administration Act 1900, 16, 19, 220
Māori Lands Administration Amendment Act 1901, 219
The Māori Messenger, 198, 200
'Māori orientation', 101
Māori Patriotic Committee, 19
Māori Purposes Fund Control Board, 19
Māori Reserved Lands Amendment Act 1997, 150, 215
Māori seats, 212
Māori Social and Economic Advancement Act 1945, 59, 65
Māori Soldiers' Trust, 157
Māori Trust Board, 210
Māori Trustee, 134, 255, 289, 291–2
forcible sales to, 108
and Hereheretau Station, 160
loans from, 17
as sponsor, 143
Māori values, 249, 251
Māori War Effort Organisation, 45–6, 59
Māori welfare officers, 66–7
Māori women, xvi–xvii, 39, 45, 46, 48, 55, 63, 78
in Ahuwhenua Trophy, 45–6, 52, 74, 228–9
and Young Farmer award, 288
Mapu, James Waitaringa, 64, 69

Marae Mānuka dairy unit, 163
Maraenui, 29, *53*
Maranga Station, 243–4, 246
field day, *246*
Marginal Lands Act 1950, 130
Marks, J W, 110
Marotiri Farm Partnership, 144
Marsden, Samuel, 8–9
Marshall, B B, 113
Martin, C R and P R, 136
Mason, H G R, 52, 54
Massey, William, 17, 220
Matariki Partnership, 151–2
Matchitt, T F, 106
Matene, Tau Iwarau, 83
Matenga, T, 89, 91
Matenga, Wananga, 101
Mateparae, Jerry, *171*, 239
Matheson, James, 286
Matua, Henare, 15
Mauriohooho, Mita, 86
Mauriohooho, Robert, 150
Mauriohooho, Wiremu, 89, 91, 103, 105, 107, 150, 232
Mawhera Incorporation, 258, 260–1
field day, *260*
McAllister, Alec, 99
McConachie, M D, 115
McDonald's Restaurants, 264
McGregor, K A, 73, 232
McIntyre, Duncan, *112*
McKenzie, A, 130
McLean, Donald, 10
McMillan, M, 88
McMillan Bull, H, 33, 97–104, 116
McNaught, J D, 129–30, 132, 134
Meat and Wool New Zealand, xvii, 145, 150, 157
Meat Board, xix, 167
Meat New Zealand, 143
Mercer, A D, 59–61
merino, 147
methane emissions, 274
Mihaere, Jack, *275*

Mihaere, P, 29
Mihinui, Roku, *171*, 172, 184
milk solids, 150, 155–6, 163, 166, 240–1, 249–50, 252, 259–60, 275, 278
    Bledisloe on, 171
milking, 3-in-2, 277
milking platforms, 240–1, 250–1, 260–1, 279–80
milking sheds, 54, 80, 83
Mill, Whare, 61
Minginui Nursery, 267
Minister of Native Affairs
    and Ahuwhenua Trust Board, 27, 175
    first, 210
    Ngata as, 221
    under Liberal government, 215, 217–19
Ministry of Primary Industries, 143, 267, 280
Miraka Dairy Company, 173, 182, 244
Te Miringa te Kakara marae, *179*
missionaries, 8–11, 198
Mitchell, R, 90
Mitchell, Tai, 33
Moana, Howard, 127
Te Moananui, Tareha, 212
Moawhango, 5
Moekino, 153
Mohi, M R, 134
Moke, Whare, 52–3
Montgomery, K M, 52–4, 70–1, 230–1, 234
Moore, Sam and Kate, 292
Morehu, M, 79
Morgan, Quinn, *292*, 329
Morikau Station, 157, 161
    field day, *160*
Morris, J H, 130, 134
Morris, James, 125
mortgages, 14, 21–2, 102, 161, 219, 245
Morunga, E, 88
Morunga, W, 91
Motu Valley, 263–4
Mourea, 29, 227
Murphy-Peehi, Whatarangi, 151
Murray, J R, 88–92, 104–16, 121–3, 126

N

Naera, Wiremu Matene, 64–5
Te Nahu, J, 117
Naki, Katerina, 18
Napier, 4
Nash, Walter, 161
Native Affairs Committee, 22, 218
Native Circuit Court Act 1858, 216
Native Committees Empowering Act 1883, 215
Native Districts Regulations Act 1858, 216
Native Land (Validation of Titles) Act 1983, 216
Native Land Act 1862, 211
Native Land Act 1888, 215
Native Land Act 1909, 19, 220
Native Land Act 1931, Section 522 of, 22, 34
Native Land Amendment and Native Claims Adjustment Act 1929, 22, 39, 191, 227
Native Land Amendment and Native Land Claims Act 1923, 21
Native Land Court, 13, 15–16
    and Ahuwhenua Trophy, 38, 45–6
    establishment of, 212
    and land dispossession, 204, 209, 215–16, 221
    legislation on, 217, 219–20
    Māori resistance to, 212–14
    Ngata and, 18–19, 21
    reforms to, 217
    see also Māori Land Court
Native Land Purchase Act 1892, 216
Native Land Purchase and Acquisition Act 1893, 217
Native Land Purchase Board, 217
Native Land Purchase Ordinance, 10
Native Land Rating Act, 219
Native Lands Administration Bill 1886, 215
Native Lands Administration Bill 1898, 217–18
Native Lands Consolidation Commission, 22, 191
Native Lands Validation Act 1892, 216
Native Purposes Act 1935, 215
native reserves, 214–15
Native Rights Act 1865, 212
Native Territorial Rights Bill 1858, 12, 210–11
Native Trust Office, 16, 21–2
Native Trustee, 22, 215, 221

neatness and cleanliness, 33, 35–8, 40, 72
Nelson, James, 83, 86
Nepia, Huinga, xvi, 38, 45, 48, 228
New Zealand Dairy Group, xx
New Zealand National Fieldays, 138
New Zealand Settlements Act 1863, 211
New Zealand Transport Agency, 174
Newdick, R N, 122
Newell, Cyril, 60
Ngā Kupu Ōra Māori Book Awards, 191
Ngā Puhi, 8, 33, 213, 289
Ngā Ruahine, 240
Ngā Whenua Rāhui, 3, 162, 178, 257
Ngaheu, Harepaora, 288, 289, 290
Ngāi Tahu, 210, 260, 291
Ngāi Tahu Farming Limited, 248, 250
    field day, 251
Ngai Tukairangi Trust, 269–70, 272–4, 273
Ngakauroa Farm, 239, 241, 243
    field day, 242
Ngakoma, Wati, 275
Ngamoki, Te Ara, 69
Ngamoki, Paul, 69
Ngata, Apirana, ix, 18, 19
    1952 memorial hui, 160
    and Ahikōuka A6B, 152
    and Ahuwhenua, xi, xv, 27, 30, 45, 61
    and dairying, 17
    and land reform, 19–23, 27, 39, 69, 108, 181, 209–10, 222, 278
    and Māori farming, 4–5
    political career, 15
    on Royal Commission, 220
    and Second World War, 59
Ngata, Paratene, 18
Ngāti Apakura, 291
Ngāti Awa
    and agriculture, 8
    land confiscations from, 211
    Nikoras and, 155
Ngāti Awa Farms Ltd, 239, 241, 243
Ngāti Awa Group Holdings, 241
Ngāti Hinemihi, 146

Ngāti Hineuru, 176
Ngāti Kahungunu
    farmers from, 59
    land losses of, 10, 147, 213
Ngāti Kauhanganui Repudiation Movement, 15
Ngāti Konohi, 158
Ngāti Maniapoto, 12, 213, 291
Ngāti Manunui, 146
Ngāti Maru, 277
Ngāti Pāhauwera, 289
Ngāti Parekawa, 156
Ngāti Porou, 5, 14, 152, 170, 291
    farmers from, 61
    Ngata and, 18, 20–1
Ngāti Porou Tapuaeroa farm project, 152
Ngati Pukenga, 174
Ngāti Rangi, 3, 18
Ngāti Raukawa, 153, 162
Ngāti Raukawa ki Wharepūhunga, 149–50
Ngāti Tama/Maniapoto, 155
Ngāti Tamaterā, 277
Ngāti Tara Tokanui, 277
Ngāti Tūhourangi, 27
Ngāti Tūramakina, 146
Ngāti Tūwharetoa, 6, 146, 153, 156, 162, 173, 218
    livestock holdings, 202
    Rauhoto hapū, 166
Ngāti Wāhiao, 155
Ngāti Whātua, 163, 168, 213
Ngatirongo Trust Farm, 278
Nikora, Dean and Kristen, *154*, 155, 183
Nikora, Paki, *275*, 276, 278
nitrogen management plan, 163
Niwa, Joseph, 107, 111, 113, 231
Nokohau, Hugh, 99
Nokohau, James, 98
Norman, James, 88
North Auckland District, 33–4, 40, 49
Northland, 8
    Ahuwhenua Trophy in, xviii–xix
    Crown land acquisitions in, 12
    land consolidation in, 21
Nuku, Tamihana, xvii, *xviii*, 144, 176

O

O'Connor, Damien, *193*, 258, 261–2, 269–70, 274–6, *275*
Ōhinemutu, xv, 30
Ōhura, 97
Ōkuhaerenga dairy unit, 162
Oliver, James, 88
Olliver, N, 93
Ōmaio, 67, 231
Omana, T, 54
Omapere Taraire E & Rangihamama X3 Ahu Whenua Trust, 253–5
 field day, *255*
Oneroa, Charles, 29
Onewa Patu, 33
Onuku Māori Lands Trust, 258–60
Opouriao Dairy Company, 17
Te Oranga o te Taiao, x
organic farming, 275, 278–9
Ormsby, G, 111
Ormsby, J, 106
Ormsby, K P, 115
Ormsby, N J, 121–2
Ormsby, R, 111
Orr, Darren, 268
Ōtakanini Tōpū Incorporation, 168–9
 field day, *168*
Otama Marere Orchard, 269, 272, *273*
Otanepae Station, 163, 169
Otene, Bunny Meihana, 52–3, 67, 233
Otimi, J T, 111
Ottewill Silversmiths, *268*, 270
overgrazing, 81, 89

P

Padlie, Mathew, 88
Paetahi, Mete Kingi, 213
Paewai, Hepa, *133*
Paewai brothers, 132, 137–8
Pah Hill Station, 3, 4, *151*, 152, 182
Paikea, Paraire, 45
Pākarae Whāngārā B5 Partnership, 157, *158*, 183
 field day, *xx*, *157*, *281*

Pākehā farmers
 and Māori farmers, 70, 202, 204, 230
 and Māori land, 16–17, 21, 214, 219
 Māori working for, 198
Paki, E, 90
Pākihiroa Station, 168, 171
 field day, *170*
Paku, H, 91
Paora, Haukino, 69
papakāinga, 200, 220, 272
 on Morikau Station, 161
 neatness and cleanliness of, 33, 37
 pre-colonisation, 6–7
papakāinga housing, 255
Paraone, Chad, *243*
Paraone, Henare, 52, 54, 61–3, *62*, 232
Parata, Selwyn, 170
Parekarangi Trust, 138, 154–6
Parekawa, 153
Parengarenga Incorporation, 246–7
Parker, Bernie, 99
Parker, R, 91–2
Parliament
 finalists announced at, 239, 243, 252, 257, 261–2
 Māori seats in, 212, 217
Parsonson, Ann, 8
paspalum, 61, 69, 71
pastoralism, 213
pasture control, 61, 67, 76, 114
pasture growth, 74, 77, 80, 90, 101, 110, 112–13, 115
pasture improvement, 18, 36, 70, 81, 88, 106, 116
pasture management, 136, 149, 163, 173, 178, 280
 Allo on, 70
 Arnold on, 83, 86
 Hall on, 123–5, 127, 129
 McNaught on, 130
 Murray on, 113
 Simmonds on, 134
 Taylor on, 77, 80
 Walker on, 35
pasture production, 71, 103, 166, 171, 266, 279
pasture quality, 40, 149, 227, 260, 264

pastures, Flowers' report on, 73
Patu Nui A Aio, 6
Te Patukirikiri, 277
patupatu, 200
Paua Station, 243-4, 246-7
    field day, 247
peaches, 10, 12, 199
Peachey, G, 125
Peachey, W P, 134
Peehi, Whatarangi and Christian Murphy, 144
Peni, Hardie, 178
Pepper, Matt and Louise, 163
Pepper, Nick, 163
Pere, Wi, 15
Pere, Wiremu, 15-16, 182, 215, 218
Perry, Ian, 125
personal development, 266, 293
Peters, C, 113
Peters, Winston, 276
Peterson, John, 88, 93, 100
Petricevitch, George, 100, 106-7
Petrie, Hazel, 7-8, 10
PGG Wrightson, 143, 150
Phelan, Michelle, 287
phosphate, 65, 79, 85, 105, 260
pig clubs, 46
pig husbandry, 38
pig production, 38, 40-1, 52, 70, 73, 76, 83
piggeries, 53-4, 60-1, 67, 71, 74, 78, 102
pigs, preventing from wandering, 201-2
Pine, Tawake, 218
pine trees, 105, 107, 175, 178, 250, 280
Pioneer Battalion, 19
plantations, 6, 8, 49, 105, 130, 147, 153, 200
planting, Māori and European methods of, 199-200
ploughs, 5, 8-9, 21, 98, 198-200
Pohia, Edward Clayton, 80-1, 83
Pohio, C, 125
Pohio, E, 233
Polack, J. S., 199
Polaris, 143
Pooley, Mathew, 290, 291

Potaka brothers, 233
potash, 73, 76-7, 79
Potatau Te Wherowhero, 210
potatoes, 8-11, 83, 198-9, 203
pou, 210
Pou, Eru Mako, 71, 233
Te Pou a Kani Farms, 146-7, 148, 182
Te Pou Tutaki/FitzRoy's Pole, 211
Pouarua Farms, 274-8
    field day, 277
Poverty Bay, 6, 220
pōwhiri, xvi, xx, 59, 160, 170, 176, 179, 189, 247, 255
Priest, Andrew, 248
Primary ITO, 278, 287, 292
problem soils, 11
Project Rerewhakaaitu, 260
Puaka, Iti, 82
Public Trustee, 21, 214-15, 221
Public Works Act 1894, 217
Te Puea Hērangi, 62
Puhipi, Riapo, 37
Te Puke, 29
Puke, Hare Jr, 99
Pukehika, Hori, 37
Pukepoto Farm Trust, 253, 257
    field day, 256
Puketawa Station, 253, 255, 257
    field day, 256
Pukunui, Tame, 52
pumice soil, 59-60, 79, 89, 99-100, 146-7, 158, 166-7
Te Puni Kōkiri, 175
    as sponsor, 143, 150, 267
Puriri, Hohaia, 78, 233
Purua, Ben, 292
Putauaki Trust, 239, 241-2

R
ragwort
    Allo on, 88
    Arnold on, 83
    control of, 60, 64, 73

McMillan Bull on, 100
　　Murray on, 111, 115, 117, 123
　　Taylor on, 73, 76, 78–80
Raharuhi, Jack, *288*, 289
Raharuhi, Pat, 35, 79
railways, 12
Rakaia Incorporation, 248–50
　　field day, *250*
Rakaupai, Kohika, 69
Rakiraki, John Puahau, 6
Rameka, Toby, 173
Rānana, 3
Te Ranga a Kauika Pā, 3
Rangatira 8A 17 Trust, 162, 166–7
　　field day, *166*
Te Rangi Development Scheme, 246
Rangiaowhia, 12
Te Rangitake, Wiremu Kingi, 5
Rangitihi Pā, 259
Rangitīkei, 5, 10
rapa maire, 200
Rare, Sam, 83
Raroa, Parekura, 90–2
Ratahi-Pryor, Enid, *239*
Ratana, Iriaka, *106*
ratstail, 98, 105
Te Rauparaha, Tamihana, 210
Raupatu Station, 137
Rawiri, W B, 111
Rawson, W E, 22
Re:Gen, 143
Read-Jones, Carl, *261*
reaper binder, *108*
recycling, 250
Reddy, Patsy, *193*, *258*, 259, 270
Rees, W L, 15
Rees, William, 13
Reeves, Paul, *114*
regional field days, 144–5
regional finals, 129–30
　　2003, 143–4
　　2004, xv, 144–5
　　2005, 145–9
　　2006, 149–50
　　2007, 151–4
　　2008, 154–7
　　2009, 157–62
　　2010, 162–7
　　2011, 168–71
　　2012, 171–5
　　2013, 175–8
regrassing, 89, 144, 156, 209, 245, 280
Rehabilitation Board, 234
Rehu, D, 86
Rei, Kiriwaitingi, *239*
Reid, John Joseph, 98, 234
Reid, M, 74, 229
Reid, Wiremu, 286–8, *287*, 293, 329
rents, statutory, 150, 215
reproductive performance, 174, 278
Resident Magistrates, 201, 212
Retemeyer, Monte, 124
returned soldiers see ex-servicemen
Rewa, Herewini, 76, 233
Rewa, Waka, 82
Rewharewha, Whareparoa, 38
Richmond, Christopher, 210
Riini, J, 125
Riini, Sonny, 99
Rika, H, 93
Rika, Heke, 99
Rika, P T, 83
Ripaki, Simon, 100
riparian planting, 163, 257, 278
Roa, Kristy Marai, *290*, 291, 293
Roach, H, 107, 115, 123
Roberts, Mere, 199
Rodda, T E, 61, 63–4, 232–3
Rogers, Tapuae, 74, 80, 230
Romdale sheep, 246, 289
Rondon, Alan, *259*
rongoa, 156
Rongomaipapa Development Scheme, 59, 233
Rongonui, T T, 111, 113

root crops, 6, 48, 53–4, 71, 79, 127, 200
roto-cutter, 92
Rotorua-Atiamuri Road, 27, 40
Rowe, H, 115
Royal, Desmond, 80, 85, 89, 234
Royal, T, 80, 234
Royal Commission 1891, 216
Royal Commission 1907, 220
Te Rua o Te Moko Ltd, 239, *240*
    field day, *241*
Ruakituri Valley, 244–5, 257
Rūātoki
    and Ahuwhenua Trophy, 29, 34
    dairying in, 17
Ruatōria, dairying in, 17, 19
Rudland, Gina, *xvii*, 143
Ruha, Shelley, 244
Ruki-Fowlie, Dylan, 289
rūnanga, 170, 201, 210
rush growth, 90–1, 137, 147
Russell, James, *248*, *249*
Rutledge, G and R, 98
Ruwhiu, T, 111
ryegrass, 53–4, 71, 83–4, 154, 158, 161, 169, 278

S

Samuels, Dave, 270
sandy loam, 161, 251, 265
Satyanand, Anand, *162*
Savage, Albert, 93, 99, 103
Savage, John, 52, 65, 69, 71, 78, 231
Savage, Michael Joseph, 34, 171
scab, 5
Scholefield, Richard, *267*
secondary growth, 36, 80, 115
section rotary sheds, 250, 277
Seddon, Richard, 182, 217–18
Sentry Hill Redoubt, *213*
Sergeant, Charles, 40
Servicemen's Rehabilitation Board, 81
Severne, Charlotte, 291–2
Sewell, Sonny, 144

Seymour, Abe, *xxi*
shareholders
    communication with, 152, 156, 181, 266
    dividends for, 150
    pride of, 182
sharemilking, 149
    Amoamo family and, 136
    and Himiona Farm, 241
    and Kapenga M Trust, 172
    and Mawhera Incorporation, 261
    and Ngakauroa Farm, 243
    Nikora family and, 155
    and Tahu a Tao farm, 249
    and Tataiwhetu Trust, 278
    and Tewi Trust, 251
    and Tunaphore B2A, 280
    and Young Farmer award, 286
Sharples, Pita, *xviii*, xx–xxi, 157, 191, *239*
Shearer, Aloma, *253*
shearers, *84*, *132*, *135*, 198
Sheath, Gavin, *xxi*
Shedlock, Winnie, 76, 229–30
sheep
    droving, *122*, *138*, *139*
    shearing, *72*
sheep and cattle competition
    in 2005, 145–9
    in 2007, 151–4
    in 2009, 158–62
    in 2011, 167–71
    in 2013, 175–8, 285–6
    in 2015, 243–7, 288
    in 2017, 252–7, 289
    in 2019, 261, 291
    held alone, 138
    problems with low numbers, 79, 86, 88, 97, 99, 101, 107, 109, 113–14, 121, 125
sheep dipping, *116*
sheep farming, 3–4, 8
    capital for, 17
sheep grazing, 10, 63, 79, 213
Shelford, Sam, 76–7, 234

shelter, 231
    Allo on, 70, 88
    Arnold on, 65, 82–3, 86
    Blake on, 67
    Glanville on, 52
    McKenzie on, 130
    McMillan Bull on, 99, 103
    Mercer on, 61
    Murray on, 89–92, 105, 107, 115
    Rodda on, 64
    Taylor on, 48–9, 74, 76–7, 79
    in Taylor's points system, 72
    Walker on, 38, 40
shelter belts, 65, 227
Shepherd, J F, 54, 231
Shepherd, Tau, 103
Sheppard, Tau and Robert, 83
Sherlock, Winnie, 74
Sidney, Te Moana, *286*
silage, 163, 250–1
    Allo on, 88
    Arnold on, 86
    McMillan Bull on, 103
    Mercer on, 61
    Montgomery on, 71
    Murray on, 89, 109
    Rodda on, 63
    Shepherd on, 54
    Taylor on, 48, 74
    Walker on, 41
silt loam, 149, 161
Simeon, Clarence, 99, 103
Simeon, G, 111
Simeon, Rangiharuru, 99, 107
Simmonds, J G, 132, 134
Skinner, C F, 65
small farmers, 22–3
Smiler, Kingi, xviii, 157, 167, *169*, 191, 193, 266
    and 2014 competition, 239–40
    and 2015 competition, *244*
    and 2016 competition, 247–9
    and 2017 competition, 252, 254
    and 2018 competition, 258–9
    and 2019 competition, 262–3
    and 2020 competition, 266–7, 269–71
    and 2021 competition, 274, 277
    and human element, 181
    and Te Pou a Kani Farms, 147, 182
Smith, Jordan, 175, 285, *286*
Smith, Robert, 104
soil health, 272
soil management, 99
soil testing, 77
South Auckland, 34, 49, 117
South Island, 10
    acquisition of Māori land in, 10, 12
    Crown land acquisition in, 210
    participation in Ahuwhenua Trophy, 116, 249
    pre-colonial cultivation in, 5
    sheep farming in, 4
South Island Dairy Association, 17
Stainton, Manu, 74, 232
State Advances Corporation, 102, 133
steamships, 11
Steedman, Jack, 89, 101–2, 109
Stevens, Mihi, 65, 69, 80, 90, 103, 229
Stewart, C J D and A F, 134–5
Stewart, J R and S C, 130
Stewart, R, 99, 106
Stirling, Mihi, 69
stock bloat, 147
stock quality, 12, 122, 136, 227
    Arnold on, 83
    Simmonds on, 134
    in Taylor's points system, 72
    Walker on, 35, 38, 40
stock rearing, 5
stock replacement, 72, 83
Stockman, Diane, 166
Stokes, Evelyn, 12
Stoney Creek Block, 100
Stout-Ngata Commission, 19
strategic planning, 145, 148, 159
subdivisions, 134, 136–7, 152, 158, 166, 169, 178, 247, 264
    Allo on, 70

    Arnold on, 81
    McMillan Bull on, 101
    Montgomery on, 54, 71
    Murray on, 89–90, 92, 105, 107, 109, 111–13, 115
    Taylor on, 74, 76–7, 79
    of tribal lands, 19
    Walker on, 36, 38
Subritzky, Arthur, 104
Subritzky, James, 88
succession planning, 163, 178, 272
supplementary feed, 230, 278
    Allo on, 70
    Arnold on, 81–3, 85
    Murray on, 91, 111, 115
    Taylor on, 76–7, 80
    in Taylor's points system, 72
sustainability strategies, 178
sustainable farming, xviii, 146, 153
sustainable management, 16, 170
Sutton, Jim, 145
Suzuki as sponsor, 150
swamps, 69, 76, 80, 82, 86, *100*, 109, 115, 130, 137, 234
Swinton, James, 29–30
Swinton, Roka, 30
Swinton, W J, 83, 86, 89
Swinton, William, 29, 30, 209

T

Tahau, P H, 134
Tahu a Tao farm, 248–50
Tahuri, John, 84, 105, 107
Taia, G, 80
Te Taiao Framework, 274
Taikato, John, 127
Tainui, 39, 92, 98–9, 210, 216, 220
Te Tairāwhiti, 16, 106, 222, 262
Tairāwhiti Land Development Trust, 160
Taiwhanga, Rāwiri, 9, 197–8
Takamoana, Karaitiana, 15
Takapau dairy unit, 163
Takuira, Hohepa, 70
tall fescue, 64, 91–2, 93, 101, 103, 130

Tamahori, Heta, 169
Tāmaki, Tiaka, 61, 64–5, 70, 230
Tamati, Edward Rongomaire, 103, 105, *106*, *114*, *128*, 231
Tamihana, Wiremu, 12
Tana, J, 117
tangihanga, 155, 176, 266, 272
taonga tukuiho, 253
Tapa, Robert Tanginoa, 61, 63
Tapara, Areka, 104, 117
Tapsell, Homman, 269
Tarakaiahi, 153
Taranaki
    Ahuwhenua Trophy winners from, 69
    dispossession of Māori land in, 10, 17, 214–15
Tararewa Station Trust, 144
Tarawera eruption, 7, 155, 172, 259
Tarawera Station, 175–6, 178, 265–6
Tareha, Kurupō, 17, 197
taro, 6, 199, 267
Tataiwhetu Trust, 237, 275–8
Tatana, Ratahi, 82, 86
Tau, Raniera, *253*
Taua, John, 88
Tauhara Moana Trust, 171, 173
    field day, *173*
Taupopoki, Mita, 27
Taurua, George, 107, 109, 123
Tawhai, Hone Mohi, 213
Tāwhiao, Tūkāroto Matutaera Pōtatau Te Wherowhero, 213
Taylor, C R, 47–8, 55, 71–4, 76–80, 228–32, 234
Teepa, Whareauahi, *279*
Tewi Trust, 248, 251
    field day, *252*
thistle, 85, 134–5, 200
Thomas, Kawati, 93, 98, 101
Thompson, David S, 115
Thompson, G, 70, 231
Thompson, Joseph William, 79, 102, 105
Thompson, R J, 38
tikanga Māori, 145, 170, 178, 260, 278
Tikitiki, 34–5, 38, 61

timotimo, 200
Tiopira, Noti, 78, 233
Tiratu Station, 132, 137–8
Tiroa E Trust, 151
Tiwha, John, 70, 231
tobacco, 199
Tohe, Pierre, 249
Toi, P, 91
Toi, W, 115
Tokaanu, 90–2, 97
Tokomaru Bay, 20, 144, 220
Tolaga Bay, 6, 220, 291
top dressing, 54, 71–4, 137
    aerial, 73–4, 81–3
    Arnold on, 86
    heavy, 91–2
    McMillan Bull on, 98
    Murray on, 90–1, 109, 111
    phosphatic, 63–5
Topia, Percy, 76
topography, difficult, 68, 70, 86, 124, 234
Tōrere, 29
Toroa, Paul, 69, 71, 230
tractors, 82, 85–6, 88, 125, 150
    accidents with, xvii, 110
training farms, 90, 198
Treaty of Waitangi, 12
    pre-emption provision of, 210–11, 216–17
trees, native, 40, 65, 156, 163
trespass, 201–2
tribal committees, 18, 45, 59, 212, 215
Trust Order, 153, 266
Tūaropaki Trust, 151, 153
Tuhaere, Paora, 213
Tuhakaraina, Irihapeti, 69, 230
Tūheitia Potatau Te Wherowhero VII, 244, 276
Tuhiwai, Finnisha, 291
Tūhoe, 6
    and agriculture, 8
    land confiscations from, 211, 217, 221
Tūhoe Pūtaiao, 278
Tūhourangi, 144, 155, 172
Tumata, B, 115

Te Tumu Paeroa, 289, 291–2
Tunapahore B2A Incorporation, 275–6, 279–80, 282
    field day, 279
Te Ture Whenua Māori Act 1993, 144, 153, 161
Tūria, Tariana, 3, 223
turnips, 53, 199, 251, 278
Tuuta, Dion, 182, 183
Tuuta, Jamie, 243, 288–9

U

undershot, 73, 76
under-stocking, 89, 101
United Party, 221
Te Uranga B2 Incorporation, 175–6, 181
    field day, 176
urbanisation, Māori, 39, 45, 66
Te Urewera, 6, 21, 200, 217, 221
Urewera District Native Reserve Act 1896, 217
Te Uri O Hau Settlement Trust, 163
Te Uru Rākau, 267

V

Validation Court, 216
Van Heuckelum, Marty, 261
variance reporting, 265–6
Vercoe, Frank, 278
Vercoe, Ruhi, 38, 79
The Village Goldsmith, 268
volunteering programmes, 246

W

Waaka, Watchman, 82, 90
Waewaetutuki 10 – Wharepi Whanau Trust, 171, 174, 174
Wahawaha, Rapata, 5
Waiapu, 5, 14
Waiapu Farmers' Co-operative Company, 18
Waiapu Farmers Store, 37–8
Waiariki and Bay of Plenty area, xvi, 29
    acreages of crops, 199
    livestock holdings in, 202
Waiariki District, 52
Waiheke Island, 202

Waihi, Kopua, 63–4, 71–2, 230
Waihi Pukawa Trust, 146–7
Waikato, land confiscations in, 17
Waikato River, and horticulture, 198
Waimate, model farm at, 198
Waimate Mission Station, 9
Waiomatatini, 18–19, 151–2, 220
Waipaoa Cadet Training Trust, 291
Waipapa 9 Trust, 162–3, 168–70, 184
Waipiro block, 18
Wairangi, 153
Wairarapa
    acquisition of Māori land in, 10
    Māori farmers from, 4, 17
Wairarapa Moana Incorporation, 10, 146–7, *148*
Wairuru Women's Marae Committee, 69
Waitangi, Treaty Grounds, 28
Waitangi Tribunal, 214
Waitara, 5
Waititi, Hirinia, 228
Waititi, John, 69
Waititi, Moana, 69
Waititi, Sid, 69
Te Wake, Joseph, 90–1, 93, 99, 101, 104, 106
Walden, E and R, 115, 117, 122, 124–5, 127
Walden, Wayne, xvii, 143
Walker, C, 34–5, 37–8, 40, 45–7
Walker, Tangaroa, 175, *285*
Walker and Hall, 268
Wallace, Hannah, *287*, 288, 329
Walters, Wano, 149
wānanga, 178
Warbrick, James, 155
Warbrick, Te Whitu Rareata, 69
Warrington, Norman, 38, 103
wastelands, 200–1, 219
Watene, W, 91, 117, 125
water, reticulated, 147, 154, 159
water piping, 70–1, 81
water quality, 175, 247, 278
water races, 111–12
water reticulation, 70, 82, 135, 159, 246

water supplies, 136–7, 147, 149, 152, 156, 158, 169, 230, 272
    Allo on, 88
    Arnold on, 81, 83
    fouling, 102
    Hall on, 128
    McMillan Bull on, 102
    Murray on, 89, 106, 109
    in points systems, 49, 65, 72
    Taylor on, 74, 76, 79–80
waterways
    fencing of, 154–5, 159, 163, 175, 255
    protecting from stock, 103
Watkinson, Tuhi, *248*
weed control, 36, 41, 53, 104–5, 113, 115, 161, 200, 227
The Weeping Woman, *212*
Wells, Charles, 52
Wells, Charles Berry, 99–100, 113, 115, 134
Wereta, Tumanako, 153
Wesleyan Church, 198
West Coast Settlement Reserves Act 1881, 214–15
Wetere, Koro, 138
Wetere, R, 111
Te Wetini, 12
wetlands, 156, 222, 264
    creation of, 278
    protection of, 155, 163, 170, 173
    restoration of, 272
Whaitā, 153
Whaitiri, Meka, *258*, 259, 276
Te Whakaari, 288
Whakapahi, Hekeawai see Chadwick, John
whakapapa, 170, 182, 240
    evidence of, 146
    of organic things, 199
Whakarewarewa, 27, 155
Whakatōhea, 29, 211, 227
Whakaue Farming Ltd, 153
Te Whānau a Apanui, 66–7, 271–2
Whānau a Pākai, 61
Whānau a Taupara, 16
Te Whānau a Te Ehutu, 271

whānau trusts, xviii–xix, 22–3
whanaungatanga, 288
Whanganui, acquisition of Māori land in, 10, 213
Whanganui Regional Museum, 144
Whangara Farms, 158–9, *206*, 262, 264–5
    field day, *264*
Te Whare, Kira, 69
Wharekura, Joe, 38, 54, 59, *60*, 228, 233
Wharepi Whanau Trust, 174
Wharerau, Ben, 99
Whata, Aperehama, 90, 98, 105
Te Whata, W, 93
wheat, 8–12, 198–200, 202–4
Whenua Kura Māori training and leadership programme, 250
White, John, 200
Te Whiwhi, Matene, 210
Wi Hongi, Hiria, 228
Wi Pere Trust, 152
Wi Tako Ngatata, 12
Wikaira, M and D, 80
Williams, Henry, 8
Williamson, Josh, *243*
Wilson, Cheyenne, 290–1, 293
Winiata, *xix*
Winstone Pulp, 3
winter feed, 227, 278

    Allo on, 88
    Arnold on, 81, 83
    Dempster on, 29
    Glanville on, 52
    Mercer on, 61
    Montgomery on, 53–4, 71
    Murray on, 89
    Rodda on, 63
    Shepherd on, 54
    Taylor on, 48, 73, 78–80
    Walker on, 35, 38, 40, 46
wintering, 130, 135, 147, 158, 163
wool production, 89, 98, 121
wool quality, 105, 107
wool sales, 72–3
woolsheds, *20*, 53, *72*, 122, 132, 147
    Arnold on, 81
    Murray on, 105, 116
    Taylor on, 79
Wordley, J T, 91
Wright, Rawson, 101–2
Wright, Rawson (son), 163
Wright, Wikitoria, *102*

Y

Yates, Bowman, 71, 115
Yorkshire fog, 107, 109, 111, 113–14